METHODS IN KARST HYDROGEOLOGY

INTERNATIONAL CONTRIBUTIONS TO HYDROGEOLOGY

26

Series Editor: Dr. Nick S. Robins
Editor-in-Chief, IAH Book Series
British Geological Survey
Wallingford, UK

INTERNATIONAL ASSOCIATION OF HYDROGEOLOGISTS

Methods in Karst Hydrogeology

Nico Goldscheider
Centre of Hydrogeology, University of Neuchâtel, Switzerland

David Drew
Geography Department, Trinity College Dublin, Ireland

CRC Press
Taylor & Francis Group
Boca Raton London New York

CRC Press is an imprint of the
Taylor & Francis Group, an **informa** business

Cover photo: Rural scene at the La Bo village spring in northwestern Vietnam. The karst spring is used as a drinking water source for the people and for animals, for washing clothes, bathing, as a fishpond and for irrigation (photo: Nico Goldscheider, 2004).

CRC Press
Taylor & Francis Group
6000 Broken Sound Parkway NW, Suite 300
Boca Raton, FL 33487-2742

First issued in paperback 2019

© 2007 by Taylor & Francis Group, LLC
CRC Press is an imprint of Taylor & Francis Group, an Informa business

Typeset by Charon Tec Ltd (A Macmillan Company), Chennai, India

No claim to original U.S. Government works

ISBN-13: 978-0-415-42873-6 (hbk)
ISBN-13: 978-0-367-38898-0 (pbk)

Library of Congress Cataloging-in-Publication Data
Methods in karst hydrogeology / edited by Nico Goldscheider & David Drew.
   p. cm.
   ISBN 978-0-415-42873-6 (hardcover : alk. paper) 1. Hydrology, Karst – Research – Methodology.
I. Goldscheider, Nico. II. Drew, David (David Phillip)

GB843.M48 2007
551.49072 – dc22   2006102475

**Visit the Taylor & Francis Web site at
http://www.taylorandfrancis.com**

**and the CRC Press Web site at
http://www.crcpress.com**

CONTENTS

8   Tracer techniques

Ralf Benischke, Nico Goldscheider & Christopher Smart

PREFACE

One of the UN Millennium Development goals is to halve, by 2015, the proportion of people without sustainable access to safe drinking water. Most of the planet's non-frozen freshwater is groundwater stored in aquifers, while lakes and rivers represent a much smaller percentage. Groundwater is not only more abundant than surface water, but the geological environment also provides some degree of natural protection against contamination. Therefore, the exploitation and wise management of groundwater resources represents a key step towards achieving this goal.

In many areas of the world, karstified carbonate rock formations hold important groundwater resources that are vital for the freshwater supply of many regions and big cities. Although the natural quality of this groundwater is often excellent, karst aquifers are more vulnerable to contamination than other aquifer types. Therefore, karst aquifers require special protection. At the same time, karst aquifers have unique hydrogeological characteristics, such as the presence of conduits, which allow groundwater to flow rapidly over large distances. Because of their specific nature, karst aquifers also require specifically adapted investigation methods.

This is the first textbook that provides an introduction to various groups of methods that can be applied to study karst aquifer systems. It was written by an international team of 21 authors from 8 countries, including scientists and practitioners: in particular, scientists who are often involved in applied projects and practitioners who have an interest in scientific research. We are grateful to all authors who contributed with their specific expertise to the chapters of our book. Working on this book was not always an easy task and it took a long time until the work was accomplished, but it was an honour and pleasure for us to collaborate with this distinguished group of experts.

Nico Goldscheider & David Drew

LIST OF AUTHORS

Bartolomé Andreo-Navarro, Dept. of Geology, Faculty of Sciences, University of Malaga, 29071 Malaga. Spain

Timothy D. Bechtel, University of Pennsylvania and Enviroscan, Inc., 1051 Columbia Ave., Lancaster, PA 17603, USA

Ralf Benischke, Joanneum Research, Institute of Water Resources Management, Hydrogeology and Geophysics, 8010 Graz, Austria

Frank P. Bosch, Institute for Geophysics, University of Münster, 48149 Münster, Germany

Robert E. Criss, Department of Earth and Planetary Sciences, Washington University, St. Louis, MO, USA

M. Lee Davisson, Flow and Transport Group, Lawrence Livermore National Laboratory, Livermore, California, USA

David Drew, Department of Geography, Trinity College, Dublin 2, Ireland

Nico Goldscheider, Centre of Hydrogeology (CHYN), University of Neuchâtel, 2009 Neuchâtel, Switzerland

Chris Groves, Hoffman Environmental Research Institute, Western Kentucky University, Bowling Green KY 42101, USA

Marcus Gurk, Institute of Engineering Seismology and Earthquake Engineering (ITSAK), Thessaloniki, Greece

Philipp Häuselmann, Swiss Institute for Speleology and Karst Studies (SISKA), 2301 La Chaux-de-Fonds, Switzerland

Daniel Hunkeler, Centre of Hydrogeology (CHYN), University of Neuchâtel, 2009 Neuchâtel, Switzerland

Pierre-Yves Jeannin, Swiss Institute of Speleology and Karst-studies, SISKA, 2301 La Chaux-de-Fonds, Switzerland

Attila Kovacs, Centre of Hydrogeology (CHYN), University of Neuchâtel, 2009 Neuchâtel, Switzerland

Neven Kresic, Malcolm Pirnie, Inc., 3101 Wilson Blvd., Suite 550, Arlington, VA 22201, USA

Jacques Mudry, Department of Geosciences, University of Franche-Comté, 25030, Besançon, France

Martin Sauter, Geoscience Centre, University of Göttingen, 37077 Göttingen, Germany

Chris Smart, Department of Geography, University of Western Ontario, London, Ontario N6A 5C2, Canada

Heinz Surbeck, Centre of Hydrogeology (CHYN), University of Neuchâtel, 2009 Neuchâtel, Switzerland

William E. Winston, Department of Earth and Planetary Sciences, Washington University, St. Louis, MO, USA

Stephen Worthington, Worthington Groundwater, Dundas, Ontario, Canada

CHAPTER 1

Introduction

Nico Goldscheider, David Drew & Stephen Worthington

1.1 THE PURPOSE OF THIS BOOK

The main aim of this book is to provide an introduction to the most relevant investigation methods and techniques that can be used to study karst hydrogeological systems, including modern approaches to interpreting and modelling the data obtained. Studying karst aquifers is not only scientifically challenging, but it is also important for humanity. Ford & Williams (1989) estimated that 25% of the world's population drink water from such resources. Although this number is probably an over-estimate, karst groundwater constitutes a crucial freshwater resource for many countries, regions and cities around the world. At the same time, karst aquifers are commonly highly vulnerable to contamination and are impacted by a wide range of human activities (Drew & Hötzl 1999). Bakalowicz (2005) noted that karst waters are often avoided as a resource if possible because of the perceived difficulties in exploitation and their high vulnerability. For example, around the Mediterranean, karst water resources are still under-exploited, and huge quantities of potential drinking water drain via submarine springs into the sea, even in regions with a shortage of freshwater. However, karst waters can be developed if general rules, specific to karst, are followed. The techniques presented in this book can assist in better exploiting and protecting these resources.

The book is intended to be useful for practitioners dealing with karst groundwater resources, their protection, management and exploitations. Also postgraduate students, PhD students and senior scientists doing their research in the field of karst hydrogeology will find much useful information about available techniques. The methods presented range from basic to sophisticated, so that the reader can select techniques that are adapted to the purpose, and to the availability of time, technical and human resources.

The methods described can be used for a wide range of applications, such as the delineation of spring catchment areas, the siting of pumping wells, the monitoring of water quality, the assessment (and resolution) of chemical and microbial contamination problems, and vulnerability mapping. Problems related to karst and groundwater may also arise during construction works such as tunnels or reservoirs. Although this book does not specifically deal with engineering aspects as this is done in other books (e.g. Milanović 2000), the methods described can also assist in solving hydrogeological problems that occur during construction works.

The techniques described in this book are also useful for scientific studies that aim to achieve a better understanding of water movement and the behaviour and transport of matter in the different components of karst hydrogeological systems. It is hoped that the book will also provide a basis for scientists and technicians working on the development of new and better investigative methods – and there is still much to do! Of course, much more can be said about each method than can be done within the framework of this book. Another goal of this book is thus to provide useful references for further reading.

1.2   WHY KARST AQUIFERS REQUIRE SPECIFIC INVESTIGATION TECHNIQUES

Karst aquifers require specific investigation techniques because they are different to other hydrogeological environments such as fractured and granular aquifers. They are always liable to surprise – "expect the unexpected", as Ford & Williams (1989, 2007) put it in their textbooks on karst geomorphology and hydrology. These and other books (e.g. Bonacci 1987, Dreybrodt 1988, Klimchouk et al. 2000, White 1988) provide a detailed introduction to karst landscapes, the evolution of caves and groundwater flow systems, while this book focuses on the available investigative methods and techniques. Some important characteristics that distinguish karst aquifers from other aquifers and hence require particular methods for their investigation are highlighted in the following paragraphs.

1.2.1   Evolution of the aquifer

Karst aquifers are not unchanging; they evolve with time as the CO_2 in the flowing water dissolves the carbonate rock, enlarging a proportion of the initial fractures into conduits and caves. The orientation and extension of the flow system and conduit network may change with time; conduits may collapse or be filled with sediments; saturated (phreatic) conduits may transform into unsaturated (vadose) conduits or vice versa. The importance of this temporal evolution in practical terms is particularly evident in coastal karst systems, where lower sea levels in the past may have initiated the formation of a karst drainage system that is far below the present sea level but which may still be partly active. In more general terms, the temporal evolution of karst aquifers means that there may be unexpected groundwater flow paths and drainage outlet points, which cannot be predicted on the basis of the present topographic and hydrologic setting. Taking into account landscape history during hydrogeological investigations can help to predict such flow paths and drainage points, which is crucial, for example, in establishing appropriate spatial sampling strategies.

1.2.2   Spatial heterogeneity

All aquifers are heterogeneous, but if a well is drilled into a sand and gravel aquifer, groundwater will nearly always be encountered, whereas the extreme heterogeneity of karst makes it difficult to drill a successful well. Groundwater flow may occur in large quantities in conduits and caves, yet there may be massive, unproductive rock only a few metres away (Fig. 1.1). Heterogeneity is a major issue in karst hydrogeological investigations. On one hand, there are techniques that aim to identify this heterogeneity, such as geophysical methods that allow potentially water-bearing fractures or conduits to be localised. On the other hand, heterogeneity also means that all types of interpolation and extrapolation are problematical if not impossible in karst. In particular, it is difficult to draw potentiometric

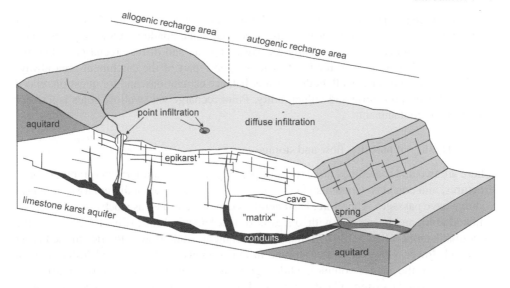

Figure 1.1.   Schematic illustration of a heterogeneous karst aquifer system characterised by a duality of recharge (allogenic vs. autogenic), infiltration (point vs. diffuse) and porosity/flow (conduits vs. matrix).

maps on the basis of water-level measurements in wells or piezometers, and then to use these maps to predict the direction and velocity of groundwater flow unless there are measurements from a large number of wells. Other methods, such as tracer tests, are thus often more appropriate to determine flow direction and velocity.

1.2.3   Hydraulic conductivity-scale effect

This effect is a direct consequence of heterogeneity. The hydraulic conductivity determined in rock samples is often very low and is mainly determined by pores and micro-fractures. Pumping tests sample larger rock volumes and thus often give higher conductivity values, reflecting the influence of macro-fractures. Finally, the behaviour of the entire aquifer system reflects the very high conductivity of the conduit network (Kiraly 1975). The practical consequences are apparent: although laboratory measurements and pumping tests can provide valuable information on the hydraulic properties of the bulk rock and the near-well zone respectively, the obtained conductivity values cannot be transferred to the entire karst aquifer system. The hydraulic behaviour of the entire karst aquifer is strongly influenced by the presence of solutionally enlarged fractures and conduits. Numerical models of karst aquifers thus need to include specific high-conductivity elements that represent the conduits and lower-conductivity elements that represent the less karstified rock volumes.

1.2.4   Duality of recharge and infiltration

Recharge into karst aquifers may either originate from the karst area itself (autogenic) or from adjacent non-karst areas (allogenic); the water may either infiltrate into swallow holes or dolines (point recharge) or diffusely into fissures in the rock, sometimes through overlying soils (Fig. 1.1). Allogenic recharge often infiltrates via swallow holes, while autogenic recharge is often more diffuse, although flow concentration in the epikarst layer may funnel the infiltrated water towards some shafts and conduits. Other aquifer types, particularly granular aquifers, mainly receive autogenic and diffuse recharge, while the described duality is specific for karst. The allogenic part of the catchment area always needs to be considered as well, both for water balance estimations, and for question related to groundwater protection and vulnerability. Point recharge via swallow holes is a major pathway for contaminants.

1.2.5   Duality of porosity, flow and storage

There are two and sometimes three types of porosity in a karst aquifer: intergranular pores, fractures, and conduits. Conduits may range from cm-wide solutionally enlarged fractures to huge cave passages. Therefore, karst aquifers can be described as a network of conduits embedded in, and interacting with, a matrix of less karstified rock (Fig. 1.1). Flow in the conduits is rapid (often > 100 m/h) and often turbulent, while flow velocities in the matrix are much lower. However, water storage in the conduits is often limited (often a small percentage of the aquifer volume), while significant storage may occur in the matrix, and in other parts of the system, like the epikarst. The existence of conduits is the big difference and the big problem compared with other aquifer types, which only contain intergranular pores and/or fractures. Consequently, all karst hydrogeological studies need to consider the conduits. Speleological surveys make it possible to directly observe water flow in caves. However, accessible caves are not always present and in any case, only represent a fraction of the total conduit network, mostly the inactive part, while tracer tests allow flow and transport to be studied also in the non-accessible but active parts of the network.

1.2.6   Temporal variability

Karst aquifers are often characterised by a rapid and strong reaction to hydrological events, such as storm rainfall and snowmelt. The karst water table may rise rapidly (often within hours) and dramatically (sometimes more than 100 m); the discharge of karst springs may also vary through several orders of magnitude within a short space of time. Likewise, the water and quality may display a dramatic variability, for example, in the content of suspended mineral particles, organic carbon and bacteria, including pathogens. Flow routes and relative discharges may be stage dependent to a much greater degree than in other rocks. This temporal variability requires specially adapted sampling and monitoring strategies. For example, it is inadequate to assess karst groundwater quality on the basis of water samples that are taken at regular, widely spaced time intervals. Instead, event-based sampling strategies need to be employed, i.e. significantly shorter sampling intervals during high-flow events. For the same reason, continuous monitoring devices are particularly useful in karst systems: for example the monitoring of discharge, temperature, electrical conductivity, but also of parameters such as turbidity and organic carbon, which indicate the bacteriological water quality and can help to optimise sampling strategies (Pronk et al. 2006).

1.3   OVERVIEW OF METHODS USED TO STUDY KARST AQUIFERS

Karst aquifers not only require specific investigation techniques, they also offer unique opportunities to observe the flowing water inside the aquifer, i.e. in caves (Fig. 1.2).

Figure 1.2. Diving expedition in a water supply karst spring in Florida. The divers carry a cave radio (below), and a team at the surface follows their route using a receiver (above). The cavers re-emerged at a trash filled sinkhole. Several lessons can be learned from this example: 1) conduits are crucial for groundwater flow and contaminant transport in karst aquifers, 2) there might be large conduits even when there are no surface karst landforms, 3) karst aquifers do not only require specific investigation techniques but also offer unique opportunities to study groundwater flow (Wes Skiles/Karst Productions, Inc.).

In fact, speleological techniques are the only group of methods that are specific for karst. All other methods presented in this book can also be applied to other hydrogeological environments but may require specific adaptations when applied to karst – or the other way round: they were initially developed for karst and were later transferred to other environments.

For example, tracing techniques, were first applied in karst flow systems, where they are still the most powerful tools to identify underground connections. Nowadays, they are increasingly used in all other types of aquifers, although the relatively low flow velocities in granular aquifers usually limit the use of tracers over long distances.

Other methods, like pumping tests and piezometric maps, work very well in granular aquifers, but the heterogeneity of karst often limits the application of such approaches. However, some karst aquifers are only accessible via wells and piezometers, which can yield valuable information when the experimental and interpretation techniques are adapted accordingly.

The groups of methods that are available to study karst hydrogeological systems correspond to the main chapters of this book.

The geological framework defines the external limits and internal structure of an aquifer and thus always needs to be considered when studying karst hydrogeology. Geomorphological observations can also deliver information concerning the groundwater flow system.

Speleological investigations make it possible to study the conduit network, and to directly monitor and sample dripping waters and groundwater (Fig. 1.2). All other methods presented in this book can also be applied inside caves, e.g. tracer tests in caves. However, caves are not always present or have not been explored and are usually not representative for the entire system.

Hydrological techniques essentially aim at characterising and quantifying recharge into, flow within, and discharge from karst aquifers, i.e. establishing the water budget. Due to the high temporal variability of karst systems, special adaptations are necessary.

Hydraulic techniques include the construction of piezometric maps and hydraulic tests in wells. As mentioned above, due to the specific characteristics of karst, such methods require special care and adaptations when applied to karst. However, especially in confined aquifers or coastal aquifers without accessible springs, such techniques may be a preferred choice.

Hydrochemical and microbiological investigations are primarily done to assess water quality and study contamination problems. However, hydrochemical parameters can also be used as natural tracers for the origin of the water, residence times, water-rock interactions, and mixing. The monitoring techniques employed need to be adapted to the high variability of karst waters.

Isotopic parameters can also be used as natural tracers. They may provide valuable additional information on the altitude of the recharge area, the origin of the water, groundwater age and residence times, and much other information. Isotopic techniques are particularly useful when they are combined with hydrochemical and tracing techniques.

Tracer tests using fluorescent dyes or other substances are often the only method that is able to provide clear and detailed evidence about underground connections in karst groundwater flow systems. Tracer tests are indispensable to delineate spring catchment areas and to study all type of flow and transport processes.

Geophysical techniques make it possible to see into the earth without digging or drilling (but can also be applied in boreholes). In karst hydrogeology, such techniques are particularly useful to identify fracture zones and other structures below a sediment cover, and to localise appropriate sites to drill pumping wells. Geophysics can also be used for many other investigations, such as determining overburden thickness, which is crucial for vulnerability assessment. Although most geophysical methods can also be applied to other environments, the localisation of conduits by microgravity or other techniques, such as cave radio (Fig. 1.2), would be an example of a karst-specific application.

Mathematical models are powerful tools to simulate groundwater flow and contaminant transport in granular aquifers, and can also be used to make reliable prediction, for example on the impact of water abstraction or contaminant release. In karst aquifers, such model-based predictions are generally problematic (see section 1.4). This is not only because turbulent flow in vadose and phreatic conduits is difficult to model, but mainly because the network of conduits and fractures is rarely sufficiently well known. Nevertheless, models can assist in solving a variety of practical problems in karst hydrogeology if the model is adequately adapted to the nature of karst and if the user is aware of its limitations. Furthermore, models can contribute to a better understanding of specific flow and transport phenomena in karst aquifers (e.g. conduit-matrix interaction), and speleogenesis.

1.4   WHAT CAN GO WRONG – AN EXAMPLE FROM WALKERTON, CANADA

The case of contaminated water at Walkerton provides a graphic example of what can go wrong when the karstic nature of carbonate aquifers with their concomitant rapid groundwater flow is not appreciated.

Walkerton is a rural town in Ontario, Canada, with a population of some 5000 people. In May 2000 some 2300 people became ill and seven died from as a result of bacterial contamination of the municipal water supply. The principal pathogens were *Escherichia coli O157:H7* (a pathogenic strain of *E. coli*) and *Campylobacter jejuni*. Subsequent epidemiological investigations indicated that most of the contamination of the water supply had occurred within hours or days at most after heavy rain.

In the summer and autumn of 2000 a hydrogeological investigation was carried out for the town of Walkerton. This included the drilling of 38 boreholes, surface and downhole geophysics, pump tests, and the testing of numerous samples for both bacteriological and chemical parameters. Three municipal wells had been in use at the time of the contamination. The aquifer at Walkerton consists of 70 m of flat-bedded Paleozoic limestones and dolostones, which are overlain by 3 to 30 m of till. A numerical model of groundwater flow (using MODFLOW) indicated that the 30-day time of travel capture zones extended 290 m from Well 5, 150 m from Well 6, and 200 m from Well 7. These results suggested that if a groundwater pathway was implicated in the contamination then the source must have been very close to one of the wells.

A public inquiry (The Walkerton Inquiry) was held to investigate the causes of what came to be known as the Walkerton Tragedy. During the inquiry the question was raised as to whether the aquifer might be karstic and thus have rapid groundwater flow. The original hydrogeological report had not mentioned the possibility of karstic groundwater flow. Subsequent investigations by karst experts found that there were many indications that the aquifer is karstic (Worthington et al. 2003, Smart et al. 2003). These included a correlation between bacterial contamination in wells and antecedent rain, demonstrating rapid recharge and flow to the wells; localised inflows to wells which video images showed to be solutionally-enlarged elliptical openings on bedding planes; the presence of springs with discharges up to 40 L/s; rapid changes in discharge and chemistry at these springs following rain; and rapid, localised changes to electrical conductivity in a well during a pumping test.

All these tests strongly suggested that the aquifer is karstic, but the most persuasive test was tracing testing. Earlier numerical modelling had suggested that groundwater velocities

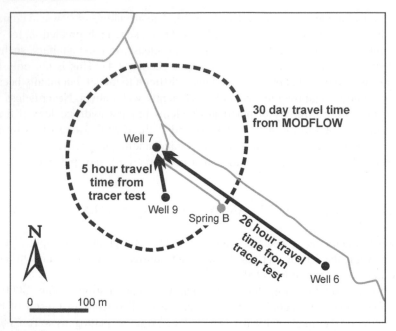

Figure 1.3.   Trajectories and travel times for tracers injected in monitoring wells #6 and #9 and recovered in pumping well #7, showing velocities >300 m/day, compared to a 30 day capture zone for well #7 predicted using MODFLOW (after Worthington et al. 2003).

were typically in the range of a few metres per day, but tracer testing demonstrated that actual velocities were some one hundred times faster (Fig. 1.3). This demonstrated that the source for the pathogenic bacteria could have been much further from the wells than the earlier modelling had indicated.

The experience of Walkerton underlines the importance of taking a precautionary approach when carrying out hydrogeological investigations in carbonate aquifers. Theory indicates that carbonate aquifers, and especially unconfined ones, will become karstified; the wise course is to assume that they are and carry out appropriate investigations.

CHAPTER 2

The geological and geomorphological framework

Nico Goldscheider & Bartolomé Andreo

2.1 INTRODUCTION

Geological and geomorphological investigations are indispensable when studying karst hydrogeology. It is impossible to understand a karst aquifer system without considering the lithology of the different geological formations, the stratigraphy, fracturing, fault and fold structures, as well as the topography, the karst landforms and the landscape history.

A lithological analysis comprises studying the mineralogical composition and texture, which affect the karstifiability, porosity and permeability of the rock. The lithostratigraphy of a rock sequence can subsequently be interpreted as a hydrostratigraphy, which means subdividing the rock sequence into different aquifers, aquicludes and aquifuges.

Water movement in hard rock aquifers mainly occurs in a network of fractures and bedding planes. In carbonate rocks, some of these tectonic and stratigraphic separation planes are enlarged by karstification to form a network of cavities and conduits. The spatial orientation and initial aperture of the fractures thus control the hydraulic properties and anisotropy of the karst aquifer. Faults with important displacement may either limit the aquifer or hydraulically connect different aquifers. Fold structures also influence the geometry of the aquifer and the underground drainage pattern. Groundwater flow often follows the synclines, while anticlines may act as local groundwater divides.

In karst areas, groundwater flow is largely independent of topography but is often guided by geological formations and structures. Groundwater flow often runs across under valleys and mountain ridges. Nevertheless, topography is important, as it defines the hydrologic base level and thus guides the regional orientation of the drainage pattern. The catchments of sinking streams originating on non-karst formations can also be delineated on the basis of topography. Karst landforms, for example dolines, often constitute zones of preferential infiltration and provide indications on the karst development and hydraulic properties of the aquifer. Karstification evolves over time, and the climatic and topographic conditions change over time, which may result in changes in base level and flow directions. The present underground drainage pattern of a karst system can often only be understood by considering its landscape history.

Geological and geomorphological information is available in the form of geological maps, cross-sections and literature. Aerial photos and satellite images provide important information and are particularly useful while studying large and remote areas with scarce vegetation, e.g. arid and high mountain areas. Additional fieldwork is often necessary. This chapter discusses geological aspects that are crucial for groundwater movement in karst areas and describes how geological information can be translated into karst hydrogeology.

2.2   MINERALOGY, LITHOLOGY AND STRATIGRAPHY

2.2.1   Minerals that form karstifiable rocks

The mineralogical composition of a rock determines to a large extent its karstifiability (chemical solubility) and thus its hydrogeological properties. Minerals that form karstifiable rock are sometimes referred to as "karst minerals" although this term does not correspond to any mineralogical classification. Karst minerals belong to three classes: carbonate minerals (e.g. calcite, dolomite), sulphate minerals (gypsum, anhydrite) and halide minerals (halite).

Carbonate minerals are composed of the carbonate anion (CO_3^{2-}) and different bivalent cations (Füchtbauer 1988). Limestone consists of calcite, the most important karst mineral. It is the low-pressure modification of $CaCO_3$ and primarily forms by bio-chemical precipitation from seawater or freshwater. With increasing water temperature and Mg/Ca ratio, an increasing amount of Mg^{2+} is built into the crystal lattice, although it has a smaller radius (78 pm) than Ca^{2+} (106 pm). The solubility of calcite increases with increasing Mg content (Davis et al. 2000). Other chemical impurities in the crystal lattice also influence mineral solubility and thus the karstifiability of the limestone. During diagenesis, high-Mg-calcite transforms into low-Mg-calcite. Aragonite is a metastable modification of $CaCO_3$, which sometimes forms in seawater and transforms into calcite during diagenesis.

Dolomite ($CaMg(CO_3)_2$) has a similar structure to calcite but half of the Ca is replaced by Mg. In specific environments, dolomite precipitates as a primary mineral, e.g. in the evaporite facies. However, it most often forms as a secondary mineral during diagenesis, when calcite reacts with Mg-rich pore water. Dolomitisation leads to a reduction of the crystal volume and thus increases the intergranular porosity of the rock by 5–15% (Ford & Williams 1989). Due to the scant connectivity between the newly created pores, the permeability does not increase as much as the porosity. Dolomite may also retransform into calcite (dedolomitisation).

Carbonate rocks also contain non-soluble minerals, the most important of which are clays, predominantly illite, and different forms of quartz (Llopis 1970). The karstifiability of carbonate rocks generally decreases with increasing content of such mineralogical impurities. The dissolution of limestone often leads to the formation of residual clays, which may occur as soils and cavity infillings. Calcarenite rocks contain detritic quartz grains, while micro- and cryptocrystalline quartz (chert and flint) forms by chemical precipitation during diagenesis and occurs in nodules or strata.

Sulphates and halides precipitate from over-saturated water under arid climatic conditions (evaporite facies). The former consist of the sulphate anion (SO_4^{2-}) and different cations. Gypsum ($CaSO_4 \cdot 2H_2O$) and anhydrite ($CaSO_4$) are the most relevant sulphates. Gypsum often transforms into anhydrite during diagenesis under increasing pressure and temperature. Anhydrite retransforms into gypsum when it comes in contact with groundwater, which results in increasing volume. Halite or rock salt (NaCl) and all other halide minerals are highly water-soluble. The described transformation and dissolution processes can also affect overlying carbonate rock formations and cause fracturing, which favours karstification.

2.2.2   Karstifiable rocks

Rocks that are chemically soluble are sometimes referred to as "karst rocks". The most important karst rocks are carbonate rocks, though evaporite rocks, such as gypsum, anhydrite and salt, are also karstifiable. Under tropical climatic conditions, dissolution phenomena may also develop in rocks composed of quartz and silicate minerals. This book focuses on the hydrogeology of carbonatic karst aquifers. Carbonatic rocks are all type of sedimentary or metamorphic rocks that consist of more than 50% of carbonate minerals. Carbonate rocks *sensu stricto* are biogene and chemical sedimentary rocks.

There are different classification schemes for carbonate rocks. The simplest and most useful for hydrogeology is based on the bulk mineralogical composition, which is crucial for the karstifiability and hydraulic properties of the rock: Carbonate rocks that predominantly consist of calcite are classified as limestone; rocks that are made of the mineral dolomite are called dolomite or dolostone. Mixtures of both minerals are referred to as dolomitic limestone (limestone with 10–50% of dolomite) or calcitic dolomite (dolomite with 10–50% of calcite). Dolomitic rocks are often less karstifiable than limestones. Carbonate rocks that contain 10–50% of sand or clay are named arenaceous (sandy) or argillaceous (clayey, marly) limestone or dolomite. The karstifiability of these rocks decreases with increasing sand and clay content. When this content exceeds 50%, the rock is classified as calcareous sandstone, calcareous shale or marl. Calcareous sandstones may act as fractured aquifers, while shale and marl formations generally form aquicludes.

Several more sophisticated classification schemes have been proposed for carbonate rocks, often on the basis of the fundamental work done by Folk (1959) and Dunham (1962). These classifications are crucial for petroleum geology, but only some aspects are relevant for karst hydrogeology. Three criteria are considered of particular importance: the size and type of particles or mineral grains, the presence and type of material (matrix, cement) that fills the voids, and the texture of the rock, i.e. the relation between the particles and the matrix or cement.

The carbonate particles include, among others, fossils and fossil fragments, grains of faecal origin (pellets), rock fragments, and ooids, i.e. spherical, sand-sized grains consisting of concentric $CaCO_3$ layers. The voids between the particles may partly be empty and thus constitute the intergranular porosity. In most carbonate rocks, however, the voids are partly or fully occupied by matrix or cement. The term 'matrix' is used to describe the microcrystalline calcite mud (micrite) that formed during the sedimentation of the rock. The cement consists of sparry calcite crystals (sparite) that formed during diagenesis. The primary porosity (for the definition see below) is negatively correlated with the micrite content. The highest porosities can be observed in limestones that are bound by organisms, e.g. reef limestones. The porosity of sparitic rocks is often 5–10%, while that of micritic rocks is less than 2% and sometimes close to zero. Karst aquifers may also develop in clastic sediments that predominantly consist of carbonatic components in a carbonatic matrix, such as sandstone and conglomerate (Göppert et al. 2002).

When limestones and dolomites undergo metamorphism, they are transformed into calcitic and dolomitic marbles respectively. The compaction and recrystallisation associated with metamorphism leads to a significant increase of the crystal size and to a reduction in primary porosity down to less than 1%. Therefore, marbles are often less karstifiable than their sedimentary equivalents, although there are examples of well-developed karst phenomena in marble (e.g. Williams & Dowling 1979).

2.2.3   Influence of lithology upon karstifiability

The lithology of a carbonate rock affects its porosity, permeability and karstifiability. The chapter on Hydraulic Methods provides detailed information on the porosity and permeability of different unconsolidated and consolidated sedimentary rocks. It is possible to distinguish two different types of porosity and two corresponding types of permeability: primary porosity formed during the genesis of the rock; and secondary porosity developed later. Primary porosity includes the small voids between the crystals, grains and fossil fragments (intercrystalline, intergranular or interstitial porosity). Secondary porosity includes all kind of fractures and karst conduits. In karst hydrogeology, the intergranular pores and the small fissures are often generically termed as matrix porosity, in contrast to the conduit porosity. The primary porosity is relatively high in reef limestones, arenitic and ruditic limestones, tufa and travertine, while it is small to negligible in many other limestones and marble. The matrix porosity is important for groundwater storage, as well as for the transport and attenuation of contaminants (Zuber & Motyka 1994).

The conduit porosity depends on the type and degree of fracturing and on the karstifiability of the rock. The karstifiability is controlled by three key factors (after Lamoreaux & Wilson 1984, Dreybrodt 1988, White 1988, Ford & Williams 1989, Morse & MacKenzie 1990):

- The bulk mineralogical-chemical composition: rocks made of halide and sulphate minerals are more karstifiable than carbonate rocks, which have a low solubility in pure water but dissolve readily in acid solutions. In nature, the dissolution of carbonate minerals is controlled by the presence of CO_2 in the water.
- The mineralogical purity of the rock: small percentages of non-soluble components, such as clay or silt, significantly decrease the karstifiability of carbonate rocks; rocks that contain more than about 25% of such impurities are commonly not karstifiable.
- The degree of diagenetic compaction and cementation, which determines both the intergranular porosity and the hardness of the rock. In slightly cemented or non-cemented carbonatic sediments, there is diffuse flow through many small pores and groundwater will rapidly become saturated. Cavities will not form in soft carbonate rocks or will not persist. In highly cemented and mechanically strong rocks, flow is restricted to fractures, which will be enlarged by chemical dissolution to form karst conduits.

Other lithological factors that are reported to influence karstifiability include crystal size (small crystals dissolve more readily than larger ones, which can be observed most clearly when calcite veins form ridges on limestone pavements), chemical purity of the crystals (traces of magnesium in calcite crystals increase solubility) and the content of organic matter in the limestone, which favours the growth of heterotrophic bacteria that produce CO_2 and thus reinforce karstification (Dreybrodt 1988, Gabrovsek et al. 2000). The most intense karst development is commonly observed in massive and thick-bedded limestones, while dissolution phenomena are usually more dispersed in thin-bedded rocks (details see below).

2.2.4   From lithostratigraphy to hydrostratigraphy

After characterisation of the lithology and hydrogeological properties of each formation, the lithostratigraphic sequence can be translated into hydrostratigraphy (Fig. 2.1). In some

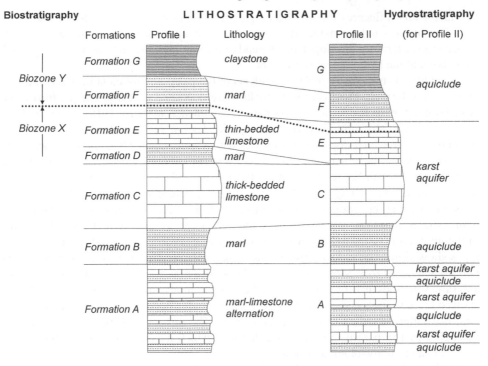

Figure 2.1.   Translation of lithostratigraphy into hydrostratigraphy. In some cases, several formations (G + F, C + E) can be combined in one hydrogeological unit; in other cases, one formation
(A) can be further subdivided into several hydrogeological units. Lateral facies changes and thin-
ning of geological formations must be taken into consideration. Biostratigraphic boundaries often
do not coincide with formation boundaries. The "karst aquifers" shown in this diagram are potential
aquifers only; the actual degree of karstification depends on various other aspects, such as fracturing
and hydrological framework.

cases, hydrostratigraphic profiles can be directly recorded in boreholes, e.g. by means
of hydraulic double packer tests (e.g. Muldoon et al. 2001). This section focuses on the
hydrogeological interpretation of lithostratigraphic information, as biostratigraphy is of
little use for hydrogeology, particularly in diachronous sedimentary sequences (Göppert
et al. 2002).

It is often possible to combine several adjacent formations into one hydrogeological unit.
For example, adjacent marl and claystone formations can be combined into one aquiclude
(G + F in Fig. 2.1), while adjacent carbonate rock formations may form one karst aquifer
(C + E). On the other hand, it is sometimes necessary to subdivide one formation into
different hydrogeological units. For example, it might be possible to divide a formation
consisting of limestone and marl layers into a sequence of different karst aquifers and
aquicludes (A). In some cases, even a thin limestone bed within a thick marl formation
may form a locally important karst aquifer. Different lithologies within one carbonate
rock formation might explain differences in its karst development and hydraulic behaviour.
White (1969) proposed a classification of carbonate aquifer systems, which is mainly based
on hydrostratigraphy. For example, thin beds of karstifiable rock between impervious strata
are called "sandwich flow type". It is important to note that the lithology is only one

key factor that influences the hydrogeological properties of a rock formation. Carbonate rock formations can be considered as potential karst aquifers, while the actual degree of karstification (aquifer development) depends on other aspects, such as fracturing, rock exposure and the hydrological framework.

The study of stratigraphy includes analysis of the paleo-environment in which the sediments were formed. Most carbonate rocks were deposited on carbonate platforms that covered large areas in epeiric seas on passive continental margins. Different facies belts can be distinguished: subtidal, peritidal, evaporite, lagoon, reef and slope facies. The transition between reef and slope often coincides with a facies change from massive limestone to breccia, while the transition between reef and lagoon corresponds to a facies change from massive to well-bedded limestone (Füchtbauer 1988). Such facies changes may have a strong impact on karst hydrogeology, e.g. a marl aquiclude separating two karst aquifers at one place may be absent at another place (D in Fig. 2.1). Seiler & Hartmann (1997) demonstrated that facies change from bedded limestone to massive reef dolomites in the Franconian Alb (Germany) has a major influence on hydrogeology, although both rock types show similar karst development. More than 150 tracer tests were carried out, about half in each facies zone. Most tracer tests in the bedded limestone gave higher recovery rates and flow velocities than the tracer tests in the reef dolomites. Furthermore, the groundwater showed higher nitrate and pesticide concentrations in the bedded limestones, although the land-use pattern in quite similar in the entire area. The authors explained these differences by higher matrix porosity in the reef dolomites, which means enhanced conduit-matrix exchange and microbiological activity.

Stratigraphic flow control can be defined as the degree to which groundwater flow is controlled by the stratification. The degree of stratigraphic flow control depends on the contrast in hydraulic conductivity between the aquifer and the intervening aquicludes, on the thickness of the aquicludes, and on the degree of faulting (Goldscheider 2005, Fig. 2.2). Facies changes may also limit the degree of stratigraphic flow control. For example, reef bodies crossing bedded marl and limestone formations may allow for groundwater flow across the stratigraphic sequence (Batsche et al. 1970).

2.3   GEOLOGICAL STRUCTURES

2.3.1   Folds

Folds influence the geometry of an aquifer, the groundwater storage and the underground drainage pattern. It is thus important to map and analyse the fold structures, including their amplitude and wavelength, their lateral continuity and the plunge of the fold axes.

The degree of karstification and the hydraulic conductivity are often assumed to generally decrease with depth. However, this is not always so in folded rocks. Extensional stress and open fractures predominate in the outer parts of folds, while compressional stress and closed fractures occur in the inner parts. Consequently, the karstification often increases with depth in synclines, while the opposite pattern is found in anticlines (Rodríguez-Estrella 2002).

Fold structures often have a major impact on the underground drainage pattern, mainly in shallow karst aquifer systems with a high degree of stratigraphic flow control. In contrast, the influence of the folds is limited in deep karst aquifer systems and/or in aquifer systems with a low degree of stratigraphic flow control, e.g. a sequence of different limestone

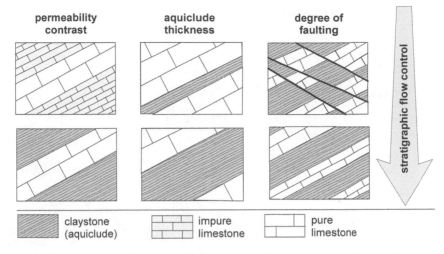

Figure 2.2. Dependencies of stratigraphic flow control. Upper row: groundwater flow across the stratification is possible, when the hydraulic conductivity contrast is low, when the aquicludes are thin and/or when there is intensive faulting. Lower row: flow is likely to follow the stratification, when the conductivity contrast is high, the aquicludes are thick and there is little faulting.

formations without intervening thick and continuous marl layers (Zötl 1974, Bögli & Harum 1981, Andreo 1997, Goldscheider 2005, Herold et al. 2000, Jeannin et al. 1995).

In karst systems with a high degree of stratigraphic flow control, different drainage patterns can sometimes be observed in the zones of shallow and deep karst (Fig. 2.3). In the zone of shallow karst, where the base of the karst aquifer is above the hydrologic base level, groundwater flow occurs near the base of the aquifer on top of the underlying aquiclude. As a consequence, the plunging synclines form the main underground flow paths and the anticlines act as local water divides. Surface streams often flow along anticlines, where the karst aquifer was eroded, or along synclines, where it is confined by the overlying aquiclude. In the zone of deep karst, the synclines are often entirely saturated with groundwater, which may overflow the anticlines. Flow across the fold can thus be observed in this zone (Goldscheider 2005).

2.3.2   Faults

Faults are of major importance for karst hydrogeology. In structural geology, faults are defined as fractures in the rocks of the earth's crust along which movement has occurred or still occurs (Davis & Reynolds 1996). The displacement at faults may range from centimetres to hundreds of kilometres. Small faults contribute to the fracture porosity of the aquifer, together with tectonic joints and stratigraphic bedding planes (see below). Karst conduits and cave systems often follow the network of faults, joints and bedding planes. Faults with high displacement often define both the external limits and the internal geometry of a karst aquifer system (Audra 1994, Häuselmann et al. 1999, Toussaint 1971). The hydrogeological effects of faults depends on three aspects: the aperture width of the fault plane, the presence and permeability of the filling material, and the fault displacement (Herold et al. 2000). Closed faults may form a hydraulic barrier by juxtaposition of an aquifer and

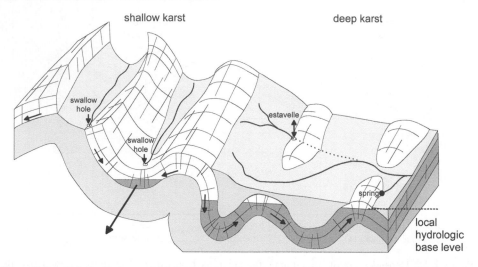

Figure 2.3.   Relation between fold structures, surface streams and groundwater flow in a karst system with strong stratigraphic flow control. In the zone of shallow karst, synclines form the main flow paths, and anticlines act as local water divides. In the zone of deep karst, flow across the folds is possible (Goldscheider 2005).

an aquiclude (Fig. 2.4a). Closed faults may hydraulically connect different aquifers if the fault displacement exceeds the thickness of the intervening aquiclude (b). Open faults act as a hydraulic barrier if the fault filling consists of impermeable material, e.g. clay, mineral veins or highly breccified and re-crystallised rock (c). Open faults may cause hydraulic short circuit across the entire hydrostratigraphic sequence if the fault is empty or if the filling consists of highly permeable material, e.g. coarse-grained fault breccia or karstified calcite veins (d). The hydrogeological characteristics of faults are heterogeneously distributed. A fault may thus act as a barrier at one place and connect two aquifers at another place.

Large sub-horizontal thrusts with high displacement rates can be observed in fold and thrust belts. The thrust sheets are also referred to as nappes. Nappe tectonics may lead to a multiplication of the hydrostratigraphic sequence (Wildberger 1996).

2.3.3   Joints and bedding planes

Joints are fractures without significant displacement, resulting from brittle rock deformation, mainly due to tectonic stress. Bedding planes separate the individual beds of sedimentary rocks (Pollard & Fletcher 2005). These tectonic and stratigraphic separation surfaces are crucial for groundwater flow and karst development in carbonate rocks. In structural geology, joints are further classified according to their genesis and properties. Release joints form parallel to the topography, often along the bedding planes, when the load of the overlying rock masses is removed due to uplift and erosion. This type of joint favours the development of sub-horizontal karst conduits and caves. Near steep cliffs, release joints and the corresponding caves can also be sub-vertical (Henne et al. 1994). Systematic tectonic joints always follow preferential directions, as they form normal to the minimum principal stress. They are often vertical and normal to the bedding planes but can also be inclined and have a sinuous shape. With increasing depth, the joints and

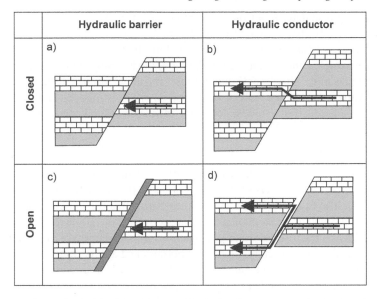

Figure 2.4. Hydrogeological function of normal faults in a limestone-marl sequence. Dependent on the displacement, closed faults may form a hydraulic barrier (a) or connect different aquifers (b). The presence and permeability of the filling material decides whether an open fault acts as a hydraulic barrier (c) or connects several aquifers (d). Similar schemes can be set up for other types of faults.

bedding planes are more and more closed. However, open fractures also exist in great depth, mainly in an extensional stress field (Van Der Pluijm & Marshak 2003), which enables the development of deep karst groundwater flow systems.

In hydrogeology, all planar discontinuities like bedding planes, joints and faults are often collectively referred to as fractures. The hydrogeological function of a three-dimensional fracture network depends on several aspects: the spatial orientation and extension of the separation planes, their density, i.e. the number of fractures per length unit, the degree of connectivity between adjacent fractures, the aperture width, the roughness, and the presence and hydraulic properties of fillings, such as clay or mineral veins (Witthüser 2002).

Joint frequencies, extensions and aperture widths in carbonate rocks can vary over several orders of magnitude and seem to be correlated with the thickness of the beds. Thin-bedded limestones are often densely fractures but the fractures tend to have small spatial extensions and apertures. Groundwater circulation is thus often dispersed and dissolution phenomena are not very pronounced. In thick-bedded and massive limestones, joints are often more widely spaced, continuous and have larger aperture widths, which favours the infiltration and circulation of water and thus karstification. Joints that cross many beds are called master joints; they often extend to a depth of several tens of metres and control the orientation of master conduits (Dreybrodt 1988, Ford & Williams 1989). The intersections lines between different separation surfaces are particularly relevant for karst hydrogeology. Sub-horizontal conduits often follow the intersections between bedding planes and joints, while shafts and dolines are often located at the intersections of two vertical joints. Massive carbonate rock formations (e.g. reefs) that suffered little or no tectonic stress can be nearly unfractured.

The fracturing also determines the anisotropic and strongly heterogeneous character of karst aquifers. Kiraly (1975) found that the hydraulic conductivity of karstified carbonate rocks depends on the scale of investigation (see Introduction). The hydraulic conductivity of small rock samples that are studied in the laboratory mainly reflects the small fissures and intergranular pores (if present) and is thus very low. Pumping tests in karst aquifers give higher values due to the effect of larger fractures. Investigations at a catchment scale give the highest conductivity values, due to the influence of the karst conduit network.

2.3.4   Fracturing studies

A fracturing study is usually performed at two scales: on the basis of aerial photos and in the field. Subsequently, the two sets of data are compared (Grillot 1979, Razack 1984, López Chicano 1992). The first step is to identify linear structures (lineaments) on aerial photos or satellite images and draw them on a transparent overlay or GIS layer. Lineaments often represent faults and fractures or other geological structures. However, lineaments can also represent anthropogenic structures, land-use borders and a wide range of other features. Therefore, care is needed while interpreting lineaments. The directions of geologically relevant lineaments are grouped into angular intervals in order to determine the frequency of each interval. The spatial density and length of the lineaments are also of interest. The zones with the greatest fracturing density are often those with the most intense karst development.

In the field, the orientation (strike and dip), the spacing and the aperture width of the fractures can be measured at natural rock outcrops, in quarries, on karrenfields or in caves. The interpretation of the field data should be addressed within the framework of a tectonic analysis that makes it possible to determine the stress ellipsoid for each deformation phase (López Chicano 1992, Andreo et al. 1997). For this purpose, it is necessary to take measurements of other structural elements such as stylolites, calcite veins and other filling material (breccia, clay), and striations on fault planes. Such interpretation allows for the differentiation of the fractures into those that favour karstification and groundwater flow, and those that act as hydraulic barriers or have little influence on karst hydrogeology.

The comparison between the results obtained from the analysis of aerial photos and field studies does not always provide a good agreement. In some cases, there is no relation at all, which is a problem of scale: Large regional fault-lines are often not represented in small rock outcrops, while systematic tectonic joints may not display on an aerial photo. A detailed fracturing study can reveal the tectonic history of the area, i.e. the variation of stress in space and time. It also helps to better understand the surface and underground karst development and might provide indications on preferential groundwater flow directions. Therefore, the results obtained from fracturing studies can also be integrated into distributive groundwater models, as they might indicate the location and orientation of high-permeability zones where no direct speleological information are available (see Speleology and Modelling chapters).

2.4   GEOMORPHOLOGY

2.4.1   Classification of karst landforms and geomorphological mapping

Numerous authors describe and classify karst landforms and discuss their importance for hydrogeology, e.g. Bögli (1980), Milanovic (1981), Jennings (1985), Bonacci (1985),

White (1988), Ford & Williams (1989) and Kranjc (1997). Karst features are usually classified into exokarstic (subaerial), epikarstic (subcutaneous) and endokarstic (subterranean). Geomorphological maps mainly show exokarst features, which are, however, surface expressions of epikarst development and aquifer structure.

The epikarst is defined as the uppermost zone of exposed karst rock, in which permeability due to fissuring and karstification is higher and more uniformly distributed than in the rock mass below. The thickness of this zone ranges between decimetres and tens of metres. The main hydraulic functions of epikarst are water storage and concentration of flow (Fig. 2.5) (Klimchouk 1997, Mangin 1975).

Karst features can also be subdivided into destructive or dissolutional landforms that predominantly result from carbonate dissolution, and depositional or constructive forms that form by carbonate precipitation. Another classification distinguishes negative (concave) and positive (convex) landforms. It is also possible to classify karst landforms on the basis of their state of development, e.g. into initial and mature landforms, and forms of decay.

Karren are small-scale (cm–m), concave dissolutional landforms, which can further be classified according to their size, shape and genesis. Karrenfields often indicate zones of intense epikarst development, important groundwater recharge and high vulnerability.

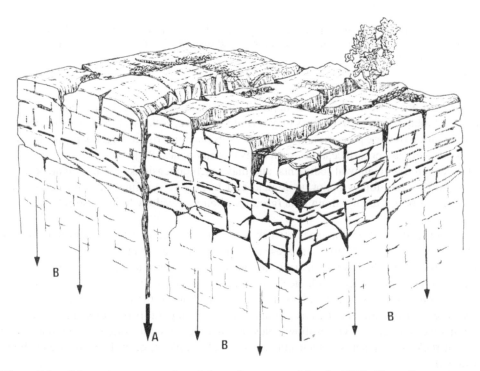

Figure 2.5.   Schematic representation of the epikarst zone (Mangin 1975). The epikarst concentrates infiltrating water toward vertical shafts (A). Another portion of the stored water contributes to diffuse recharge (B). Exokarst features, like karren, are surface expressions of the epikarst zone. However, epikarst may also be present when no exokarst features are visible. Both the exokarst and epikarst development reflects the network of sub-vertical tectonic joints and sub-horizontal bedding planes.

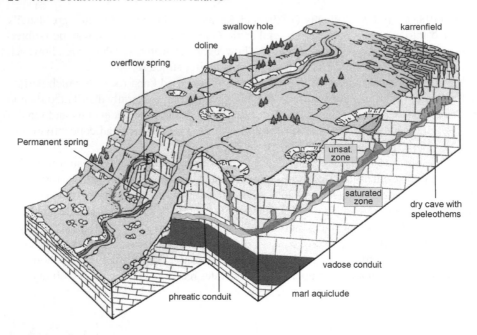

Figure 2.6.   Block diagram of a karst landscape from the French/Swiss Jura Mountains with some typical surface and underground karst landforms and their relation to the hydrogeological system (Schaer et al. 1998).

Dolines are closed, sub-circular depressions, ranging in size between metres and hundreds of metres (Fig. 2.6). Dolines may form by solution, collapse, piping, subsidence or a combination of these processes. Similar processes may give rise to larger depressions, with progressively larger surface areas, such as uvalas and poljes. Poljes have a flat valley floor, often covered with alluvial sediments, and may reach dimensions of many square kilometres. Poljes often follow major geological structures, like synclines or grabens. Dolines, uvalas and poljes usually drain underground, often via swallow holes, also referred to as ponors.

Although the surface drainage network is usually not well developed and often entirely absent, various types of valleys, gorges and canyons are present in some karst areas. Some valleys formed by fluvial erosion in the past but then dried up due to increasing karstification. Karst valleys often reflect geological structures, such as syncline troughs, fault lines and stratigraphic contacts. Karst valleys that are situated above the regional hydrologic base level are often dry. Surface streams that may locally or temporarily be present in such valleys often sink underground via swallow holes. Karst valleys are thus often discontinuous and end at rocky escarpments at their upper end (pocket valleys) or at their lower end (blind valleys).

The main tools for geomorphological mapping are topographic maps, aerial photographs, satellite images and field studies. Large-scale geomorphological maps (e.g. 1:50,000) help to better characterise an entire karst aquifer system, while local-scale maps are useful for detailed hydrogeological site investigations. The Working Group "Survey and Mapping" of the "Union Internationale de Spéléologie" (UIS), in cooperation with the

Karst Commission of the International Geographical Union (IGU), proposed a legend for the mapping of geomorphological and hydrological karst features, which is suitable for computer-aided cartography. The complete list of mapping symbols can be downloaded from: www.sghbern.ch/surfaceSymbols/symbol1.html

2.4.2 Relations between karst landforms and hydrogeology

Karst landforms can reveal much hydrogeological information, although it is not possible to conclude unambiguously from the surface karst development as to the deeper karstification and groundwater flow system. Confined karst aquifers generally show no surface karst landforms but may still have a well-developed underground drainage network. The presence of thick soils and dense vegetation may also hide the extent of karstification. In the Arctic zone and in areas that have been covered by ice sheets in the Pleistocene, glaciers have often removed the epikarst and exokarst landforms and/or buried the landscape under moraines, while deeper karst conduits might still be present. On the other hand, karren develop quite rapidly on limestone outcrops but are not always associated with or connected to, an important karst aquifer. In some cases, karst landforms have formed in the geological past (paleokarst) but are not active any more and thus not indicative of the present hydrogeological situation. Despite these limitations, the mapping and analysis of karst landforms can be recommended as useful tool in karst hydrogeology.

Landforms related to point recharge and discharge, i.e. swallow holes, estavelles and large springs, are the most reliable indicators for the presence of a karst aquifer with an active conduit network. Dolines are often considered as "diagnostic karst landforms", although some dolines might be paleokarst phenomena that are no longer connected to an active karst aquifer.

The spatial orientation of karst landforms may reflect the fracturing and thus provide indications on the geometry of the underground conduit network. Dolines and potholes are often situated at the intersection of two fractures. Aligned dolines and dry valleys often follow master joints or faults, which may also constitute preferential directions of cave development and groundwater flow (Ford & Williams 1989). However, the relation between surface landforms and the hydrogeological system is rarely straightforward.

Geomorphological and topographic observations help to delineate recharge areas and to identify infiltration processes. Areas formed of karstified carbonate rocks, with or without soil cover, represent zones of autogenic recharge. Very little evapotranspiration is to be expected on bare karstified limestones (i.e. karrenfields), and recharge may exceed 80% of the precipitation. The presence of a soil and vegetation cover significantly increases evapotranspiration, so that the proportion of recharge is much lower. However, the percentage of recharge obviously also depends on the overall climatic conditions, such as annual precipitation, type of precipitation (rain or snow), rainfall intensity, and temperature.

Highly variable infiltration conditions can be observed at dolines, poljes and dry valleys. Dolines often indicate the presence of vertical shafts at depth, which drain the perched aquifer of the epikarst zone. In some cases, dolines serve as swallow holes for surface waters. Dolines may also be filled with sediments and drain water diffusely though these sediments or laterally via the karstified margin of the doline. Poljes and other karst depressions often drain via swallow holes, which may transform into springs during high-water conditions (estavelles). When non-karstifiable geological formations overlie the karst aquifer, the predominant infiltration processes depend on the hydraulic properties of these formations.

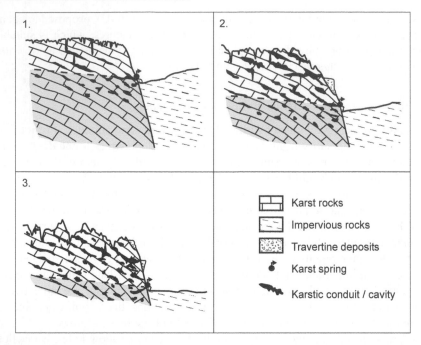

Figure 2.7.   Schematic illustration of the evolution of a karst aquifer system. Perched travertine deposits make it possible to identify paleo-springs and reconstruct landscape development. The conduits and springs get systematically older and inactive with increasing altitude but may be re-activated during flood events.

Diffuse infiltration and subsequent percolation can be observed on highly permeable formations. Concentrated surface runoff may occur on low permeability formations, which may thus constitute the catchment areas of sinking streams, i.e. zones of allogenic recharge. Such catchments can be delineated on the basis of topographic and geomorphological criteria.

The underground drainage pattern of karst systems cannot always be completely understood solely in terms of the present hydrological conditions. In many areas, regional uplift and/or deepening of the valleys results in a lowering of the regional hydrologic base level. Previously phreatic karst conduits thus transformed into vadose conduits and finally dried up, while a new generation of karst conduits developed in a greater depth. Both the conduit system and the springs thus get progressively younger with increasing depth. Older and higher conduits may be re-activated during flood events and give rise to overflow springs. Perched travertine or sinter deposits indicate the location of paleo-springs (Fig. 2.7). Their age can be determined using geochronological methods, e.g. U-Th dating. The topographic position, composition and age of perched travertine systems helps in the reconstruction of the evolution of the karst hydrogeological system, which is important for the understanding of the present-day aquifer functioning (Cruz Sanjulián 1981, Martín-Algarra et al. 2003).

In other cases, regional subsidence and sedimentation results in the burial of the karst aquifer so that the conduits may cease to be functional. Some of the old and deep conduits may still remain active and result in unexpected groundwater flow directions. Such old conduits may also be reactivated during the re-emergence of the karst system (see Speleology chapter).

2.5 SUMMARY

The first and indispensable step in karst hydrogeological investigations is the character-isation of the geological and geomorphological framework. This generally includes the interpretation of existing geological literature, maps and sections, as well as data acquired from fieldwork. Observations in caves and geophysical surveys may provide additional geological information.

The lithostratigraphy and the geological structures define the external limits and inter-nal geometry of the karst hydrogeological system. The geological setting also determines whether a karst aquifer is shallow or deep, confined or unconfined, and controls the extent of the unsaturated and saturated zones. Bedding planes, fractures, faults and folds often predetermine the initial preferential orientation of karst development and groundwater flow, though this may however, change during geological time. Landscape history thus needs to be considered for hydrogeological interpretations. Geomorphological and topographic obser-vations may also reveal much hydrogeological information, and help to identify infiltration processes, and to delineate zones of autogenic and allogenic recharge.

CHAPTER 3

Speleological investigations

Pierre-Yves Jeannin, Chris Groves & Philipp Häuselmann

3.1 INTRODUCTION

A method that sets karst hydrogeology apart from those methods practiced in other terrains is that a part of the groundwater flowpaths – the larger conduits – can sometimes be directly explored and mapped by speleologists. In combination with water tracing, this can be useful to get a more complete overview of the locations of underground flows and to determine key sites for the collection of water samples, either directly or through the placement of monitoring wells. For example, direct mapping of caves can identify underground stream confluence locations that can only be approximated through tracing, which can allow for sampling that represents entire sub-groundwater catchments. For example, in Slovenia, the confluence of the Pivka and Rak underground rivers has been found in Planinska Jama. Both rivers have huge and independent catchment areas, one being mainly allogenic (derived largely from surface drainage from Flysch deposits) and the other being mainly autogenic (derived from precipitation directly onto a limestone surface). In Milandre Cave (Jura, Switzerland) three main tributaries feed the underground river. Their respective catchments have been delineated quite precisely with the aid of more than a hundred tracing experiments. Access to the cave made it possible to sample each tributary separately (Fig. 3.1). Within Mammoth Cave (Kentucky, USA), discovery and mapping of the confluence of the Logsdon and Hawkins Rivers deep in the cave system allowed the placement of monitoring wells 100 m upstream from the junction, allowing easy sampling of their 25 km² and 75 km² catchment areas, respectively. This was especially important to understand agricultural impacts on groundwater quality in this cave, which has been designated not only as a US National Park, but also by UNESCO as an International Biosphere Reserve and World Heritage Site (see also chapter 4).

The three-dimensional nature of the subsurface karst plumbing networks means that conduits that are above the water table and thus dry during base flow conditions can become active flow paths as the water table rises during storm events. In some cases, this can result in modification of groundwater drainage basin boundaries, as normally dry conduits at various levels higher in the aquifer may direct water across what were divides under drier conditions, and to springs draining formerly adjacent basins.

For example, during relatively dry conditions the Echo River Spring within the Mammoth Cave System has very good water quality as its waters are derived from forested land within the highly protected Mammoth Cave National Park. As the stage rises, however,

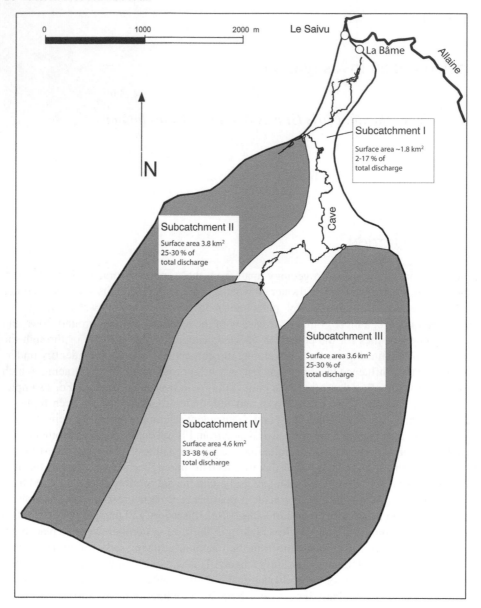

Figure 3.1.   Map showing delineation of the Milandre underground stream catchment area and its subcatchments.

flood activated conduits direct water to the spring from the adjacent Turnhole Bend Basin to the west, which drains large areas of agricultural land outside of the Park, and faecal bacteria concentrations have been measured to increase by tens of thousands of times within only a few hours (Meiman & Ryan 1993). Direct exploration and survey of underground conduits can be useful in helping to understand the hydrologic dynamics of such time-variant conditions.

Caves allow observation not only of water (including, for example, head, discharge, chemistry, isotopes), but also of the geological setting in a detailed way. Aquifer boundaries as well as lithologic or structural features that may exert influence on groundwater flow, such as major faults or folds can be identified with precision (e.g. Häuselmann 2002). Characteristics of the rock (chemistry or lithology) and of the fracture patterns can also be documented.

Caves are natural, inexpensive, and extensive "boreholes". It is worth using them!

Because the hydraulic conductivity of karst conduits is generally thousands to millions of times higher than that of the rock matrix, an identification of the conduit network geometry is essential for the understanding of groundwater flow. Flow at a regional scale is controlled by the conduits, which in turn control flow in the surrounding rock matrix. The conduit network is the key factor in determining flowpaths in karst aquifers, and where possible hydrogeological investigations in karst systems should aim at assessing the position of the major conduits.

Cave systems are typically complex, and some caves are more important than others in influencing contemporary hydrogeology. Three categories of conduit systems can be distinguished. One can define "speleological conduit networks" as the total length of conduits explorable by cavers. In a mature karst, typical speleological conduit density ranges between 10 and 50 km per km^2 (Jeannin 1996, Worthington 1991).

"Karst conduit networks" are defined as the total length of the existing conduits within a karst massif, i.e. the length of conduits wider than about 1 cm in diameter (as defined by Ford & Williams 1989). The expected density is in the order of several hundreds of km per km^2. It is thus clear that the accessible conduits only represent a fraction of all karst conduits.

These conduit systems are made of both old, fossil and recent, active conduits. Obviously the currently active ones are the most important for understanding the contemporary hydrogeological situation. Active conduits are defined as the "flowing conduit network". This system includes all sizes of conduits. Therefore, during the evolution of an overall karst conduit network, a succession of flowing conduit networks has been active. The presently active flowing conduit network is the latest one of the series. The estimated conduit density of flowing conduit networks, based on speleological observations and confirmed by hydrogeological models, is in the order of 2 to 15 km/km^2. It should be kept in mind that the part that can be explored (the speleological network) commonly does not correspond to the part, which is of direct interest in hydrogeological investigations (the flowing conduit network). Observations made in accessible caves may or may not be representative of the complete system. There can thus be some danger in extrapolating conclusions drawn from the accessible passages to the behaviour and characteristics of the whole aquifer.

Thus, caves are not always present, not always accessible, and not always representative of the present flow system. However, in some cases a significant proportion of the underground water may flow in the explorable conduits, there, the cave data can be extremely important.

Many methods applied in caves are adaptations of methods described elsewhere in this book (e.g. tracer tests), while others are unique for caves (e.g. exploration and mapping of the vadose (unsaturated) and phreatic (saturated) conduit network, measuring drip waters, and analyses of cave sediments). This chapter focuses on the techniques unique for caves.

3.2   USE OF CAVE SURVEY DATA FOR HYDROGEOLOGY

In order to use survey data, caves have to be mapped. In this section, we first explain how caves are mapped, and how the data can be used for 3-D modelling and precise well emplacements. Then, we explain how the cave survey can be interpreted.

3.2.1   Cave mapping

Determining the location, lateral and vertical extent of a cave is done by exploring and mapping the cave. Cave survey is normally carried out by cavers concurrently with exploration of the cave. Standard cave survey methods are conceptually similar to those used in normal surface surveys, with modifications to accommodate the conditions found in caves. Surveying equipment must be compact and robust enough to withstand the often wet and muddy conditions encountered there, particularly when mapping underground streams and rivers. A standard survey technique is used by cavers in general, however there are some regional and international variations in the methods employed. From the cave entrance, a series of points (or stations) is measured and in some cases labelled within the cave. For the line between each survey station the distance is measured with a tape or a laser distance-meter, the slope with a clinometer, and an azimuth with a compass. Normally the dimensions of the conduit at the survey station are also measured or closely estimated to determine conduit width and height at the site. With these methods, under conditions typically encountered underground, it might be possible to survey several hundreds of meters in a day in large conduits, or only a few tens of meters in constricted passages. Under good conditions, the accuracy of this type of survey can be within 1% of the surveyed length. Under these conditions, the positioning errors of surveyed cave conduits would be on the order of 10 m per km of surveyed line from the entrance. Some cavers, however, may survey less precisely, and as a consequence, errors of 5 to 10% may be observed in some systems.

This degree of accuracy may be not sufficient for particular purposes, for instance well placement. One method to improve the accuracy is to use theodolites, although the tubular and often chaotic pattern of caves, not to mention the less than favourable physical conditions, do not lend themselves to convenient survey with a theodolite. This technique can also be very time consuming. Depending on the effort dedicated to these measurements, though, the precision ranges between 0.1 and 0.01% of the distance to the cave entrance.

Another way to determine an exact underground location is to use electromagnetic positioning systems (see section 3.2.3 below).

Several specialized computer programs have been written to process and plot the raw cave survey data in order to produce a map, including *Toporobot* (Heller 1983), *Visual Topo* (David 2006), and *Survex* (Betts 2004) which are popular in Europe, as well as *COMPASS* (Fish 2004) and *Walls* (McKenzie 2004), widely used in North America. These programs use trigonometric relationships to convert the polar coordinates of the raw data to three-dimensional Cartesian coordinates. The programs also correct for small survey errors using loop closure algorithms. They can plot plan views and longitudinal sections of the caves at any scale as well as 3-D sketches. They can also export data to usual 3-D formats and plot survey stations with the respective wall position in plan view and section. Increasingly, programs are designed to export files in formats that can be directly imported into Geographic Information Systems (GIS) software that offers powerful data management, analysis, and visualization capabilities.

Such computer-generated maps (usually line plots or volumetric plots) are not, in isolation, adequate to describe caves and to report observations. Cave surveys have to be systematically complemented by cave cartography.

As with surface topography, the cave maps are required to represent the main characteristics of karst conduits. To produce such maps, a running sketch is drawn along the line of survey. The cave survey and passage dimensions collected at each station make it possible to draw the sketch to scale. Because a cave is a 3-dimensional entity, the plan view alone is not sufficient to represent a cave. Therefore a profile along the conduit (called a longitudinal section) is added, as well as a series of cross-sections normal to the passage.

The programs plot the survey points, which provide a reference on which to draw in the positions of the cave walls and other features using standardized symbols (Fig. 3.2). These symbols are defined at an international level by the *Union Internationale de Spéléologie* (UIS). Drawing of cave maps (cave cartography) is not a simple, straightforward process, especially for complex cave systems. It is suggested that experienced cave cartographers are contacted if a map is needed.

A written description of the mapped cave is another important source of information in association with maps and longitudinal sections. Information about cave biology, flooding conditions, genetic observations, and other time-dependent information that cannot be drawn on a map, is reported in this way. Unfortunately, in many cases only a plan view of the cave is available. However, one has to consider that the trilogy: map – longitudinal section and cross section – description, should represent a standard documentation for all cave systems.

3.2.2   Cave modelling in 3-D

Cave survey data make it possible to represent cave networks in three dimensions (3-D) on computers. This is often the only way to visualise the geometry of cave networks. Shown together with surface topography and the main elements of the geological setting, this type of image can provide a very clear understanding and visualization of the cave system under investigation (Fig. 3.3). Unfortunately for book-readers the 3-D effect is most apparent when the user moves across the 3-D-scene. For hydrogeology this type of 3-D representation can be very useful in order to understand the conduit development in a particular region.

More spectacular visualization techniques became available at the end of the 20th century and continue to evolve. Laser scanners and computer 3-D visualisation make it possible to represent a cave in 3-D with a very high degree of precision. 3-D laser scanners allow for a detailed and precise survey of the walls and floor. By scanning the cave from many viewpoints, it is possible to locate cave walls with millimetre precision. Photographs of the walls (or floor) can then be projected onto those virtual surfaces, producing an almost exact reproduction of the cave, and allowing any virtual analysis without damaging the real cave (Fig. 3.4). At present, this technique requires specialists and a high level of technology, and is not useful for most applications in hydrogeology. The techniques are applicable mainly in large and easily accessible voids.

Another technique is to use a 2-D laser scanner and to scan many profiles at 0.3 to 1 m intervals. The profiles are positioned in 3-D and interpolated to make a reasonably accurate model of the conduit. These scanners are less fragile than the 3-D type and the technique requires less precision in the positioning of the laser itself than for the 3-D

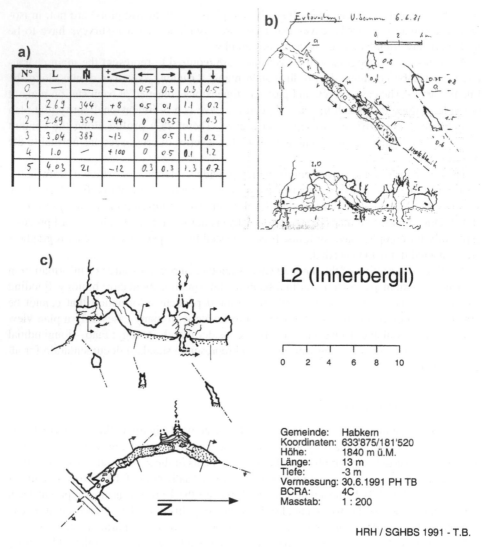

a)

N°	L	⩘	±<	←	→	↑	↓
0	–	–	~	0.5	0.3	0.3	0.5
1	2.69	344	+8	0.5	0.1	1.1	0.2
2	2.69	354	-44	0	0.55	1	0.3
3	3.04	387	~13	0	0.5	1.1	0.2
4	1.0	–	+100	0	0.5	0.1	1.2
5	4.03	21	~12	0.3	0.3	1.3	0.7

b) Entdeckung: U. Sommer 6.6.81

c)

L2 (Innerbergli)

0 2 4 6 8 10

Gemeinde: Habkern
Koordinaten: 633'875 / 181'520
Höhe: 1840 m ü.M.
Länge: 13 m
Tiefe: -3 m
Vermessung: 30.6.1991 PH TB
BCRA: 4C
Masstab: 1 : 200

N ⟶

HRH / SGHBS 1991 - T.B.

Figure 3.2.   Cave L2 at Innerbergli (BE, Switzerland); a) survey data, b) field sketch, c) final map. Note that a plan view as well as a longitudinal section and three cross sections are presented. The written description accompanying that map is not shown here.

laser scanner. Therefore, this technique is more adaptable to the cave environment and can provide meaningful results.

3.2.3   Precise cave positioning and well placements

A useful product of cave surveys is to make it possible to undertake water sampling at specific locations of importance within underground rivers, either directly by accessing the aquifer for sampling, or by placement of monitoring wells. Direct sampling, while useful in many circumstances, can be limited. Storm pulse sampling, for example, might

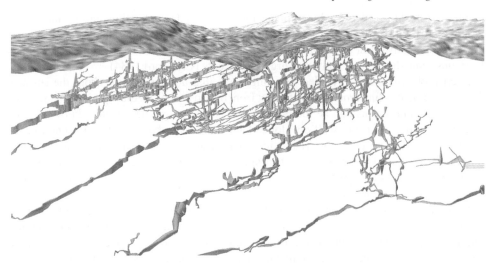

Figure 3.3.   3-D image of the Siebenhengste area (Switzerland). The land surface above (wavy surface) is underlain by a complicated maze of cave passages. Because of the perspective view, a scale cannot be given. The width of the main cave maze is around 3 km.

Figure 3.4.   3-D view of an underground chamber. This 3-D model has been obtained by combining 3-D and 2-D lasers canner techniques (by ATM3D/SISKA 2006).

be impossible if the active conduits flood to the ceiling, and travel within the cave might also be made difficult by the presence of volatile contaminants affecting the cave atmosphere. Passages beneath urban areas have been shown to be contaminated at times with fumes from gasoline, diesel fuel, and other organic solvents (Crawford 1984).

Such problems can be avoided by using wells that allow sampling from the surface. Their placement is fundamentally different from that in porous media aquifers where broad plumes of contaminants can be intersected by wells down-gradient from a known source of contamination. In many well-developed karst aquifers, tributary networks tend to concentrate flow into larger order streams in a manner analogous to surface stream systems. Sampling of a karst aquifer must therefore take place within active conduits, and wells must intersect those flows. Wells that miss an underground river conduit laterally by even a few meters can be totally ineffective.

Several methods exist for determining the location of such wells. In many cases traditional cave surveys must be improved, because drilling a well into an underground river generally requires a high degree of precision, often within a meter or two. Cave radios, which are actually magnetic induction transmitters, can be used from a cave location with a receiver on the surface to determine a surface point directly above (Glover 1976, Gibson 1996, 2001, France 2001). A transmitter antenna in the cave is placed horizontally, and an amplifier introduces a very low frequency signal into the antenna. A signal within the receiving antenna is induced by the resulting magnetic field. By varying the position of the receiver it is possible to identify the geometry of the magnetic field and in particular the lateral position of the transmitter below. The depth between the two antennas can also be determined by a number of methods, including measurement of the signal strength and geometrically using the shape of the magnetic field generated by the transmitting antenna.

The expected precision of these systems is in the order of 0.5 to 1% of the cave depth below the surface. For shallow caves (less than 50 m), the positioning is thus quite precise. Usually these systems can be applied up to a depth of 200 m.

There are times when it may be desirable to drill a monitoring well into an underground stream that is not accessible to humans. In that case, the survey and radio techniques are not feasible. There are evolving geophysical methods for use in karst settings (e.g. Butler 1984, Crawford et al. 1999, McGrath et al. 2001, Paillet 2001, Hoover 2003) including those used for location of subsurface cavities. When combined with test drillings, they can sometimes allow well placement for groundwater monitoring within stream conduits. Geophysical methods useful for karst investigations, discussed in more detail in Chapter 8 of this volume, are associated with seismic refraction, electrical resistivity, electromagnetic terrain conductivity, ground penetrating radar, and gravity. None of the existing methods really make it possible to detect a cavity located deeper than 20–30 m, unless the conduit volume is large.

3.2.4   Interpretation of cave surveys

Five main types of interpretation of interest for hydrogeology can be derived from cave maps:

The conduit pattern (giving insight into possible recharge conditions). This was discussed in detail by Palmer (1991). It was shown that conduit patterns give a basic insight not only into the dominant type of porosity, but also about the type of recharge that prevailed during the genesis of the cave (Fig. 3.5).

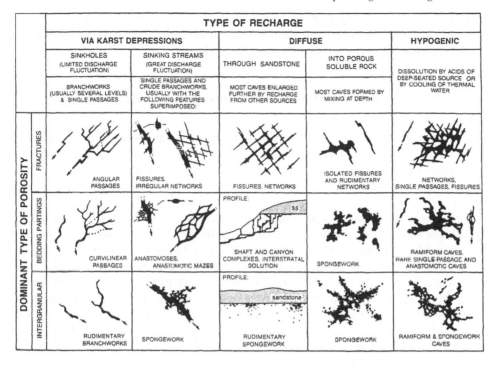

Figure 3.5.   Interrelationship between cave patterns, type of porosity, and type of recharge (from Palmer 1991, by permission).

The position of the known active conduits and their characteristics. Information about major active conduits is usually clearly visible on cave maps, but indications about high water flow conditions might not be apparent. If additional information is present in a written description, valuable insight into the karst groundwater behaviour might be gained. For example, observations in St. Beatus Cave (Switzerland) proved that vadose rivers may bifurcate; rivers may cross each other (due to an impervious layer between), and even might flow across "aquiclude" layers (Häuselmann 2005). These types of information are not accessible from surface observations, and can only be obtained from detailed subsurface observations (see also section 3.1).

The position of known infiltration inlets. Section 3.4 presents investigations that can be carried out at infiltration sites.

The characteristics of present and past flow conditions in known conduits. These can often be derived based on three major indicators: conduit morphology (conduit cross-section), micromorphology, and cave sediments. A good cave map-section-description can provide sufficient information to distinguish between conduits created under vadose and phreatic conditions, and to reconstruct the major phases of the evolution of flow conditions. In some cases cave maps may allow an interpretation of flow direction and discharge assessment. In many cases they can be helpful in locating sites for more detailed observations. However, direct investigations and measurements in the caves are often necessary to provide data relevant to a particular hydrogeological investigation.

The position of the expected active conduits and their characteristics. Existing speleological data can be used in order to generate an unknown karst conduit network and to extrapolate it in space. In most karst hydrogeological studies, an understanding of the positions of the main flowpaths are desired. This is especially important when trying to make a numerical model of karst systems. See below.

3.2.5   The position of the expected active conduits and their characteristics

3.2.5.1   *Statistical methods*
The principle is to describe the distribution of known conduit networks in space and to generate the positions of unknown conduits with the same distribution. This method does not typically provide good results because of the inherent difficulty of defining the basic characteristics of the conduit networks, particularly the connectivity (Jaquet & Jeannin 1993).

A random walk technique has been applied with more success (Jaquet 1995). This method includes inputting the dissolution process in a simplified manner into a statistical generator. The principle is to use a grid of pathways (in 2-D or 3-D) and to inject particles at the upstream part of the grid. At each time step the particles move forward to a neighbouring node using one of three (in 2-D) or five (in 3-D) possible pathways. The karst process is included by assuming that the probability of using a given pathway increases according to the number of particles that have already travelled through this particular flowpath. The number of particles is assumed to represent the size of the conduits. This process induces a converging structure of the generated networks, which is quite realistic. This method provides interesting results, however it should be validated with comparisons to speleogenetic models and real systems, which has not been accomplished so far.

3.2.5.2   *Fractal methods*
Caves have been shown to display some fractal characteristics (Curl 1986, Jeannin 1992, 1996, Kincaid 1999). Several geometric relations that have been quantified including network structure (typical fractal dimension of 1.67 for karst networks), tortuosity (typical fractal dimension of 1.09 for karst conduits), and conduit size (the fractal dimension should be close to 2.79). For the generation of conduit networks using fractals, one can use a "Y" structure with a 60° opening angle (corresponding to a fractal dimension of 1.76) and iterate 3 to 10 times in order to get a branching system. As each segment will be straight, a tortuosity generator can be applied to each segment as well as a size generator. As pure fractals lead to highly symmetric and regular networks, a random process has to be added in order to get images that appear more similar to natural karst networks. Unfortunately, it is very difficult to relate simulated model networks developed using such fractal methods to observed data.

3.2.5.3   *Empirical methods*
In some cases, conduit networks have been extrapolated empirically. For example, in a modelling approach, Kovacs (2003) expanded the Milandre conduit network up to an entire aquifer scale, applying a similar pattern as the one of the known cave. He then verified that the virtual system produced consistent hydraulic responses. A similar approach had been previously used by Mohrlok et al. (1997) and Mohrlok & Sauter (1997). This method is time consuming, mainly because it is often only constrained by flow model results, i.e. the geometry is improved by an iterative "trial and error" process.

3.2.5.4 *Speleogenetic models*

There are now many models, which include the coupling of flow and calcite dissolution in order to simulate conduit growth (Groves & Howard 1994a,b, Howard & Groves 1995, Siemers & Dreybrodt 1998, Kaufmann & Braun 2000, Sauter & Liedl 2000, Palmer 2000, Dreybrodt & Gabrovsek 2000, Dreybrodt et al. 2005). The respective models use different solution schemes (for example: finite elements, finite differences, discrete fractures, dual-porosity, and matrix-fractures), leading to generally comparable results. These models are the most complete ones. They provide interesting results that can be useful for understanding constraints on aquifer evolution, however they still remain quite distinct from observed conduit networks. Also, the sensitivity to some parameters appears to be quite extreme, and, unfortunately, it is difficult to estimate them in the field. These models should be applied to real and well documented cave systems in order to provide comparisons between model results and existing field data. This has not been seriously attempted up to now.

3.2.6 Cave genesis and identification of phase development in cave networks

In some hydrogeological investigations of karst aquifers it may be desirable to obtain an understanding of the evolution of the aquifer through time by identifying conduit systems that operated in the past (possibly under different environmental conditions) but are now inactive.

Usually, cave systems are the result of a succession of active conduit networks that evolve through time (Klimchouk et al. 2000). It has been demonstrated that phreatic conduit systems may develop within a few thousand years if recharge is sufficient (Palmer 2000), especially if initial openings are wide enough (>0.1 mm). In many mountainous karst areas those conditions are fulfilled, and active conduits normally are located close to the valley bottoms. In mountain ranges it can be assumed that uplift is more or less compensated for by the entrenchment of valleys, therefore new conduits have to form each time the valley floor is entrenched further, previous (higher-level) conduits become only temporarily active, and later are eventually permanently abandoned by the water. Most of the caves explored by cavers are part of those dry (fossil) systems. This simple model may be more complex if, for any reason, the base level was raised at any time in the past (Audra et al. 2004).

Observations of conduit morphology, major flowpaths, flow directions, scallops, and sediment in large cave systems can make it possible to reconstruct in detail the flow conditions prevailing in the conduit network over millions of years of conduit evolution. It may be possible to distinguish several paleoflow systems and to characterize flow conditions in each of them. In some cases the respective systems and their levels can be dated using speleothems (U-Th) or quartz grains (Al-Be) (Granger et al. 2001, Stock et al. 2004, 2005, Häuselmann & Granger 2005). For this type of study 3-D visualisation of cave networks is very useful for identifying phases in cave genesis.

This work requires detailed observations in caves and thus much fieldwork. However, the potential for paleo-environmental reconstruction, including paleo-hydrology, is very high. An instructive example is given by Häuselmann (2002), where the genesis of the Siebenhengste-Hohgant cave system (Switzerland) was elaborated and partially dated, giving information on valley incision, glacial advances and retreats, and adaptation of the flow systems within the cave. Other examples include a study of the speleogenesis of the Picos de Europa (Spain) (Fernandez-Gibert et al. 2000), the Agujas Cave system (Rossi et al. 1997), and the Mammoth Cave System and adjacent Green River Valley (Granger et al. 2001).

Figure 3.6.   To the left, a picture of a typical shaft (photo R. Wenger); in the middle is shown a typical meandering canyon (photo R. Wenger); to the right, a phreatic conduit (photo P. Morel). It is clear that the passage to the right has formed while the valley bottom was above or close to the conduit level.

3.2.7   Availability of cave data for hydrogeologists

Cavers have produced a large volume of work by surveying, mapping and documenting caves. By far, most of this work has been done on a voluntary basis and published in caving journals. The UIS has a list of national caving societies, which can provide addresses of cavers active in any part of their country (www.uis-speleo.org).

3.3   INVESTIGATION METHODS SPECIFIC TO THE CAVE ENVIRONMENT

3.3.1   Analysis of conduit morphology

The analysis of conduit morphology (conduit cross-section) makes it possible to characterize past and present flow conditions in conduits. There are three main types of conduit cross-sections, each of them characterising a given type of flow when the conduit is (or was) active (Fig. 3.6). Vertical **shafts** (or pitches/pits) form in the unsaturated zone by water cascading vertically down through fractures, continuously enlarging them. The diameter of shafts is typically significantly larger than the conduits feeding and draining them. **Meanders** (or meandering canyons) are high and narrow passages entrenched by vadose streams. **Phreatic conduits** usually have an elliptical cross-section (expanded either horizontally, oblique, or vertically). They form in the saturated zone or slightly above (epiphreatic zone, where floodwaters still reach the passages). Sometimes the transition where a meandering canyon (vadose passage) changes into a phreatic conduit (saturated zone) can be found. This transition point defines an ancient water table.

In phreatic conduits, the cross-section of the passages is often quite constant along the flowpath. This sometimes makes it possible to follow paleo-flowpaths for several kilometres.

Micromorphological features of cave passages may be important. For example, scallops can be used to determine the direction of past water flows (Ford & Williams 1989).

3.3.2   Analysis of cave sediments

Sediments offer a wide range of methods for determining the paleoflow direction and velocity. In addition, the presence or absence of certain sediments may also indicate the discharge present at the time of sediment deposition.

Dripstones and flowstones usually only form above the water table. In addition, their presence usually suggests rather low discharge.

Detrital sediments offer a wide range of paleoflow information. The geometry of elongated pebbles that form an imbrication, or of sand ripples and dunes, indicate the direction of flow. The position of sediments can also give directional information. For instance, fine sediments tend to accumulate upstream of a constriction in the cave passage. The constriction causes the flow rate to increase, thus the passage lacks sediments, whereas the passage enlargement farther downstream decreases the flow rate, and coarser sediment is re-deposited.

White (1988) described mechanisms for fluvial sediment transport. The size of the sediment particles found on the floors of conduits can be used to assess a flow velocity according to the Hjulstrom (1939) diagram (bed-load). Discharge rates during the active phase in a phreatic conduit can therefore be estimated considering the conduit cross section. This method is quite useful for the interpretation of cave sediment profiles, making it possible to estimate the evolution of discharge rate through time.

In some cases, values obtained by such indicators can be validated by direct measurements of the flow velocity (Jeannin 2001).

3.3.3   Water tracings within caves

Often, caves allow access to active conduits. In some cases, however, they cannot be followed due to constrictions or to the presence of water-filled passages. Water tracing with fluorescent dyes and other substances within the cave systems can identify locations of, and flow conditions within, underground flowpaths that cannot be directly observed and surveyed (see chapter 8 in this volume). Moreover, it might be useful to subdivide the total response from surface to spring into discrete portions within the caves, in order to fully characterize each portion of the flow system. For example, this has been extensively applied in Milandre Cave, where each section of the underground stream could be characterized (flow velocity, dispersivity, e.g. Hauns et al. 2001). In Mammoth Cave, the technique has also been used (Meiman et al. 2001).

3.4   MONITORING WATERS IN CAVES

3.4.1   Introduction

This section outlines methods specific to hydrological measurements in caves. Other methods of flow measurement, which are not specific for caves, are described in chapter 4. Interpretation methods for head and discharge measurements in underground streams or in (epi)phreatic conduits is also presented there. This section focuses on the characteristics of drip and seepage water, i.e. of infiltration through the unsaturated zone.

Shallow caves (typically between 2 and 50 m below the surface, but sometimes deeper) can provide useful observation points for the characterization of flow and transport in the infiltration zone of karst aquifers. Classically, this zone is divided into three subsystems, each having their own characteristics: soil, epikarst (or subcutaneous zone, Williams 1983, Perrin et al. 2003a,b, Groves et al. 2005), and vadose zone (White 1988).

Soil may be absent, thin, or as thick as several meters. Epikarst is characterized by a layer of relatively highly weathered limestone at or near the soil/bedrock interface, having an increased porosity and permeability compared to the massive limestone below (Williams 1983, Klimchouk 1997). The vadose zone often contains vertical shafts conducting water from the epikarst down to the main karst conduit network. Flow through the fissured limestone surrounding shafts is slow compared to shaft flow, but seems to represent a significant part of the base flow.

Caves that allow observations in the infiltration zone are usually fossil conduits stretching more or less horizontally. They intersect fissures and shafts, making it possible to characterize flow regimes and water chemistry of the respective water inlets. This gives direct access to information about recharge, which is not accessible in other types of aquifer rocks without drilling tunnels. In many cases, this type of observation makes it possible to identify and precisely locate the origins of contamination.

3.4.2   Characterisation of flow in the infiltration zone

3.4.2.1   *Site selection and instrumentation*
There are two extreme types of flow regimes of the water inlets into caves in the vadose zone (Smart & Friederich 1986, Perrin et al. 2003a):
1) *Vadose flow* with very large variations of discharge, i.e. almost no flow at low water and up to several hundred l/s during floods;
2) *Seepage flow* with highly buffered responses to recharge, i.e. no peaks and only an annual cycle with discharge varying by a factor of 2;
3) *Intermediate or combined responses.*

Before installing instrumentation in a cave, it is important to decide what type of water inlet is relevant for the purpose of the study.

Vadose flow is difficult to measure with a high degree of precision mainly because of its great variability (usually 2–3 orders of magnitude). The first difficulty is to collect the complete amount of water into the gauge station. The best way is to fit a "cover" to the conduit walls. In some cases a gulley has to be constructed in order to collect the water correctly. Prior to setting up such a collecting system, it is advised to visit the site once during very high water conditions.

The gauging station may need to be quite complex. For low discharge a rain gauge station (tipping buckets) usually covers the range of observed discharge rates. However, under medium to high discharge conditions the flow rate is much too high to measure in this way. Only a weir-type system can measure such high rates. For drip water measurements at the Milandre cave site (Switzerland), Perrin (2003) developed a gauging station consisting of a vertical, perforated cylinder that is open at the top and in which a pressure probe records the water level variations that reflect variations of the flow rate. At another vadose flow epikarst monitoring site in Cave Spring Caverns, Kentucky (USA), Groves et al. (2005) used a tipping bucket mechanism that clearly missed some of the falling water at the site, especially during the higher flow conditions, but showed regular, relative relation to the flow conditions. In order to relate the measured flow rate to the actual discharge rate, they periodically measured the entire discharge of the water using a tarp and timing the flow into a container with known volume. They then used this information to construct a rating curve that related discharge to the measured rate of the tipping bucket at any time (also see chapter 4).

Seepage flow is easier to measure because of its smaller variations. The collecting system is basically the same. A rain gauge station (tipping buckets) is usually adequate.

In some cases it may be of interest to measure discharge rate variations from one single stalactite. Drop counting methods have been developed for this purpose (e.g. Yonge et al. 1985, Genty & Deflandre 1998, Tooth & Fairchild 2003, Musgrove & Banner 2004, see also addresses at the end of this chapter).

3.4.2.2 *Interpretation methods*
Classical hydrological methods can be applied to such data, including use of water budgets, recession analyses, variation coefficients, correlation and spectral methods, wavelets, and others.

The first interpretation might be to classify a flow inlet of interest into either a vadose, seepage, intermediate or conduit flow type. Conduit flow is similar to vadose flow but with a discharge rate larger than 1 l/s at high water stage. Seepage flow is assumed when no peaks are visible besides the annual cycle. Conduit flow is assumed to be in a quick relation with the ground surface (i.e. can be quickly polluted, but will not accumulate pollution). Seepage flow is assumed to have the opposite behaviour.

The shape of curves usually shows that larger inlets are more a combination of vadose and seepage than a pure vadose flow with a larger discharge rate.

Water budgets are used to reconstruct the size of the catchment area and to give a first idea of where the water comes from (e.g. White 1988, Perrin 2003, Ray & Blair 2005).

This model (Fig. 3.7) gives a possible interpretation of the reservoir cascades explaining measured responses of seepage, vadose and intermediate flow in the infiltration zone of Milandre Cave. ST, VI, EN, SO, EC, AM, BU, FA, RO refer to various observation points in the cave (from Perrin & Kopp 2005).

Further interpretation methods can be applied if quantitative interpretations are required. Input-output signal analysis methods will clearly show that the relation between the rain and the output discharge rate is neither linear nor stationary (Perrin 2003, Perrin & Kopp 2005). In most cases the interpretation is further enhanced by comparing the flow response to signals given by one or more solute transport parameters.

3.4.3 Characterisation of transport in the infiltration zone

Investigations of infiltration water are often conducted in order to characterize water quality (Plagne 2000). For example, in relation to groundwater protection zones, it is important to know if soils and epikarst protect groundwater from a given pollution risk.

The characterization of water quality in the infiltration zone often focuses on understanding the flow organization (e.g. Fig. 3.8). Classically, two cascading sub-systems are assumed: soil and epikarst. Each sub-system has storage and a fast flow component. The epikarst has a perennial outflow at its base and a temporary outflow at a given height. Boundary conditions of the infiltration zone are given by input and output water characteristics (discharge rate and tracer concentration). The significance of the respective components can be understood because tracers have specific reactions or transformation

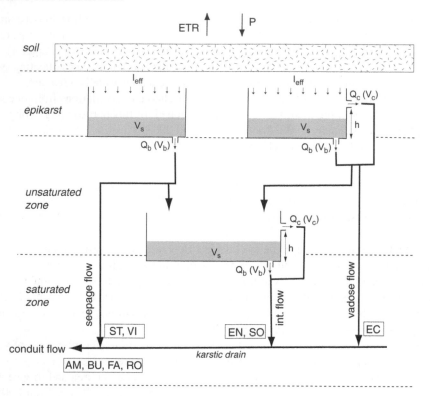

Figure 3.7.   Possible interpretation of flow responses; for explanation see text (from Perrin 2003).

within the sub-systems (e.g. TDS, nitrates, stable isotopes, radon or TOC have highly contrasting behaviour in soil, epikarst or shafts). Therefore, it is possible to characterize the dominating elements of a water inlet by studying its hydrographs and chemographs (c.g. Fig. 3.8, Perrin 2003, Perrin et al. 2003b).

However, as infiltration in karst is highly heterogeneous in space (from one water pathway to the other) and in time (depending on the flow conditions), it is not straightforward to extrapolate the results at catchment scale.

The selection of water inlets depends on the aim of the investigation. In a first approximation, water inlets with strongly dampened flow reactions can be considered as displaying a more stable water quality. However significant exceptions are known. It is recommended to sample one or two flood events with an hourly time interval. Continuous measurements of some parameters (temperature, conductivity and possibly others) are recommended. Water inlets with strong discharge variations may be expected to display strong chemical/isotopic variations as well. Again, exceptions are known and in those cases it is absolutely necessary to sample/measure tracers at short time intervals. As it is expensive and time consuming, it is better to sample those for only two or three flood events, rather than to sample on a weekly or monthly time step. The latter method may miss important variations. The application of electronic data loggers in these kinds of studies has been a major technical development in the last decade, and continues to evolve. These methods are discussed in chapter 4. Alternatively, "integrated sampling" systems can be used for tracking the "average" behaviour of the water outlet.

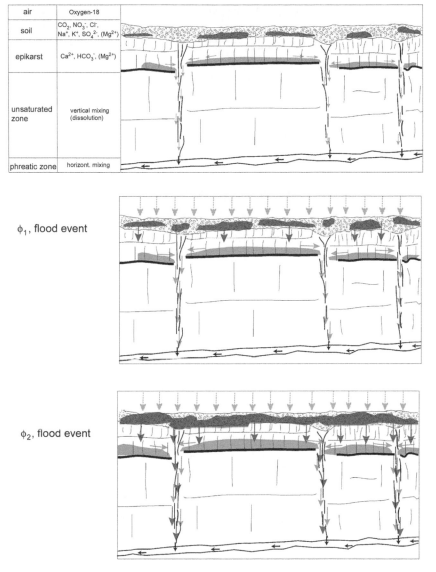

Figure 3.8.   Conceptual model of flow and transport in the unsaturated zone of karst systems (from Perrin 2003). See detailed explanations in the text.

Water sampling and measurement techniques are highly dependent on the tracer. For dissolved gases (e.g. radon, CO_2, and oxygen) it is important to sample as close as possible to the "spring point" at the cave wall or roof and to keep the samples sealed. Conservation and calibration problems have to be given serious consideration before starting measurement in caves, because some of these problems are exacerbated, especially when electricity is not available.

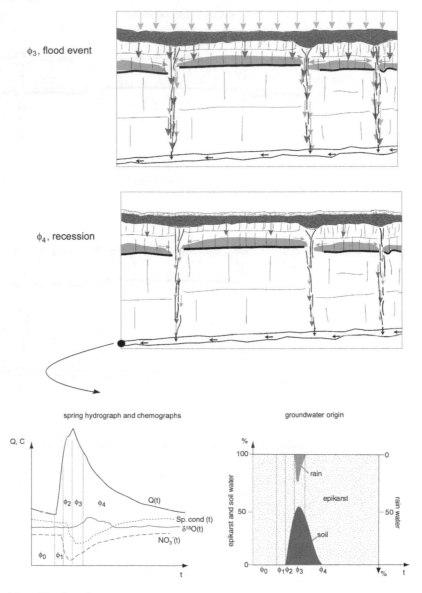

ϕ_3, flood event

ϕ_4, recession

spring hydrograph and chemographs

groundwater origin

Figure 3.8.   (Continued)

Each natural tracer has a specific behaviour. Therefore interpretation of the measured signals has to be adjusted. At least three main groups of interpretation approaches can be distinguished (adapted from Perrin 2003):

1. Conservative tracers: The best example is the ^{18}O isotope, which comes from rain and flows through the system with (theoretically) very little exchange with the medium. The response of this tracer in the cave compared to the input concentration characterizes the mixing between fresh infiltration water and water previously stored in the system. Observations made by Perrin et al. (2003a) showed that ^{18}O responses are always highly dampened, even in water inlets presenting strong and quick reactions. This is

assumed to be due to the mixing of fresh infiltration water with water stored in soil and epikarst. This storage seems to be as significant as 100 mm (or more) in temperate karst systems.

2. Other conservative tracers are anthropogenic, such as salts with chloride (Cl^-), Bromide (Br^-) and Potassium (K^+) and under certain conditions fluorescent dyes (e.g. fluorescein). The use of these tracers in test-sites can confirm the mixing of infiltration with large amounts of storage water (typically on a long duration input of a dye tracer, recovery concentration in a cave 15 m below the injection surface is in the order of 20% of the input concentration). Nitrate is quite conservative, although under some conditions significant concentrations of background nitrate are present, having been created in the soil and epikarst by the oxidation of ammonium.

3. Reactive tracers: Most chemical tracers are reactive, as they may be transformed by one or more processes (e.g. adsorption and/or participation in reactions) along the flowpath. Fortunately some naturally occurring chemical species that can be used as tracers are typically found in a given karst catchment sub-system. For example radon or CO_2 occur naturally in the soil, but more calcium and bicarbonate (contributing to the total dissolved solids (TDS) and conductivity) are introduced from the epikarst. Therefore, comparison of radon, CO_2, TDS, and conductivity responses to a recharge event makes it possible to distinguish inputs of soil and epikarst waters (Savoy 2002).

A detailed study of these respective tracers made by Perrin (2003) lead to the conceptual model of flow and storage in the unsaturated zone of karst systems presented in Fig. 3.8. Five phases can be differentiated:

In low-water conditions (Φ_0), the system is fed by waters stored within the epikarst. Discharge and chemistry are stable. Water is still stored in the soil, but held by the capillary barrier.

During phase Φ_1 rainfall has started and soil water is pushed down into the epikarst, which pushes further epikarst water into the system. Therefore, the system is mainly fed by epikarst water, i.e. the discharge rises, but the chemistry remains stable. If the rain stops, discharge decreases and no chemical signature is visible.

If the rain continues, the system reaches phase Φ_2, in which soil water bypasses the epikarst reservoir and directly reaches the karst conduits. In this situation, the system is fed by epikarst and soil water. If soil water has spatially heterogeneous characteristics (e.g. in polluted regions), these parameters will display concentration changes. Usually soil water is highly unsaturated with respect to calcite, i.e. dissolution-related parameters including calcite saturation indices and dissolution rates will be impacted. Stable isotope values will remain constant because soil and epikarst water usually have very similar values.

If the rain continues, the system may attain phase Φ_3, in which the soil's capacity is attained, i.e. some freshly infiltrated water bypasses the soil and epikarst reservoirs and directly reaches the main conduit system. This corresponds to strong flood events. In this situation, stable isotope values will change, because the isotopic concentration of rain water usually differs from the values from soil and epikarst.

Phase Φ_4 corresponds to recession after the flood event.

In this model, the water storage in the phreatic system is considered to play a negligible role in water chemistry. This is supported by field observations (Perrin 2003, Perrin et al. 2003b). The flow conditions in the unsaturated zone, especially the role of the soil and epikarst storages, is assumed to be dominant.

3.5 SUMMARY

The utility of the cave-based methods described in this chapter, in relation to measurements at springs and boreholes, may be summarised as follows:

Drilling of monitoring wells in karst areas is expensive and often ineffective, as the probability that the wells will not intersect the flowing pathways is high. Drillholes may therefore have limited value for the delineation of catchment areas or for assessing flow velocities (Worthington 2003). In contrast, in many cases caves can provide access to perennial rivers, giving access to major flowpaths for sampling and measurement.

Springs often offer very useful information, but these measurements represent an average of the whole catchment area. Caves that provide access to multiple rivers and tributaries allow sampling that provides information about sub-catchments. Moreover, the presence of active caves may give sufficient information about the flowpaths, so that water-tracing experiments (needed for that information without caves) may be omitted in some cases. In other cases, even when underground rivers are accessible, other areas within the aquifer are not, and a combination of direct travel and dye tracing provides a source of integrated data.

Caves are accessible to humans, observable, and quantifiable; wells are not directly accessible and observable; most information in boreholes has to be interpreted from indirect data. Caves can offer the only place where groundwater circulation can be observed and continuously monitored in two or three dimensions.

The map of an active cave provides information about the actual pathways of the water. Observations in caves may explain some seemingly "strange" responses of tracing experiments through direct observation of the flowpaths.

The basic geometry of a cave provides information about its genesis and thus about water flowpaths prevailing at that time. This may be useful in understanding the present situation.

Cave morphology, micromorphology, and sediments provide information about past flowpaths, but also about possible high-water level flowpaths. This information is obtained neither from springs nor from wells. Moreover, some sediments can be dated. They can thus give information about the velocity of changes in the flow system and in the outside landscape evolution.

Cave maps may inform about folds, faults, and impervious layers which might not be seen at the surface due to vegetation, covering rocks, or human construction. This is very useful for defining the geometry of the limestone aquifer.

As a general summary, it may be said that whilst hydrogeological observations in karst areas can be done without the information available within caves, if caves are present and accessible, they can offer great sources of information and should be utilised to the extent possible.

Addresses of people using cave radio
–   Center for Cave and Karst Studies, Western Kentucky University, 1906 College Heights Blvd. 31066, Bowling Green, KY 42101, USA, http://www.caveandkarst.wku.edu/
–   Swiss Institute for Speleology and Karst Studies SISKA, c.p. 818, 2301 La Chaux-de-Fonds, Switzerland, http://www.isska.ch/
–   BCRA Cave Radio and Electronics Group, The Old Methodist Chapel, Great Hucklow, Buxton, SK17 8RG, United Kingdom, http://bcra.org.uk/creg/

Address for automated drip counters
–   http://www.driptych.com/

CHAPTER 4

Hydrological methods

Chris Groves

4.1 INTRODUCTION

As the nature and behavior of karst aquifers differ significantly from those of porous media and fracture flow aquifers, so do the hydrologic methods that must be employed to study them. Because of the extreme permeability, heterogeneity, and variations in hydraulic conductivity that commonly typify well-developed karst flow systems, many techniques commonly used to investigate porous media aquifers are of little use. Simultaneously, however, there are indeed peculiarities of karst aquifers that can be exploited. In many cases, for example, it is possible to actually travel into active stream conduits within karst aquifers to map out the geometry of these systems, and even to determine locations for monitoring wells (Fig. 4.1).

This chapter reviews methods used in hydrologic investigations of karst aquifers, geared towards readers with general experience in hydrology. The focus here is on the water itself, and how to determine its location, where it is flowing to, and how fast it is getting there. While there is considerable overlap with related methodologies including hydrochemical studies, fluorescent dye tracing, and geophysics, these are treated in more detail in separate chapters in this book.

Not all caves and karst systems have evolved in the "traditional" concept associated with water and CO_2 as part of the near-surface hydrologic cycle, including those that have developed from deep-seated sulphuric acid sources (Jagnow et al. 2000) and karst systems developed by simple dissolution of gypsum and halite (e.g. Frumkin 1994, Klimchouk 1996). In this chapter the focus is on the most widespread karst flow systems, those developed in carbonate rocks and dissolved by meteoric waters.

4.2 GENERAL HYDROLOGIC CHARACTERISTICS OF KARST AQUIFER SYSTEMS

The basic unit of scale for analysis of natural drainage systems, the *catchment*, can be defined both for surface and groundwater flow. Groundwater systems defined in this way can be called *groundwater basins*. However, differences between these in karst and non-karst settings highlight one of the major challenges of karst hydrology – while groundwater flow divides in non-karst regions can often be identified by analysis of topographic maps,

Figure 4.1.   Cave stream monitoring well, Logsdon River in the Mammoth Cave System, Kentucky USA (photo Groves).

karst groundwater can flow independent of surface topography. Evolution of larger, water-bearing conduits in the subsurface by dissolution can to some degree be decoupled from that of the topography above, and leads to the first challenge in any basin-scale karst hydrologic investigation – delineating the basin. This is done by three primary methods: direct exploration and survey of cave streams, potentiometric surface mapping, and one of the most important methods in karst hydrology, groundwater flow tracing with fluorescent dyes or other materials (covered in another section of this volume). Other methods to determine the provenance of water, including analysis of environmental isotopes and flood pulse analysis (Dreiss 1983) were reviewed in some detail by Ford & Williams (1989).

Groundwater divides in karst regions can have other complications in both space and time. While these are often indicated as two-dimensional lines drawn on maps, they are perhaps better characterized as complex, three-dimensional surfaces within the subsurface (Meiman et al. 2001). In many cases they can also change through time, dependent on the position of the water table. Conduits within the subsurface that during dry conditions are above the water table can become activated as the water table reaches that position, providing new routes that can send water across the previous drainage divides.

Recharge to karst aquifers can occur in several ways. *Autogenic* recharge results from precipitation directly onto a karst surface where carbonate rocks are exposed at this surface, which then infiltrates and enters the aquifer directly. In contrast, in more heterogeneous geologic settings where karst rocks may be adjacent to non-carbonate units, headwater areas may be dominated by surface flow which collects, and then upon reaching the soluble carbonates can enter the aquifer at discrete points, as *allogenic* input. These sinking streams, variously called *swallets*, *ponors*, or *swallow holes*, can be at single points where water enters distinct holes or cave entrances, or along a losing stream reach where alluvium covers the limestone, and stream flow can be seen to diminish incrementally down stream until at some point the channel is dry. Commonly, a number of swallets may be present in series along a stream channel, with the flow either disappearing into the most upstream one, or else entering several or all of them simultaneously, depending on the volume of flow at a given moment.

Within many karst aquifers, autogenic water moves through a shallow flow zone called the *epikarst* or *subcutaneous zone* (Drogue 1980, Williams 1983, Bonacci 1987). This occurs in the upper part of the carbonate rock body containing the aquifer, immediately below the soil-bedrock interface, where relatively acidic solutions have intensively dissolved fractures to create a zone of relatively high permeability. This can be perched upon a tighter interval, as this intense dissolution tends to decrease with depth, and is connected to the separate, phreatic zone below by occasional pathways through the vadose zone. Flow through the epikarst has been shown to be an important part of the overall transport of flow within karst drainage systems (Perrin et al. 2003, Groves et al. 2005), especially with regard to contaminant transport (e.g. Recker 1992, Wolfe & Haugh 2001).

Within the main portion of a karst aquifer, flow can take place through a continuum of permeability elements, which can be grouped into three types: primary porosity associated with the original interstitial spaces of the carbonate bedrock, fractures, and solutionally enlarged conduits (White 1988). Although there is of course a variety of carbonate lithologies within which karst aquifers form, two general end-members might be considered separately. On one end are the compact, highly indurated, commonly Paleozoic-aged rocks within which many of the world's great karst aquifer/landscape systems have formed (e.g. Bonacci 1987, White & White 1989, Yuan 1991). In these aquifers, where primary porosity is commonly a few percent, the majority of flow is through fractures and conduits, and capillary flow within the interstitial spaces is quantitatively negligible. In contrast, within younger limestones, particularly in coastal areas or on carbonate islands, that may have primary porosity that can exceed 50%, water movement through both conduits and interstitial spaces may be significant (e.g. Martin & Dean 2001, Martin et al. 2002, Screaton et al. 2004). Conduit flow can dominate where conduits are present, while where conduits are absent flow through interstitial spaces can dominate. In some cases, both are important simultaneously.

4.3   A BASIC CONCEPT: WATER BALANCES

In any discussion of hydrology and its methods, including that for karst, it is worth noting that if there is a single fundamental basis for hydrologic analysis, it is the concept of the *water balance*. At the drainage system scale, water is neither created nor destroyed, and thus we can undertake water budget analyses that in many ways are analogous to the workings of a bank account: put verbally, what goes in minus what goes out equals that which is stored. In karst drainage systems we can write a simple balance as

$$\Delta S = P - (Q_{out} + ET) \tag{1}$$

where P is the precipitation (input) into the drainage system, Q_{out} is the discharge leaving, ET is evapotranspiration, and ΔS represents changes in storage, which over longer timescales are often assumed to be negligible. While these are the basic elements of any basin budget, there are other possibilities depending on the detailed field situation, generally outlined in most introductory hydrology texts (e.g. Hornberger et al. 1998). These might include, for example, human withdrawals for drinking water supplies or irrigation, or leakage through confining layers to underlying aquifers. In some karst systems moreover, discharge may leave a basin through multiple springs in distributary outflow systems, which can change over time with seasonal and storm-scale variations in aquifer water levels.

There are at least two basic reasons for posing analyses in such terms. The first is that some terms are relatively easy to measure, or at least to reasonably estimate. These typically include, for example, precipitation inputs and discharge using methods described in detail below. In contrast others, such as leakage, are more difficult or impossible to directly measure. Since water is conserved, if it is possible to solve for a particularly difficult to measure term in the balance but measures or estimates can be made for the others, that component can be determined. The other issue is that balances of key chemical or biological constituents moving through karst drainage systems are closely related to water fluxes. Other budgets of interest might include inorganic carbon and other products of dissolved limestone in geomorphic and karst evolution studies, nutrients or organic materials that impact aquifer ecosystem function and health, and contaminants introduced to a system from agricultural or other land uses.

While the concepts of the actual measurement or estimation of hydrologic budgets in karst landscape/aquifer systems share features with those for non-karst catchments, there are indeed uniquely karst aspects. Perhaps the most problematic, introduced in the previous section, is that the recharge areas for many karst aquifers, in particular for extensive areas of autogenic recharge where rainfall lands directly onto exposed or soil covered karst landscape surfaces, cannot be delineated by surface topography alone as is typically possible in non-carbonate regions. Thus, before it is possible to consider a water budget for a karst flow system the recharge area must be delineated by adding information from tracer tests, cave exploration and survey, and potentiometric surface mapping (all of which are discussed in various chapters of this book) to topographic data. The established autogenic recharge area can then be combined with pertinent allogenic contributing areas where water is entering the karst aquifer at discrete points having gathered as surface flow on adjacent non-karst rocks. As the contributing allogenic areas are generally formed as fluvial surface drainage systems, their areas can typically be determined based on topography. The precipitation input term can then be determined by combining information on contributing areas with areally weighted precipitation and discharge data, discussed in section 4.5.

Another set of related, karst-specific aspects influencing water flow and thus budgeting follows from the intimate connectivity between surface and groundwater in well-developed karst situations, which makes the discrimination of these as distinct flow components much less apparent than in other flow systems. This means that in many cases the outflow point defining the catchment providing the basis for the areal extent of a system under study is a karst spring, rather than a confluence of rivers or some other arbitrarily determined point on a surface drainage network. Flow from such springs represents the integrated total of all remaining (that is, not having evapotranspired) flow that lands as precipitation on the allogenic and autogenic recharge areas, regardless of the proportion of time any component of this flow has spent on the surface or moving through the karst aquifer underground. Of course, since many karst springs serve as water sources, understanding their flow conditions also provides one of the major motivations for undertaking water budget analyses.

Related to this is the fact that for significant areas of karst recharge, particularly in autogenic recharge conditions, water moves relatively quickly into the subsurface either at discrete points or as diffuse recharge through bare rock or soil-covered surface. Underground, there is less opportunity for evapotranspiration and thus losses can be smaller than those for nearby surface flow catchments that are otherwise in similar climatic conditions. Hess & White (1989), for example, showed that evapotranspiration is lower, and runoff is higher, for the south central Kentucky karst area than for other nearby surface catchments within the Green River basin by 10–15%. They interpreted this to result from rapid movement of water underground through sinkholes and sinking streams with consequent reduced opportunity for evaporative and transpirational losses. They also estimated that normalized base flow (a value that relates the contributing catchment area to base flow discharge) from the karst springs is almost an order of magnitude lower than for other nearby surface catchments dominated by surface flow. Under tropical conditions in peninsular western Malaysia, Crowther (1987) monitored discharge of more than 160 underground seepages, cave streams and springs in three karst systems and determined that the measured evapotranspiration was lower than evaporation pan measurements by some 17%.

There have been a number of efforts in a variety of hydrogeological and climatic settings whose goal has been to supplement traditional water balance methods, especially using geochemical information, to better understand how water is moving through karst systems. Isotopes, tracers, and quantitative models have been particularly important. While each of these is discussed in more detail elsewhere in this volume, some examples of studies that have been specifically aimed at water budgets include Brown et al. (1969), Grubbs (1995), Katz & Bullen (1996), Lee (1996, 2000), Lee & Swancar (1997), Sacks et al. (1998), Andreo et al. (2004), and Afinowicz et al. (2005).

4.4 SPRING HYDROGRAPHS

We introduce here another important aspect of karst hydrology that in ways sets it apart from more traditional hydrogeologic settings. That is the *spring hydrograph*, a graphical representation of spring flow response (as either stage or discharge through time) typically following a storm input. Hydrographs recording individual storms and longer time series that may show seasonal variations can offer important tools for probing the interior workings of an aquifer. These also provide the most important monitoring points in many karst flow systems as they represent the outlet for flow from karst aquifers, influenced by and containing information on the integrated set of flow processes in an entire catchment, including spatial and temporal distribution of recharge, as well as both surface and subsurface flow paths. Analogous to the situation in a surface river network, useful hydrographs can of course be obtained at any accessible spot along an underground river network, and key underground points at river confluences can be important monitoring locations for aquifer sub-catchments (see Groves & Meiman 2005 and Raeisi et al., in press, for examples). Springs, though, are most generally monitored as they obviously provide surface access.

Obtaining the data necessary to construct spring and other hydrographs is arguably the most important technique of karst hydrology and the pertinent methods are discussed at length in various sections below. The ideas follow methods typical in many aspects for surface drainage flows, involving the measurement of discharge with as high a temporal resolution as possible – especially critical for many karst springs because of the inherent rapidity with which flow and chemical conditions can change in response to storm inputs. As with surface streams this is commonly based on either direct continuous discharge measurements with a weir and associated water level monitoring methods, or combining continuous water level records with spot discharge measurements to create rating curves, or relations between water level and discharge. As in other areas of karst hydrology, the technology for making high temporal resolution measurements has advanced rapidly with the advent of electronic data loggers and associated equipment.

Methods of hydrograph analysis that have been developed for interpreting karst springs (e.g. Bonacci 1993, Felton & Currens 1994, Padilla & Pulido-Bosch 1995, Labat et al. 2000) and wells (e.g. Shevenell 1996) have followed and expanded on methodologies used for analysis of surface stream hydrographs. Springs issuing from karst systems tend to be amenable to these types of analysis (whereas in many other hydrogeologic settings they are not) in that they often exhibit rapid and substantial flow variation following storm inputs, as waters move relatively quickly through well-developed conduit systems (Shuster & White 1971). Interpretation of storm hydrographs can provide information about internal structure and geometry of karst aquifers, storage volumes of different aquifer components, and transport times. Details of information to be gained from the interpretation of these curves, made even more powerful by combination with other signals of storm-scale chemistry, sediment, and temperature (e.g. Grasso et al. 2003, Liu et al. 2004), or artificially-applied tracers (e.g. Smart 1988, Hauns et al. 2001, Goldscheider et al. 2003) are further discussed in other chapters of this volume. Methods by which to collect the field water flow data necessary to feed these analyses are included in the following section.

4.5   PRECIPITATION AND RECHARGE MONITORING

4.5.1   Introduction

The location and movement of water into and through karst drainage systems depends on the spatial and temporal distribution of recharge as rainfall or snowmelt, and thus the measurement of precipitation is an important component for the understanding of hydrologic balances. As in many monitoring situations, the methods employed and the ultimate quality of the data depend on a balance between the needs of a particular study and the available resources with which the data can be collected.

The fact that recharge to karst aquifers can occur in two fundamentally different ways, as diffuse, direct autogenic inputs and as concentrated allogenic inputs where surface streams enter an aquifer at a discrete points, suggests that there are generally two sets of methods for measuring these inputs. Autogenic recharge is best measured as areally weighted precipitation as would be done on surface flow catchments, and discussion of these methods follows. The flow of autogenic sinking streams, in contrast, can be measured with weirs or direct stream discharge methods, as can be done for springs. And as with springs, the ability to make high temporal resolution flow measurements of flow sinking into an aquifer has significantly advanced the ability of hydrologists to make accurate water budgets for karst aquifers.

There are a number of methods for, and complications associated with, measurement of recharge that occurs as snowfall (e.g. Dunne & Leopold 1978, Woo & Marsh 1978, Yang & Woo 1999) including the facts that snow is blown by wind into drifts of varying depths, different snow crystal configurations and packing can have different amounts of liquid equivalents per volume of snow, and that snow can melt both in the atmosphere and while on the ground. Guidelines for appropriate methodologies should be consulted for programs where this is required (e.g. Doesken & Judson 1996, Gray & Male 1981, Uccellini 1997), and the following discussion focuses on methods for the measurement of liquid rainfall.

4.5.2 Manual gages

At one end are measurements of rainfall depth at a single location using simple rain gages that can be relatively inexpensive and yet can offer precision to within a millimeter or less. This is done by having an opening that is larger in area than the cross section of the rest of the device, so that the measurement is exaggerated, and the device scaled appropriately. These can be as simple as plastic wedge-shaped gages, or deeper, multiple-walled metal gages that can help minimize evaporation between sample readings with readings taken with a dipstick. As with any rain collection device, the accuracy of data depends on the site selection. While there are difficulties with wind if the gage is out in the open, potential obstructions such as trees, buildings and other high objects must be far enough away from the gage not to block angled rainfall, preferably not closer than four times the height of the obstructions (e.g. Doorenbos 1976). The device should be on level ground, and accuracy can depend on the height of the opening.

The limitations of such devices are both temporal and spatial. Data from manual gages are generally read at the same time each day, giving the accumulated total of the previous 24 hours. To capture information on the spatial variability of precipitation, it is typical to establish a network of gages within and adjacent to the catchment or recharge area of interest. While the finer the resolution of such a network the better, this will be a function of the resources available both for equipment and labour to take the readings. Several spatial averaging procedures are available to determine an "effective uniform depth" of precipitation across the network area (Linsley et al. 1982, Wanielista et al. 1997). Commonly employed methods include taking a simple arithmetic mean as well as contouring and calculating areas and average depths between isohyets, or lines of equal rainfall. Another, the Theissen polygon method, utilizes fixed polygons drawn around each station made by drawing lines connecting the location of each gage and then constructing polygons composed of lines that bisect the connecting lines. The area of each gage's polygon is then used to provide a weighting factor that is considered when summing the total contributions of the areas. Both the isohyet and Theissen methods offer the advantage of areal weighting that takes a non-uniform distribution of gages into account, and while the Theissen polygons remained fixed as long as no new devices are added, the isohyets must be recalculated with each storm event.

4.5.3 Automatic-recording precipitation gages

Several styles of automatic gages allow measurement of time-distribution of rainfall intensities at finer resolutions. Weighing bucket gages measure the weight of accumulated precipitation that can be recorded on a rotating drum strip chart, providing a cumulative plot of accumulation, and therefore intensity, as a function of time. Similar devices are constructed using floats that move with accumulating precipitation.

Tipping-bucket gages, originally in analogue designs that record on paper charts how often a given amount of rainfall, say one millimetre, has accumulated, are now commonly used with computer data loggers. Within the device is a centre-pivoted arm with a small reservoir on either end and starting with one end up, the arm tips whenever the designated amount of rainfall has accumulated. This empties the filled bucket and puts the other in the position to catch the rainfall, and by recording the time at which the device tipped, the rainfall intensity is determined. The tips are counted and then recorded on the data logger at a programmed interval of, say, five or fifteen minutes. Of course, as with manual gages, the spatial resolution of the data within a catchment area depends on the spacing of a network of gages, which becomes considerably more expensive with automatic gauging.

4.5.4   Doppler radar

While rain gage measurements have historically been the only source for the measurement of precipitation distribution in space and time, recent development and deployment of weather radar technology has generated new sources of rainfall data. Doppler radar technology such as WSR-88D in the United States is able to obtain information on rainfall intensities and wind directions and velocities (Sauvageot 1992, Collier 1996). Rainfall intensities can be determined for circular areas as large as about 450 km radius from the radar station by measuring the reflected energy returned to the station having been reflected from raindrops. Wind speeds and directions are determined by analysing Doppler phase shifts in the reflected energy cause by rain moving towards or away from the radar station. Rainfall intensity observations can be integrated through time to produce plots of accumulated rainfall over say an hour or for the duration of a storm event.

An issue with Doppler radar that is still an area of research and development concerns the quantitative relationships used to relate reflected energy intensity and the actual rainfall rate (e.g. Koistinen 1986, Joss & Lee 1995, Legates 2000). Improved technology is being developed that combines radar data and real time calibration with ground weather station observations (Nixon et al. 1999, Legates et al. 1999).

4.6   WATER FLOW MEASUREMENT

4.6.1   Introduction

A characteristic of karst groundwater flow is that commonly water flows as streams of various sizes within conduits. Indeed, in many (but not all) ways, we can study karst water with methods drawn from those used in surface water studies. In quantifying stream flows, two of the primary concepts are *stage height*, or the level of a water surface at some point along the stream above an arbitrarily determined datum, and *discharge*, which is the volume of water flowing past a cross section of the stream per time. Common units of discharge are litres per second or cubic metres per second, and it is generally calculated by noting that discharge is the product of the stream's cross sectional area and mean velocity at the point of measurement. Stage and discharge are related in that generally, as stage increases so does discharge. Cave stream flow and its measurement at places within caves that the streams are accessible are generally analogous to surface stream channels, although when conduits become filled it is quite a different situation. It is also common practice to measure discharge at springs, where the underground flows reach the surface.

Especially compared to measuring stage height, the measurement of discharge is time consuming and sometimes difficult. Developing a relationship between these for a given stream location results in a *rating curve*, which shows discharge as a function of stage height, and once this curve has been developed it is only necessary to make stage observations, either manually or with automated methods, to obtain good discharge estimates. Following is an overview of the most common methods for measuring stream discharge, and details are discussed in numerous texts (e.g. Chow 1959, 1964, Henderson 1966, Rantz et al. 1982). The choice of method depends primarily on the volume, geometry, and accessibility, and range of values of the flow to be measured.

4.6.2   Stage height measurement

Measurement of stage height (often just called stage) is fundamental in any quantification of stream flow (Buchanan & Somers 1968), and once a rating curve has been established serves as a proxy for easy estimation of flow discharge. There are generally two types of methods, manual and automated. The simplest of these is to install a *staff gage*, which is essentially a vertically installed linear scale on which an observer can directly read the water level. The values on the scale are generally tied to an arbitrary datum, and often the gage is accurately surveyed to provide a relation to actual topographic elevations. In some cases a staff gage location is augmented with installation of a *crest gage*, which consists of a measuring stick inside of a pipe that has a perforated bottom so that water can enter, along with a supply of ground cork. As the water level rises the cork particles float and attach to the stick, the highest particles marking the highest elevation reached by the water before receding.

While these methods provide useful data, temporal resolution is limited to the frequency with which an observer can actually visit the site. This can create difficulties, especially with cave stream data, for at least two reasons: the first is that during floods stages of cave streams can rise very rapidly so that high resolution data are needed to track rapidly changing conditions, as well as the fact that during floods the passages and their staff gages can become inaccessible. Groves & Meiman (2005), for example, documented the response of the Mammoth Cave System's Logsdon River from a single storm, where stage rose 24.6 meters within 14 hours, with a maximum rate of 6.03 meters per hour, which is in line with numerous other anecdotal observations reported in expedition reports by cave explorers and surveyors.

These problems can be overcome with automated systems, which have been developed in several types. Springs and other surface flows can be measured with water-level recorders. In a commonly used type (although there are a variety of others) a weight and float are attached to a wire draped over a cylinder that rotates as the float moves up or down with rising or falling stage. The wire and float are generally housed in a stilling well situated beneath the recorder, which both stills the water surface on which the float is positioned, as well as prevents interference of the wire by wind. A pen mechanism that moves via a clock mechanism across a chart wrapped around the drum can record these stage changes.

There are other, non-mechanical devices for measuring water levels including gas-purge manometers and pressure-sensor systems (bubble gages) (Rantz et al. 1982, Craig 1983), which monitor stage by sensing pressure in a sensor placed at a fixed depth in a stream that is affected by the weight of the overlying column of water and applying a pressure-stage relation. These involve the physical exchange of gas through a piping system as a function of pressure that is then recorded with a mercury manometer or pressure transducer, although many mercury sensors have been replaced due to environmental or safety concerns with respect to mercury.

Solid-state electrical pressure transducers are now in use that also measure the height of a column of water above the sensor, and thus stage, but provide a digital signal output that can be recorded on a computer data logger with high precision and temporal resolution (Bell & Howell 1983, Freeman et al. 2004). Different versions of the sensors work by a variety of methods, shared by the ability to convert energy from one form (mechanical) to another (electrical). Commonly, this involves an electric transduction element that converts pressure induced mechanical changes experienced by the sensing probe to an electronic signal sent to a display or recording device. Other transducers respond to pressure related mechanical changes in response to an external excitation signal. The most widely used device in hydrologic studies is the strain-gage, or resistive, transducer, which exploits the property that the electrical resistance of a "strain element" within the device (consisting of wire, metal foil filaments, or silicon semi-conductor depending on design) changes as it lengthens under strain.

As the use of electronic data logging becomes more widespread in karst hydrologic investigations, pressure transducers are providing an important tool for high-resolution studies. Freeman et al. (2004), however, warn that in some cases a perceived ease of installation and use of these devices can lead to inadequate quality-assurance, and that a thorough understanding of the capabilities of an employed pressure transducer and careful attention to calibration and maintenance are critical to provide high-quality data.

4.6.3   Direct discharge measurement

Although it uses the lowest technology, direct measurement of flow with a bucket or other suitable volume measurement device can be very accurate at locations that allow its use. Generally suited for small flows, as those often found in caves, the method is useful where the entire flow can be caught in a bucket or even a tarpaulin (Fig. 4.2) for a timed interval, and then the volume of the collected water determined. The volume, say in litres, is divided by the time of collection, say in seconds, to arrive at a discharge value. Especially with vigorous flows that may fill the device quickly, its is prudent to repeat the measurement a few times to take an average value. In cave settings this method has an advantage that the equipment is robust, and at least small buckets can be easy to carry.

4.6.4   Current meters

Current meters are devices that measure water velocity. This method uses the relationship introduced above: that discharge is equal to a flow's mean velocity times its cross section area. For small to moderate flows (streams with a maximum depth of a meter of less), this is often done by wading in the stream with the current meter attached to a *wading rod* calibrated to measure the stream's depth. Because flow velocity in a stream channel is variable both in depth and laterally, wading methods involve dividing the stream into a number of slices, typically with widths from 50 cm to a meter or more depending on the stream width and desired accuracy, using a measuring tape stretched across the stream perpendicular to the flow. In the middle of each slice, the depth of the stream channel is measured with the wading rod, and then the cross sectional area of that slice approximated

Figure 4.2.   Discharge measurement for epikarst dripwater using timed measurement of a known volume of water, Cave Spring Caverns, Kentucky USA (photo Groves).

as the resulting rectangle. Multiplying that value by the velocity obtained with the current meter provides a discharge value for the slice, and in the end the individual discharges are summed to arrive at the total flow value. While this method is used most commonly for surface streams, it is very useful for karst springs, and indeed can be used to measure cave streams, limited by the size of the wading rod if negotiating small cave passages *en route* to an underground stream measuring location (Fig. 4.3).

Since velocity varies with depth within the stream, it is standard practice to make the measurement at 0.6 of the stream depth within each slice, which approximates the mean

Figure 4.3.   Discharge measurements using wading methods can be made underground, although this is somewhat limited by the portability of the equipment. This measurement is being made more than a half kilometre from the Doyel Valley Entrance to the Mammoth Cave System, but in this case relatively large passages make transport of the equipment feasible, if not particularly convenient (photo Groves).

velocity in profile assuming a logarithmic velocity profile (Chow 1959). Alternately, the measurements can be made at 0.2 and 0.8 of the depth, and averaged. Some wading rods are built with sliding, calibrated scales that allow very easy determination of the correct measuring depth. For deeper flows at very large springs or on surface rivers, the current meter can be deployed from a bridge or boat using a cable and a lead torpedo-shaped device with fins attached the to the meter that keeps it in the direct position during the measurement. Under *ideal* conditions, it is possible to get discharge measurements using current meter methods to within 5% of the true value (Sauer & Meyer 1992).

There are several types of meters commonly used to make stream velocity measurements, including those with propellers and others that use electromagnetic properties of flowing water. The propeller types come in various sizes and are designed with spinning cups, in a way analogous to those of anemometers for measuring wind velocity, that rotate with an angular velocity proportional to the mean velocity of water flow at the point of measurement. The operator wears headphones that make a clicking sound during each revolution of the propeller cups, and by counting clicks over a measured time interval, charts or formulas can be used to arrive at a velocity value.

Electromagnetic sensors utilise Faraday's Law, which states that a voltage is produced when a conductor moves through a magnetic field. The current meter produces such a field, and the faster the water (the conductor) moves past the sensor, the larger a voltage is produced. The resulting voltage is accurately measured and then calibrated to the appropriate velocity, which is displayed either digitally or with an analogue meter. Since the directions and velocities of the water particles that produce the voltage can be highly variable due to the chaotic nature of turbulent flows, the instruments are able to take numerous measurements each second, displaying an average value that is constantly updated. Electromagnetic sensors have a distinct advantage for karst studies in some conditions as the resulting electronic signals can be stored on a data logger, with the result that recording sensors can be automated, even within flooded cave passages. Groves & Meiman (2001, 2005, see also Raeisi et al., in press), for example, have made continuous discharge estimates, albeit limited by measurement with the sensor at a single location in the flow, within the Mammoth Cave System's Logsdon and Hawkins Rivers, even during floods with the conduit ceiling as much as 25 m below the water table.

4.6.5 Weirs and flumes

Weirs and flumes provide an excellent method for making continuous discharge measurements, especially for long-term monitoring projects (Ackers et al. 1978, Waniclista et al. 1997). They are in-stream structures that regulate and allow measurement of flow rates by imposing a constriction, with discharge proportional to the depth of flow as it passes through the structure. Flumes are open conduits that narrow flow, while weirs are constructed in a way that causes flow to accelerate as it moves through a constriction of a particular geometry, which may be rectangular, triangular, parabolic, or a variety of other shapes. While flumes tend to be larger, more permanent structures, weirs come in all sizes and are commonly used in karst hydrologic investigations to measure flows at springs, surface streams, and occasionally underground in cave streams. They are constructed as sharp-crested types, in which the constriction through which the water flows is constructed of a metal plate or other thin construction, and broad-crested weirs, with greater thickness. The geometry of the constriction determines the specific type. For a wide range of flows some weirs are constructed with multiple cross sections, where the water, for example, can pass through a small cross section during low flows that may be accurately measured, but be accommodated by a large cross section above during flood conditions. Such structures are often necessary in tropical and subtropical karst areas subject to highly variable monsoon-related wet and dry season conditions, as in southern Asia (Fig. 4.4).

Figure 4.4.   Weir for discharge measurement at Spring 31 of the Guilin Karst Experimental Site, near Yaji Village, Guangxi Autonomous Region, China. A pressure transducer makes continuous depth measurements upstream from the weir that are related to discharge using the appropriate relationships for the weir's three sections. Use of a compound weir is useful in this case to capture a wide range of flows that results from the monsoon climate of southwest China. This photo was taken in February, typically late in the dry season prior to the onset of the Asian Summer Monsoon (photo Groves).

Ackers et al. (1978) and Wanielista et al. (1997) provide information on different sizes and shapes of wiers and the ranges of flows that they can accommodate. While small right-angle "v-notch" weirs common for monitoring relatively small flows in karst water balance investigations can accurately measure from about a litre to over a cubic metre per second, larger weirs can be constructed, albeit with rapidly increasing expense for materials and labour with increasing capacity, for flows greater than $100 \, \mathrm{m}^3/\mathrm{s}$.

For all styles of weirs, mathematical relationships relate the height of the water flowing through a weir above its crest, or the bottom edge of the opening, to the discharge of the flow under that condition (Ackers 1978, ISO 1980, ASTM 2001). A great utility of these structures is that if a continuous stage record can be developed for the water level behind the weir, the appropriate relationship can be easily used to develop a continuous discharge record.

4.6.6   Tracer dilution

There are conditions under which it is impractical, or the resources are not available, to construct a weir. In inaccessible areas (including within caves), as well as in very shallow or high velocity flows, current meters may also not be appropriate. Another method for discharge that has particular utility for karst studies involves the measurement of an introduced, artificial tracer at a point in the flow downstream where it was introduced

(Cobb & Bailey 1965, Rantz et al. 1982). Small flows can be measured relatively accurately with these methods using common table salt and a conductivity meter, which are portable enough to easily carry through cave passages in a small pack. In other cases quantitative fluorescent dye traces, where the spring or other dye recovery location is known in advance and the passage of the dye is monitored with water samples taken at regular intervals, can provide data that with the same methods can be used to estimate discharge at the monitoring locations. Any tracer can be used that is easily soluble in water, naturally absent or present only at low levels, is not subject to adsorption, degradation or other processes in the stream that could alter its concentration, is easily detectable at low levels, and is not toxic at the concentrations used in the methods.

There are two general methods for tracer dilution, including constant-injection and sudden-injection (Cobb & Bailey 1965, Rantz et al. 1982). While each has advantages, the sudden-injection method using sodium chloride is particularly useful for relatively small flows, particularly for cave settings because of its ease and the portability of the required equipment, and is considered in detail here. A known quantity of salt is dissolved in a volume of water, or a salt solution of known volume and concentration is prepared and the background concentration of the stream measured. For small cave streams a litre of solution (or some 10s of grams of salt) can be sufficient, while several kg of salt are required for flow rates of several m^3/s. The slug is quickly introduced into the stream and a stopwatch started. Some distance downstream, far enough for the slug to become completely mixed in the stream, the conductivity is measured at frequent intervals, say every 60 seconds and noting the time for each observation. The choice of interval varies with flow conditions, and in rapid flows more frequent observations (as short as five seconds in very rapid flow conditions) may be necessary – judging this follows from trial and error, or ultimately from experience. These concentrations should be measured until the slug has passed, although in practice the tail of the time-conductivity curve can get very drawn out due to dispersion of the tracer within the stream and only a small error is introduced by ending when the conductivity returns close to pre-injection levels. From these data, discharge can be calculated using the relationship:

$$Q = \frac{M}{\int_0^t C - C_b \mathrm{d}t} = \frac{V_o C_o}{\int_0^t C - C_b \mathrm{d}t} \approx \frac{V_o C_o}{\sum_{i=1}^n (C - C_b)\Delta t} \qquad (2)$$

where Q is discharge, M is injected tracer quantity, V_o and C_o are the volume and concentration of the injection tracer slug, respectively, C_b is the background tracer concentration in the stream before injection, C is the observed tracer concentration, and t is time. The integrals in the denominators of the middle terms in (2) represent the area under the tracer concentration-time curve as it passes the measurement point, which in practice can be estimated using the summation in the right-hand term, where C is the observed concentration of each measurement and Δt the time interval between measurements. Any concentration units appropriate to the particular choice of tracer can be used (i.e. μS/cm for conductivity, fluorescence intensity for dyes, or mg/L for other tracers), as long as they are in a range in which a linear relationship exists between the concentrations and measurement values – for extremely high salt concentrations, for example, the linear relationship between concentration and electrical conductivity can break down. Within the linear range, since the units of concentration in the numerator and denominator cancel, the resulting discharge unit depends on those of the tracer slug and time intervals. Expressing the tracer volume in litres and the time intervals in seconds, for example, gives discharge in L/s.

4.6.7   Non-contact methods

Emerging technologies are using remote methods to measure stream discharges, especially in large flows that would otherwise be difficult or dangerous (Beres & Haeni 1991, Spicer et al. 1997, Costa et al. 2000). These use microwave Doppler radar signals that measure surface water flow velocities based on non-symmetrical scattering off of roughness elements on the water surface, in combination with ground penetrating radar that can measure the channel flow cross section by making high-resolution profiles of the stream surface and bed. At this time the instruments to make such measurements are large and expensive, generally suspended by bridges. So there is potentially utility for large karst springs and surface flows in karst areas, if not in-cave measurements. It is developing technology, however, and may have additional uses in the future as the methods, equipment, and procedures evolve.

Another method that can be listed in this category involves methods that use observations of stream and bed geometry to estimate the flow's mean velocity, which can then be multiplied by the flow cross-section area to estimate discharge (Dalrymple & Benson 1967, Chanson 2004). The most well known of the relationships used in such methods is Manning's equation:

$$\tilde{V} = \frac{1}{n} R^{2/3} S^{1/2} \qquad\qquad (3)$$

where $\tilde{V}$ is mean velocity in m/s, n is Manning's roughness coefficient, R is hydraulic radius of the stream (flow cross section area divided by wetted perimeter) in m, and S is the water surface slope. The roughness coefficient (Barnes 1967, Limerinos 1970, Arcement & Schneider 1989) is estimated by one of several methods to quantify the impact that bed-fluid interaction has on energy loss, and has been described (Barnes 1967) as "chiefly an art". Manning's n values have been tabulated for a wide variety of natural channel types (including the impacts of varying sediment sizes and distributions and vegetation, for example) and estimates of n values for cave stream channels can be estimated by comparison to similar, vegetation free surface channels. The n value for a particular stream location, perhaps subject to a long-term investigation, can be directly obtained by a direct measurement of the mean velocity with the current meter or tracer dilution methods discussed above, along with the channel slope and channel geometry and rearranging (3) to solve for n. Thereafter, that n value can be used with the slope and geometry data to estimate the mean velocity using (3), and if the cross section is known, the discharge. If careful measurements of channel cross section are made and a relation developed between stage heights and the associated flow cross sections, then continuous stage measurements can be used with Manning's equation to obtain continuous discharge estimates. These methods can work for cave streams, although modifications need to be made for flood conditions when a given passage gets completely filled, and is thus modeled by pipe, rather than open channel, flow under which conditions accurate measurements become more difficult (for discussion of pipe flow in karst and theoretically see White 1988, Ford & Williams 1989, Hwang & Houghtalen 1996, Munson et al. 1998, Roberson et al. 1998).

4.6.8 Measurement of cave drips

It has been noted that direct access to the interiors of karst aquifers provides opportunities to understand details of both background hydrogeochemical processes and the movement of contaminants within impacted aquifers. Much of his work has been focused on the collection of vadose zone waters, and in particular speleothem drip waters that can elucidate details of processes taking place in the soil, epikarst, and vadose zones (e.g. Holland et al. 1964, Tooth & Fairchild 2003, Bolster et al. 2006). A great deal of work has also been associated with interpretation of the details of geochemical processes associated with the precipitation of calcite in speleothems, as it has been well established that such minerals can contain paleoclimatological information. Methods are based on collecting drips into a suitable container left underneath of the flow of interest, and primarily differ depending on the desired interval over which flow rate variation is to be determined. This can vary from collection by grab samples with intervals of weeks or longer, to methods utilizing electronic data loggers that can measure rates with a resolution of minutes, including with tipping bucket rainfall gages and bucket-weir arrangements in which the depth in the bucket is proportional to flow rate, and which can be measured with pressure transducers or other water level measuring methods. For discussion see Holland et al. (1964), White (1988), Ford & Williams (1989), Baker & Brundson (2003), Tooth & Fairchild (2003), and Bolster et al. (2006) and note Fig. 4.2 in this chapter. It should also be mentioned that there is a vast and growing speleothem literature, and those papers in this group describing the results of field monitoring programs have descriptions of the various methods used.

4.7 ELECTRONIC DATA LOGGING

A revolution in the available technology for hydrologic and hydrochemical data collection in karst systems has taken place with the development and implementation of computer data loggers (Fig. 4.5). These are devices that are capable of controlling the operation of, and storing information received from, a wide variety of environmental sensors. The importance for karst comes not as much for the convenience provided by remote, automatic data collection, as for the ability to collect high temporal resolution data. Both hydrologic and chemical characteristics of karst flow systems typically vary over a wide range of timescales, including those as short as hours or even minutes in response to storm inputs (Fig. 4.6). The fact that data loggers can record data with resolution as fine as minutes (or seconds for that matter, limited only by the data storage room) essentially solves a long standing, thorny problem in karst hydrology – the evaluation of continuously changing data. In mathematical terms many useful expressions in hydrology in general, and karst hydrology in particular (see equation 2 above, for example), use integrals that call for evaluating areas under hydrographs, chemographs, and other curves. Doing so exactly is generally difficult or impossible when dealing with field data. However, the value of an integral-containing term can be approximated by replacing the integral with summed, discrete observations (see equation 2 right-hand term), and the exact value is approached as the frequency of observations increases. Using data loggers to make measurements with a resolution of minutes captures all important structures of hydrologic variation, even for karst systems. By summing such data, easily done on a spreadsheet, an evaluation of the associated integral is essentially achieved.

Figure 4.5.   Triplicate data loggers recording flow, specific conductance, temperature and pH within Cave Spring Caverns, Kentucky USA (photo Groves).

Key sensors for hydrologic measurements include those used to obtain essentially continuous discharge records, either with a pressure transducer to measure water levels in conjunction with a weir, flume, or rating curve, or with direct measurement of velocity using electromagnetic velocity sondes. Loggers are also used in high-resolution rainfall monitoring, described earlier in the chapter. Other powerful methods, described more fully in another chapter in this volume, allow related, high-resolution hydrochemical monitoring

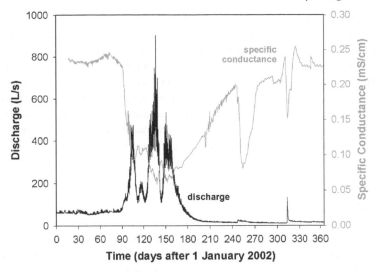

Figure 4.6.   Discharge and specific conductance of the karst spring at Tufa Falls, the exit of an alpine marble karst flow system that provides the headwaters of Spring Creek within the Sierra Nevada of Sequoia National Park, California USA (Despain et al. 2006).

of pH, temperature, and specific conductance, which in turn can be related to discrete observations of, among other parameters, calcium and bicarbonate. From these data, very useful calculations of landscape denudation rates, saturation indices, and limestone dissolution rates can be made (see Liu et al. 2004 and Groves & Meiman 2005 for an example).

An important issue with automated monitoring involves the trade off between ease of sampling once the station has been designed and implemented, and quality assurance of the data that mandates the frequency of visits for instrument maintenance and calibration. Remote sites where access may be difficult (potentially including in-cave sites) need careful planning with regard to protection of the equipment (both from the weather and vandalism), power sources, data storage, and checks for sensor drift, ideally with redundant sensors to ensure uninterrupted collection of data. Freeman et al. (2004) provide a thoughtful and thorough discussion of issues to be considered in the design and implementation of remote hydrological monitoring sites – while written specifically for the implementation of pressure transducers, much useful general information on issues such as system design, quality assurance, and instrument drift are presented.

Another challenge with high resolution data logging is that large data sets accumulate, and clear procedures must be established for the archiving of raw data, processing the data into a useful form, and storing the processed data. Each parameter measured with two-minute resolution, for example, generates 262,800 observations in a year's time. The processing of data can be done with the programming control of the data logger at the time of collection, or the raw data can be processed in the office using spreadsheet programs. In some cases a combination of real time and post collection processing can be useful. Groves & Meiman (2005), for example, used a strategy in which the data logger program recorded data hourly during static conditions and with two-minute resolution during storms where parameters varied rapidly and thus where fine-scale data were critical. This was achieved in the data logger program: when the logger queried each of four probes each two minutes, it compared a reading with the previous reading. If the changes in parameter values did not exceed a very small, programmed threshold, the data were not recorded as they were essentially the same and thus redundant. During static conditions this process was repeated until after an hour another data point was recorded, as either those in between were the same, or else the small linear drift could be determined by the difference in hourly recorded values. In contrast, during a storm when the subsequent value of any parameter exceeded the programmed threshold, the logger was programmed to begin recording all parameters every two minutes until static conditions again returned and hourly recording continued. For one year of such logging in that example, more than 90% of the data were shown to be redundant, and file sizes were thus less than 10% of what they would have otherwise been, and no significant detail was lost.

4.8   SUMMARY

In order to understand adequately the characteristics and behaviour of karst aquifers, it is necessary to quantitatively evaluate the occurrence and dynamics of water into, within, and from them. As in other terrains, surface catchments and groundwater basins serve as the basic units of investigation, and water balances can provide the basis for hydrologic analysis. However, the detailed methods used to evaluate karst water dynamics can vary from those used in traditional hydrogeology in significant ways. It is possible in some instances, for example, to enter a karst aquifer via caves to collect direct data on flow rates, chemistry and water quality, and to map out the geometry of conduit systems. In other ways, traditional monitoring and evaluation methods are similar to those used in karst settings. Networks of rainfall gauges are used to quantify autogenic recharge rates onto karst surfaces such as sinkhole plains, and traditional methods of discharge measurement such as velocity-area and control structures can be used to measure flow within cave conduits as well as from karst springs.

Methods for evaluating karst hydrology continue to evolve. Among the most important technological advances has been the implementation of electronic data logger/probe setups, which can be installed remotely throughout karst drainage systems, in both the surface and subsurface. These can record flow, temperature, chemical, and other data with frequencies down to minutes (or seconds in some applications) that capture all significant temporal variations within karst flow systems and allow evaluation of dynamics driven by both storm scale and seasonal influences.

CHAPTER 5

Hydraulic methods

Neven Kresic

5.1 INTRODUCTION

This chapter explains hydraulic and hydrogeologic parameters needed to characterise groundwater flow in a karst aquifer, as well as the methods for their determination. The parameters include: 1) porosity, effective porosity and storage capacity, 2) hydraulic head, 3) transmissivity and hydraulic conductivity, 4) groundwater velocity, and 5) groundwater flow rate. Field test methods include hydraulic borehole tests such as packer and slug tests, and aquifer pumping tests. In a true karst aquifer however, where all characteristic porosity types are developed, an accurate or just an approximate determination of all key groundwater flow parameters is the most difficult task. In addition, virtually all hydraulic test methods commonly applied in karst aquifers have been initially developed for intergranular and fractured porous media, so that the interpretation of their results in karst aquifers, where conduit and channel flows often play the most important role, is not straight-forward and requires considerable practical knowledge.

5.2 HYDRAULIC AND HYDROGEOLOGIC PARAMETERS

5.2.1 Porosity, effective porosity and storage capacity

Porosity is the most important property of rocks that enables storage and movement of water in the subsurface. It directly influences the permeability and the hydraulic conductivity of rocks, and therefore the velocity of groundwater. It is defined as the percentage of voids (empty space occupied by water or air) in the total volume of rock, which includes both solids and voids, $n = V_v/V$, where V_v is the volume of all rock voids and V is the total volume of rock. *Effective porosity* is often equated to *specific yield* of the porous material, or that volume of water in the pore space that can be freely drained by gravity due to change in the hydraulic head. Effective porosity is also defined as the volume of interconnected pore space that allows free gravity flow of groundwater. The volume of water retained by the porous media, which cannot be easily drained by gravity, is called *specific retention*. Since drainage of pore space by gravity may take long periods of time, especially in fine-grained sediments, values of specific yield determined by various laboratory and field methods during necessarily limited times are probably somewhat lower than the "true" effective porosity. Specific yield determined by aquifer testing in the field is a lumped hydrodynamic response to pumping by all porosity types (porous media) present in the aquifer. This value cannot be easily related to values of total porosity, which are always determined in the laboratory for small samples.

In general, rock permeability and groundwater velocity depend on the shape, amount, distribution and interconnectivity of voids. Voids, on the other hand, depend on the depositional mechanisms of carbonate sedimentary rocks, and on various other geologic processes that affect all rocks during and after their formation (see Chapter 2). *Primary porosity* is the porosity formed during the formation of rock itself, such as voids between the mineral grains, or between bedding planes. It is also often called matrix porosity. *Secondary porosity* is created after the rock formation, mainly due to tectonic forces (faulting and folding) which create micro and macro fissures, fractures, faults and fault zones in the brittle rock such as limestone. Sedimentary carbonate rocks may become cavernous (karstic) as a result of the removal of part of its substance through the solvent action of percolating water. Although solution channels and fractures may be large and of great practical importance, they are rarely abundant enough to give an otherwise dense rock a high porosity. Both the primary (matrix) and secondary porosity can be successively altered multiple times, thus completely changing the original nature of the overall rock porosity. In general, these changes may result in porosity decrease, increase, or altering of the degree of void interconnectivity without a significant change in the overall void volume. However, in true karst aquifers, continuing dissolution of rocks is expected to result in an overall increase in effective porosity. In general, rocks that have both the matrix and the fracture/conduit porosity are referred to as *dual-porosity* media. This distinction is important in terms of groundwater flow, which has very different characteristics in fractures and conduits compared to the bulk of the rock. Fig. 5.1 is a comparison between total porosity of main sedimentary rock types. As can be seen, limestone has the widest porosity range of any consolidated sedimentary rock.

One important distinction between the specific yield and the effective porosity concepts is that the specific yield relates to *volume* of water that can be freely extracted from an aquifer, while the effective porosity relates to groundwater *velocity* and *flow* through the interconnected pore space. Unnecessary confusion is introduced by some professionals trying to distinguish between the effective porosity for groundwater flow, and the effective porosity for contaminant transport. If the contaminant is dissolved in groundwater, its advective transport will be governed by the same effective porosity since it moves with groundwater. Diffusive transport of the contaminant is its movement due to concentration gradient and is independent of the groundwater flow. Diffusion involves the entire (total) porosity: molecules of the contaminant (and water) can move through minute pores, which would otherwise not allow free gravity flow. In conclusion, there are no two different effective porosities and it is sufficient to determine two values for any quantitative analysis of groundwater flow, or contaminant fate and transport: one for the effective porosity, and one for the total porosity.

It is widely shown that the degree of karstification, fracture density, porosity, and hydraulic conductivity all generally decrease with depth in carbonate sedimentary rocks. Older (e.g. Palaeozoic) limestones commonly have lower porosity and matrix permeability than more recently deposited ones (e.g. Tertiary) due to depth of consolidation, recrystallisation and cementation (Moore 1989). However, as the erosional basis for karst groundwater discharge is not necessarily at a static elevation and may be constantly lowered such as due to surface stream incision, the depth of karstification and the depth to water table would consequently increase in time as well. In some areas of the Dinaric carbonate platform,

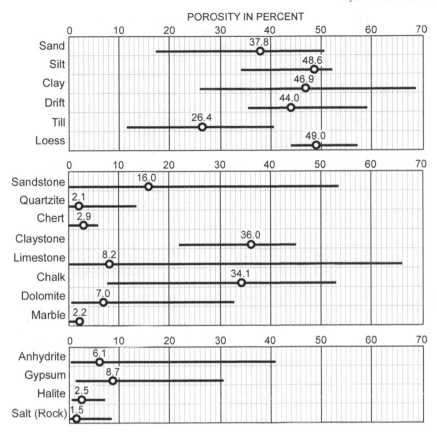

Figure 5.1. Porosity range (horizontal bars) and average porosities (circles) of unconsolidated and consolidated sedimentary rocks (Kresic 2007; porosity values processed from Wolff 1982).

depth to water table exceeds 1000 m. It is therefore not advisable to make generalisations, based on some karst literature examples, as to the expected depth of karstification and the related portion of aquifer storage capacity for any particular case.

The same general area with carbonate rocks may be subject to multiple periods of karstification depending on the depositional and tectonic history. In such cases it is possible to find very transmissive zones, together with karst conduits, at greater and varying aquifer depths and/or below overlying non-carbonates. These zones often mark position of a paleo water table where karstification intensity was the highest (Fig. 5.2).

Although storage capacity of a carbonate aquifer due to karstification generally decreases with depth, there is no simple way for quantifying this relationship: it would always be site-specific depending on many different factors including tectonic fabric, stratigraphy, sedimentology and climate. Depth of karstification can also vary widely within the same aquifer.

5.2.2 Hydraulic head

Figure 5.3 illustrates some key differences between a karst aquifer (A) and an intergranular aquifer (B), which both have the same general flow direction from the north to the south as

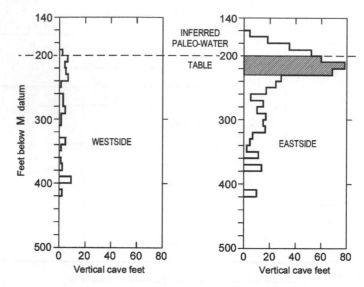

Figure 5.2.   Vertical voids ("cave-feet") in the San Andres carbonates as a function of depth below M datum, a horizon in the Seven Rivers Anhydrite. "Cave" in this example is defined as a void detected in wells which has minimum size (height) of 0.3 meters. (Craig 1988, p. 354, fig. 16.10A. Copyright Springer-Verlag 1988. Reprinted with kind permission of Springer Science and Business Media.)

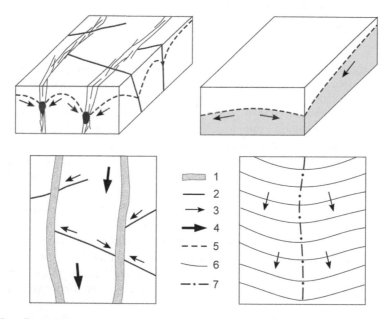

Figure 5.3.   Groundwater flow and its map presentation in a karst (A) and an intergranular (B) aquifer. 1) Preferential flow path (e.g. fracture or fault zone or karst conduit/channel); 2) Fracture/fault; 3) Local flow direction; 4) General flow direction; 5) Position of hydraulic head (water table); 6) Hydraulic head contour line; 7) Groundwater divide (Kresic 1991).

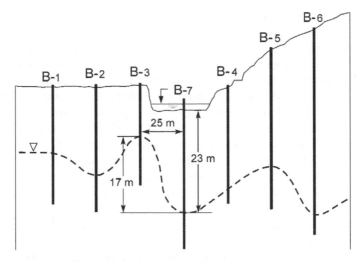

Figure 5.4.   Hydraulic head measured at a transect of piezometers perpendicular to the Trebisnjica river in eastern Herzegovina, the largest sinking stream in Europe (Milanovic 1981, copyright Water Resources Publications, printed with permission).

shown on the map view. A triplet of wells, installed anywhere in the intergranular aquifer, would reasonably accurately determine the general groundwater flow direction based on the measured hydraulic head. The same cannot be stated for the karst aquifer case where the three-well principle may give very different results depending on the position of individual wells relative to the preferential flow paths (karst conduits and/or large fracture zones). Moreover, a group of closely spaced wells may show a completely "random" distribution of the measured hydraulic head as illustrated in Fig. 5.4. One well may be completed in a homogeneous rock block, without any significant fractures and with low matrix porosity, and may even exhibit the so-called "glass effect" (no fluctuation of water table regardless of the precipitation-infiltration dynamics). A well ten or so meters away may, on the other hand, show the hydraulic head fluctuation of several meters or more.

As a conclusion, using only the hydraulic head information for the purposes of assessing groundwater flow directions and overall flow characteristics in a true karst aquifer would be insufficient. Hydraulic head measurements should be combined with dye tracing (Chapter 8) and a thorough understanding of various hydraulic factors such as flow through pipes and "non-hydraulic" factors such as geomorphology, limestone (carbonate) sedimentology, tectonics, and geologic history, including paleokarstification (Chapter 2).

5.2.3   Transmissivity and hydraulic conductivity

In quantitative terms, the transmissivity (T) is the product of hydraulic conductivity (K) of the aquifer material and the saturated thickness of the aquifer (b): $T = Kb$. It has units of squared length over time (e.g. m^2/d). In practical terms, the transmissivity equals the horizontal groundwater flow rate through a vertical strip of aquifer one unit wide. The larger the transmissivity, the larger the hydraulic conductivity and/or the aquifer thickness. However, the concept of transmissivity in karst aquifers is not straight-forward: it is obvious that the transmissive properties of conduits and channels, where the groundwater preferentially flows, are much higher than in the surrounding aquifer matrix. At the same time, the hydraulic conductivity has no real physical meaning in the case of conduit and channel flow; it is proportionality constant in Darcy's law, which was derived for intergranular porous media, and has dimension of velocity. Although the hydraulic conductivity has been used to calculate flow in all aquifer types, it should be noted that in karst aquifers it represents a lumped parameter which describes properties of an "equivalent porous medium"; in other words, the hydraulic response of all porosity types, including conduits and karst channels, is described with one parameter: an equivalent hydraulic conductivity.

The following discussion illustrates the interdependence and variability of the porosity, the effective porosity, and the hydraulic conductivity (transmissivity) of carbonate rocks. Chalk and some limestones may have high porosity, but since the pores are small (usually less than 10 micrometers), primary permeability is low and specific retention is high (Cook 2003). For example, the mean interconnected porosity of the Lincolnshire Limestone in England is 15%, while the mean matrix hydraulic conductivity is only 10^{-9} m/s (Cook 2003, after Greswell et al. 1998). The groundwater flow is largely restricted to the fractures. The aquifer hydraulic conductivity determined from pumping tests ranges between approximately 20–100 m/d, which is more than five orders of magnitude greater than the matrix hydraulic conductivity (Cook 2003). The San Antonio segment of the Edwards aquifer in Texas, the United States consists of Cretaceous limestones and dolomites that have undergone multiple periods of karstification. The average aquifer hydraulic conductivity, based on over 900 well pumping tests, is approximately 7 m/d, while the mean matrix hydraulic conductivity is approximately 10^{-3} m/d. The aquifer includes a number of wells with very high discharge rates, including one well near San Antonio with a discharge rate of 1.6 cubic meters per second (Cook 2003, after Halihan et al. 2000). This well is considered by many to be the highest yielding drilled vertical well in the world. Reported transmissivity values for Devonian limestone from the Mt. Larcom district in central Queensland, Australia range from $10\,\mathrm{m^2/d}$ in poorly fractured areas, to $3000\,\mathrm{m^2/d}$ for wells intersecting solution channels (Cook 2003, after Pearce 1982).

5.2.4   Groundwater velocity

Because of widely varying hydraulic conductivity and effective porosity of karstified carbonates, even within the same aquifer system, the groundwater velocity in karst can vary over many orders of magnitude. One should therefore be very careful when making a (surprisingly common) statement such as "groundwater velocity in karst is generally very high". Although this may be true for turbulent flow taking place in karst conduits, a disproportionably larger volume of any karst aquifer "experiences" relatively low groundwater velocities (laminar flow) through small fissures and rock matrix. One common method for determining groundwater flow directions and apparent flow velocities in karst is dye tracing (see Chapter 8). However, most dye tracing tests in karst are designed to analyse possible connections between swallow holes (sinks) and springs. Because such connections often involve conduit flow, the velocities calculated from the dye tracing data are usually biased towards the high end. For example, the results of 43 tracing tests in karst regions of West Virginia, the United States show the median groundwater velocity of 716 m/d, while 50% of the tests show values between 429 and 2655 m/d (Jones 1977). It is interesting that, based on 281 dye tracing tests, the most frequent velocity (14% of all cases) in the classic Dinaric karst of Herzegovina, as reported by Milanovic (1979), is quite similar: between 864 and 1728 m/d. 25% of the results show groundwater velocity greater than 2655 m/d in West Virginia, and greater than 5184 m/d in Herzegovina. The West Virginia data do not show any obvious relationship between the apparent groundwater velocity and the hydraulic gradient.

Various approximate calculations of flow velocity have been made based on the geometry of hydraulic features, such as scallops and flutes visible on walls, floors and ceilings of accessible cave passages (e.g. see White 1988, p. 97–98, and Bögli 1980, p. 163–164). For example, the calculated flow velocity for a canyon passage in White Lady Cave, Little Neath Valley, Unite Kingdom is 1.21 m/sec and the flow rate is 9.14 m^3/sec, for the cross sectional area of flow of 7.6 m^2 and scallop length of 4.1 cm (White 1988).

Confined karst aquifers which do not have major concentrated discharge points in forms of large springs, generally have much lower groundwater flow velocities. This is regardless of the predominant porosity type because the whole system is under pressure and the actual displacement of "old" aquifer water with the newly recharged one is rather slow. Groundwater flow velocity estimates using carbon fourteen isotope dating for the confined portion of the Floridan aquifer in central Florida (Hanshaw & Back 1974) showed that the average groundwater velocity based on 40 values is 6.9 m/y or 0.019 m/d.

5.2.5   Groundwater flow

5.2.5.1   *Introduction*

As illustrated in previous sections, interpretation of hydraulic head measurements in karst aquifers is often non-unique and should not be used to estimate rates of groundwater flow based solely on the principles established by Darcy. The Darcy's flow equation is applicable to intergranular porous media and laminar conditions only and is given as a product of the hydraulic conductivity (K), the hydraulic gradient (i), and the entire cross-sectional area of flow (A) inclusive of voids and solids: $Q = K \cdot i \cdot A$.

For a true karst aquifer, in which all three types of porosity are present, it is very difficult, if not impossible, to determine "representative" values of the three main quantities in the above equation. The hydraulic gradient determined based on the hydraulic head measurements in just several wells may be misleading as shown earlier. The same is true when selecting the most appropriate wells and times of the hydraulic head measurement (see well hydrographs in Fig. 5.5). Best professional judgment would often have to be made when deciding how to average the hydraulic gradients, in both space and time, over an area covered by monitoring wells. Once such judgment is made, however, an equally difficult task of determining the actual (effective) or "representative" cross sectional area of groundwater flow comes next. In other words, the flow takes place through all of the following: interconnected pores within the matrix porosity, micro and macro fractures, and solutional cavities, which may have almost any imaginable shape.

Except (arguably) for the application of Darcy's law in case of the matrix porosity dominated karst aquifers (which is rather uncommon), in all other situations there is a wide array of possible approaches to calculate groundwater flow rates. The simplest one is to assume that the porous media all "behave" similarly, at some representative scale, and then simply apply Darcy's equation. Although this equivalent porous medium approach still seems to be the predominant one in hydrogeologic practice, it is hardly justified in a true karst aquifer. Its inadequacy is emphasised when dealing with contaminant fate and

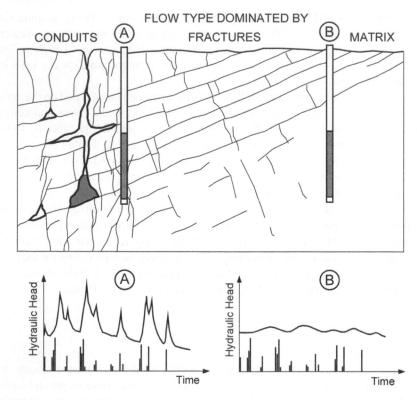

Figure 5.5.   Response of the hydraulic head measured in monitoring wells to different types of flow in karst aquifers. A: Rapid conduit flow after major recharge events and no significant storage in the matrix. B: Delayed and dampened response of aquifer matrix. Flow dominated by fractures may include any combination of these two extremes (Kresic 2007).

transport analyses where all field scales are equally important, starting with contaminant diffusion into the rock matrix, and ending with predictions of most likely contaminant pathways in the subsurface. In the analytically most complicated, but at the same time the most realistic case, the groundwater flow rate is calculated by integrating the equations of flow through the rock matrix (Darcy's flow) with the hydraulic equations of flow through various sets of fractures, pipes, and channels. This integration, or interconnectivity between the four different flow components, can be deterministic, stochastic, or some combination of the two.

Deterministic connectivity is established by a direct translation of actual field measurements of the geometric fracture parameters such as dip and strike (orientation), aperture, and spacing between individual fractures in the same fracture set, and then doing the same for any other fracture set. Cavities are connected in the same way, by measuring the geometry of each individual cavity (cave). Finally, all of the discontinuities (fractures and cavities) are connected based on the field measurements and mapping. As can be easily concluded, such deterministic approach includes many uncertainties and assumptions by default ("You have walked and measured this cave, but what if there is a very similar one somewhere in the vicinity you don't know anything about?").

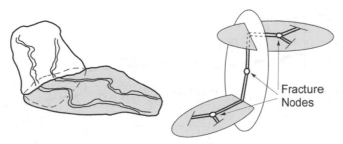

Figure 5.6. Channelling in a fracture plane and equivalent "channel model" (Cacas 1989, from Chilès & Marsily 1993).

Stochastic interconnectivity is established by randomly generating fractures or pipes using some statistical and/or probabilistic approach based on field measures of the geometric fracture (pipe) parameters. An example of combining deterministic and stochastic approaches is when computer-generated fracture or pipe sets are intersected by a known major preferential flow path such as cave. In any case, except for relatively simple analytical calculations using homogeneous, isotropic, equivalent porous medium approach, most other quantitative methods for karst groundwater flow calculations include some type of modelling.

5.2.5.2 *Flow through fractures*

Fracture aperture and thickness are two parameters used most often in various single-fracture flow equations, while spacing between the fractures and fracture orientation is used when calculating flow through a set of fractures. However, these actual physical characteristics are not easily and meaningfully translated into equations attempting to describe flow at a realistic field scale:

- Fracture aperture is not constant and there are voids and very narrow or contact areas called asperities. Various experimental studies have shown that the actual flow in a fracture is channeled through narrow, conduit-like tortuous paths (Fig. 5.6) and cannot be simply represented by the flow between two parallel plates separated by the "mean" aperture (Cacas 1989).
- Because of stress release, the aperture measure at outcrops or in accessible cave passages is not the same as an in-situ aperture. Aperture measured on drill cores and in borings is also not a true one – the drilling process commonly causes bedrock adjacent to fractures to break out thereby increasing the apparent widths of fracture openings as viewed on borehole-wall images (Williams et al. 2001).
- Fractures have limited length and width, which can also vary between individual fractures in the same fracture set. Spacing between individual fractures in the same set can also vary. Since all these variations take place in the three-dimensional space, they cannot be directly observed, except through continuous coring or logging of multiple closely spaced boreholes, which is the main cost-limiting factor.

The limited (by default) extent of fractures can be simulated with various geometric configurations, including 2D nearly orthogonal fracture traces, 3D orthogonal disks, randomly or otherwise distributed 3D plates, or in some other way. Whatever approach is selected,

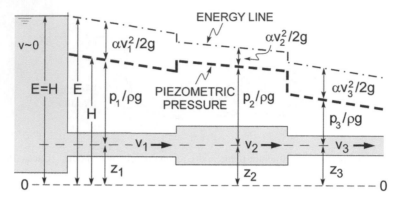

Figure 5.7.   Illustration of the Bernoulli equation for flow of real viscous fluids through a pipe with the varying cross-sectional area. E: total energy; H: piezometric pressure (hydraulic head). p_1, p_2, p_3: fluid pressure at cross sections 1, 2 and 3 respectively. ρ: fluid density. g: acceleration of gravity. v: flow velocity. z: elevation head. α: coefficient of hydraulic (velocity) loss. Explanation in text.

numeric models are arguably the only quantitative tool capable of solving groundwater flow through such complex fracture networks.

Evolution of analytical equations and various approaches of quantifying fracture flow is given by Whiterspoon (2000) and Faybishenko et al. (2000). Most common, simple analytical equations used in fracture hydrogeology are given in Chapter 10. Equations of fluid flow in non-ideal fractures with influences of various geometric irregularities, and fracture network modeling approaches are also discussed in detail by Bear et al. (1993), Zimmerman & Yeo (2000), and Faybishenko et al. (2000).

5.2.5.3   *Flow through pipes (conduits) and channels*

Flow through pipes is generally described by the well-known Bernoulli equation for real viscous fluids as illustrated in Fig. 5.7. The total energy surface (E) of flow can only decrease from the upgradient cross section towards the downgradient cross section of the same flow tube (pipe or conduit) due to energy losses. On the other hand, the piezometric surface (H) can go "up" and "down" along the same flow tube depending on the cross sectional area. The total energy surface, which includes the flow velocity component ($\alpha v^2/2g$), can be directly measured only by the Pitot device whose installation is not feasible in most field conditions. Monitoring wells and piezometers, on the other hand, only record the piezometric pressure (hydraulic head), which does not include the flow velocity component. It is therefore conceivable that two piezometers in or near the same karst conduit with rapid flow may not provide useful information for calculation of the real flow velocity and flow rate between them, and may even falsely indicate the opposite flow direction. In fact, as illustrated by Bögli (1980, p. 87) it has been shown that water rising through a tube in an enlargement passage can flow backward over the main flow conduit and into another tube, which begins at a narrow passage in the same main conduit.

There are additional complicating factors when attempting to calculate flow through natural karst conduits using the pipe approach:
1)   Flow through the same conduit may be both under the pressure and with the free surface.
2)   Since pipe/conduit walls are more or less irregular ("rough"), the related coefficient of roughness has to be estimated and inserted into the general flow equation.

3) Conduit cross section may vary significantly over short distances.
4) The flow may be both laminar and turbulent in the same conduit, depending on the flow velocity, cross-sectional area and wall roughness. The irregularities that cause turbulent flow are mathematically described through the Reynolds number and the friction factor.

Bögli (1980), Ford & Williams (1989) and White (1988) provide detail discussion on various pipe and channel flow equations and their applicability in karst aquifers.

5.3 HYDRAULIC BOREHOLE TESTS

5.3.1 Introduction

In a true karst aquifer, with heterogeneous secondary porosity (fractures and dissolutional cavities), the inflow of water to a pumped well would most likely occur from discrete intervals where the borehole intersects preferential flow paths within the surrounding rock. The presence of such intervals may be indicated by various methods of advanced geophysical logging, and their actual flow contribution may be measured and calculated using borehole flow meters ("flowmeters") and hydrophysical logging. These techniques are explained in detail in Chapter 9. The flowmeters and hydrophysical logging can be utilised in various ways, with or without pumping of the well. During well pumping packers can be used to isolate portions of the open borehole for a more precise characterisation. Finally, classic packer tests, during which water is injected under pressure into discrete intervals, can also be performed to determine permeability of the tested intervals. Simultaneous use of geophysical logging tools and flowmeters is likely the best available method for invasive characterisation of fractured and karst aquifers (Paillet 1994, Paillet & Reese 2000).

5.3.2 Packer tests

Borehole packer tests reduce or eliminate the effect that open boreholes have on the collection of hydraulic and chemical data. Borehole packers are pneumatic or mechanical devices that isolate sections of a borehole by sealing against the borehole wall. Hydraulic tests or collecting groundwater samples for chemical analyses then can be conducted on the isolated section of the borehole (Shapiro 2001). Packer testing should test both the more permeable portions of the borehole (preferential flow paths) and less permeable portions without fractures and cavities, which may be representative of the matrix porosity.

First widely applied method of pressure permeability testing was developed by Lugeon (1933), primarily for designing geotechnical grouting during construction of dams. Original Lugeon method included grouting of the tested interval before drilling to deepen the borehole for the next test interval. Because of the high cost and time requirements, this method of pressure testing is now seldom used. However, the term "Lugeon" test is still the most widely used term for describing multiple pressure tests regardless of the drilling method. Descending test is used mostly in unstable rock. The hole is drilled to the bottom of each test interval and an inflatable packer is set at the top of the interval to be tested. After the test, the boring is cased and then drilled to the bottom of the next test interval. Ascending test is suited for stable rock where the boring can be drilled to the total depth without casing. Two inflatable packers, usually 1.5 to 3 m apart, are installed on the drill rod or pipe used for performing the test. The rod section between the packers is perforated.

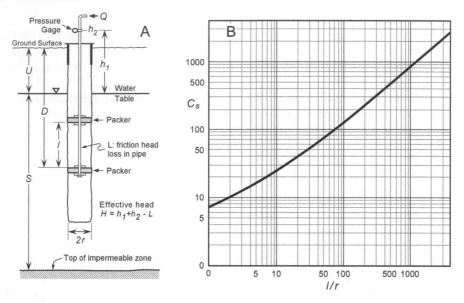

Figure 5.8.   (A) Pressure permeability test parameters for determining the hydraulic conductivity of consolidated rocks in saturated zone. l: length of test section; D: distance from ground surface to bottom of test section; U: thickness of unsaturated material; S: thickness of saturated material; h_1: distance between gage and water table; h_2: applied pressure at gage; $H = h_1 + h_2 - L =$ effective head; L: head loss in pipe due to friction; ignore head loss for $Q < 4$ gal/min in 1¼ inch pipe; use length of pipe between gage and top of test section for computations; r: radius of test hole; Q: steady flow into well. (B): Conductivity coefficients for semispherical flow in saturated materials through partially penetrating cylindrical test well (USBR 1977).

Tests start at the bottom of the hole. After each test, the packers are raised the length of the test section, and another test is performed.

The length of the test section is governed by the character of the rock, but generally a length of 3 m is acceptable. Occasionally, a good packer seal cannot be obtained at the planned depth because of bridging, raveling, fractures and cavities, or a rough hole. If a good seal cannot be obtained, the test section length should be increased or decreased, or test sections overlapped to ensure that the test is made with well-seated packers. In case of highly permeable fractures and cavities, the 10-foot (3-m) section may take more water than the pump can deliver, and no stabilisation of pressure (back pressure) can be developed. If this occurs, the length of the test section should be shortened until back pressure can be developed (USBR 1977).

Fig. 5.8A shows test parameters used to determine the hydraulic conductivity of the tested borehole interval. The hydraulic conductivity is given as (USBR 1977): $K = Q/(C_s rH)$ where C_s is conductivity coefficient for semi-spherical flow in saturated material through partially penetrating cylindrical test wells. Values of C_s are found from the graph shown in Fig. 5.8B for different values of l/r.

Pressure permeability is often expressed in Lugeon units (Lu); one Lugeon is equal to the water intake of one liter (L) per one meter of the tested interval, during time interval of one minute (min), under an injection pressure of one mega Pascal or approximately 10 atmospheres: $1\,Lu = 1\,L/min/m$ at 1 MPa. Specific permeability (q) is another measure of

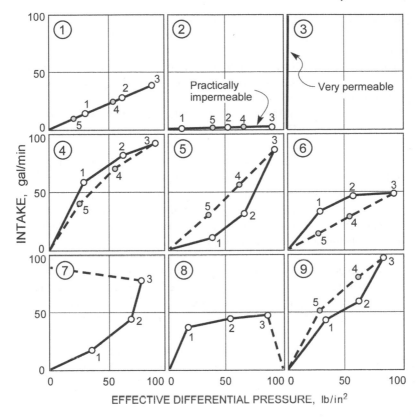

Figure 5.9. Plots of simulated, multiple pressure, permeability tests (USBR 1977).

rock permeability obtained from the pressure and water intake measurements. It is defined as volume of water injected into the borehole per one meter length, during one minute, under a pressure of one meter of water (0.1 atmospheres): $q = Q/(H \times l)$; given in L/min/m at 0.1 atmosphere, where Q is the water injection rate, H is the effective injection pressure, and l is the length of test interval.

Multiple pressure permeability test ("Lugeon test"; see Lugeon 1933) applies the pressure in three or more approximately equal steps. Each pressure is usually maintained for 20 min or so under the requirement of steady water intake. Flow readings are made at 5-min intervals. The pressure is then raised to the next step. After the highest step, the process is reversed and the pressure maintained for approximately the same time as during the first cycle. Hypothetical test results of multiple pressure tests are plotted in Fig. 5.9. The curves are typical of those often encountered. The test results should be analysed using confined flow hydraulic principles combined with data obtained from the core or boring logs. Probable conditions represented by plots in Fig. 5.9 are:

1. Very narrow, clean fractures. Flow is laminar, permeability is low, and discharge is directly proportional to head.
2. Practically impermeable material with tight fractures. Little or no intake regardless of pressure.

3. Highly permeable, relatively large, open fractures indicated by high rates of water intake and no back pressure. Pressure shown on gauge caused entirely by pipe resistance.
4. Permeability high with fractures that are relatively open and permeable but contain filling material which tends to expand on wetting or dislodges and tends to collect in traps that retard flow. Flow is turbulent.
5. Permeability high, with fracture filling material, which washes out, increasing permeability with time. Fractures probably are relatively large. Flow is turbulent.
6. Similar to 4, but fractures are tighter and flow is laminar.
7. Packer failed or fractures are large, and flow is turbulent. Fractures have been washed clean; highly permeable. Test takes capacity of pump with little or no back pressure.
8. Fractures are fairly wide but filled with clay gouge material that tends to pack and seal when under pressure. Takes full pressure with no water intake near end of test.
9. Open fractures with filling that tends to first block and then break under increased pressure. Probably permeable. Flow is turbulent.

The U.S. Geological Survey (USGS) has developed a Multifunction Bedrock Aquifer Transportable Testing Tool (BAT³), which can be used to test discrete borehole intervals by either injecting or withdrawing water. The equipment is designed to perform the following operations by isolating a fluid-filled interval of a borehole using two inflatable packers (Shapiro 2001): 1) collect water samples for chemical analysis; 2) identify hydraulic head; 3) conduct a single-hole hydraulic test by withdrawing water; 4) conduct a single-hole hydraulic test by injecting water; and 5) conduct a single-hole tracer test by injecting and later withdrawing a tracer solution. The equipment can be configured to conduct these operations with only one of the borehole packers inflated, dividing the borehole into two intervals (above and below the inflated packer). The Multifunction BAT³ is designed with two inflatable packers and three pressure transducers that monitor fluid pressure in the test interval (between the packers), as well as above and below the test interval; pressure transducers above and below the test interval are used to ensure that the borehole packers seal against the borehole wall during applications.

Sections of a borehole containing highly transmissive fractures are most easily tested by withdrawing water, whereas fractures with low transmissivity are tested by injecting small volumes of fluid. The Multifunction BAT³ is configured with both a submersible pump and a fluid-injection apparatus in the test interval to accommodate hydraulic tests that either withdraw or inject water. With this capability, the Multifunction BAT³ can estimate transmissivity ranging over approximately 8 orders of magnitude (Shapiro 2001).

5.3.3   Slug tests

Slug tests are performed in exploratory boreholes and small-diameter monitoring wells to find an approximate hydraulic conductivity in their immediate vicinity. Despite this limitation, and the fact that most methods in use cannot determine the storage properties, slug test are widely applied in groundwater studies because of the following advantages:
*   the cost is incomparably smaller than the cost of an average pumping test;
*   the time needed for its preparation and the duration of the test are short;
*   there is no need for observation wells;
*   in case of contaminated aquifers there are no issues associated with groundwater extraction and disposal;

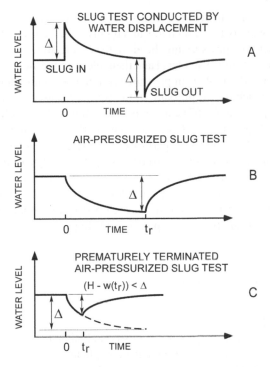

Figure 5.10. Schematic illustrating time-varying water level during (A) slug test conducted by water displacement, (B) air-pressurised slug test, and (C) prematurely terminated air-pressurised slug test, where Δ is the maximum change in water level due to water displacement or an applied air pressure, t_r is the time at which the pressurised part of the air-pressurised slug test is terminated, H is the initial water level at time $t-0$, and $w(t_r)$ is the water level at time $t-t_r$ (Shapiro & Green 1995).

- the test can be performed in exploratory boreholes (including inexpensive soil borings) that are only temporarily equipped with a screen;
- the results are not influenced by short-term or long-term changes such as variations of barometric pressure, or interference from nearby pumping wells.

In general, a slug test consists of quickly displacing a volume of water in the borehole (well) by adding or removing a "slug" and then recording the change in the head at the borehole (Fig. 5.10). Since the displaced volume of water is relatively small, the head displacement is also small and the recovery is fast (usually not more than several minutes in moderately permeable aquifers). For this reason it is necessary to take as many measurements of the head in the well as possible in order to have enough data for a graphoanalytical analysis. This is most accurately done with the help of a pressure transducer and a digital data logger, which can be programmed to take recordings every few seconds, or with even smaller intervals during the initial portion of the test. Relatively quick recovery is the main limitation of slug tests in karst aquifers with highly transmissive discrete zones.

The well known Bouwer and Rice slug test can be used to determine the hydraulic conductivity in fully or partially penetrating and partially screened, perforated, or otherwise open wells (Bouwer & Rice 1976, Bouwer 1989). Although initially developed for unconfined aquifers, the test has been successfully used for confined aquifers given that the top of the screened section is below the overlying confining layer (Bouwer 1989).

Modifications of air-pressurised slug tests offer an efficient means of estimating the transmissivity (T) and storativity (S) of both low and highly permeable aquifers. Air-pressurised slug tests are conducted by pressurising the air in the casing above the column of water in a well, monitoring the declining water level and then releasing the air pressure and monitoring the rising water level. If the applied air pressure is maintained until a new equilibrium-water level is achieved and then the air pressure in the well is released instantaneously, the slug test solution of Cooper et al. (1967) can be used to estimate T and S from the rising water level data (Shapiro & Green 1995, Green & Shapiro 1995). The total time to conduct the test can be reduced if the pressurised part of the test is terminated prior to achieving the new equilibrium-water level. This is referred to as a prematurely terminated air-pressurised slug test. Type curves generated from the solution of Shapiro & Greene (1995) can be used to estimate T and S from the rising water level data from prematurely terminated air-pressurised slug tests. The same authors also developed a public domain computer program called AIRSLUG for the analysis of air-pressurised slug test (Green & Shapiro 1995).

5.4   AQUIFER PUMPING TESTS

5.4.1   Introduction

Aquifer pumping tests are among the most costly and labor-intensive field tests in hydrogeology. It is therefore crucial to approach their planning and execution by following well-established and accepted practices and guidelines. There is an abundant published literature on the topic and numerous recommendations and publicly available guidance documents by the United States Geological Survey (USGS: Ferris et al. 1962, Stallman 1971), the United States Environmental Protection Agency (USEPA: Osborne 1993), the United States Army Corps of Engineers (USACOE 1999), and the United States Bureau of Reclamation (USBR 1977). Other excellent reference books include *Groundwater and Wells* by Driscoll (1986) which covers all aspects of well design, drilling, and testing, *Analysis and Evaluation of Pumping Test Data* by Kruseman & de Ridder (1994), and *Aquifer Testing: Design and Analysis of Pumping and Slug Tests* by Dawson & Istok (1991).

In general, aquifer pumping test consists of the following five phases: 1) Definition of test objectives; 2) Development of a focused site conceptual model; 3) Test design; 4) Data collection (recording); and 5) Data analysis. The question of defining the test objectives could simply be rephrased to "what exactly do you want to learn about the aquifer?" Is the test being conducted as part of a water budget study where the concern is defining transmissivity and storativity; or is the test part of a water supply study where the concern is specific capacity and safe well yield; is the test part of a groundwater contaminant transport study where the ultimate question is the velocity and direction of the groundwater flow? Is there any concern between the possible interconnection of two or more separated aquifers, such as a near-surface water table aquifer and a deeper artesian aquifer separated by an aquitard? A careful definition of the test objectives is therefore essential to ensure a successful test (USACOE 1999).

Although aquifer pumping tests are, among other things, performed to provide data for development of a more detailed site conceptual model, at the same time it is very important that already available hydrogeologic information be fully utilised in the test design. Ideally, this focused site conceptual model (CSM) includes locations (or their estimates) of any hydraulic boundaries that may impact drawdown measured in the pumped well and the monitoring wells. Hydraulic boundaries are contacts with less permeable formations (*aquitards*) in both the vertical and the horizontal directions, and all natural and artificial surface water drainage features near the test site including more "exotic" ones such as leaky sewers. In karst aquifers it is equally important not to oversee possible existence of subsurface equipotential boundaries such as flowing karst conduits and channels. Possible well interference due to other wells pumping in the vicinity should also be examined.

Preliminary estimates of the aquifer thickness and hydraulic conductivity, including its anisotropy, as well as of the underlying and/or overlying aquitards (where applicable), are useful in predicting possible pumping rates for the test, radius of influence, and drawdown at the test well and various distances from it (i.e. at the existing or possible future monitoring wells). These estimates and other available information on the site hydrostratigraphy integrated within the CSM are then used in the next phase – the test design. In contaminant hydrogeology every attempt should be made to delineate horizontal and vertical extent of groundwater contamination prior to any aquifer field test. This delineation should be followed by an analysis of possible impact of the proposed test on contaminated groundwater flow directions, both horizontal and vertical.

When performing aquifer tests in karst aquifers it is particularly important to dispose of the extracted water far enough from the pumping well so to exclude possibility of its percolation back to the aquifer where it may impact the drawdown. As explained further, it is also important to plan for a longer test in order to analyse possible effects of the dual porosity nature of the aquifer.

5.4.2 Data analysis

Most methods for characterising aquifer parameters in porous media, including in karst, are based on quantitative analysis of the observed data of drawdown versus the time of pumping at the well. The drawdown is measured in one or more observation wells (piezometers) and/or the pumping well itself. Because the time of pumping is explicitly included in mathematical formulas describing relationship between the rate of pumping, the drawdown and the aquifer parameters such as transmissivity and storage properties, these methods are called "transient" or time-dependent. Steady state methods of aquifer test analysis (such as method proposed by Thiem in 1906) are rarely applied in modern hydrogeology and will not be discussed. Interested readers should consult excellent work by Wenzell (1936), which describes the method in detail, including its limitations and an approximate determination of the specific yield.

Virtually all of the pumping test methods commonly applied today are in one way or another based on the pioneering work of Theis (1935) and involve some type of graphoanalytical solution during which the observed time-drawdown data are matched against theoretical model curves ("type curves"). Theis equation can be used to determine aquifer transmissivity and storage if frequent measurements of drawdown versus time are performed in one or more observation wells. Since the equation has no explicit solution, Theis introduced a graphical method, which gives T and S if other terms are known. Although the Theis method has quite a few assumptions (the aquifer is homogeneous, isotropic, and confined with an unlimited extent, horizontal impermeable base, horizontal flow lines, and no leakage), it has been shown that it often provides an approximate solution within acceptable error limits (e.g. 5–10 percent error). With some minor changes and corrections, the

Theis equation has also been applied to unconfined aquifers and partially penetrating wells (e.g. Hantush 1961a,b, Jacob 1963a,b, Moench 1993), including when monitoring wells are placed in the non-horizontal flow domain closer to the pumping well (e.g. Stallman 1961, 1965). However, hydrogeologic conditions in the field commonly prevent direct application of Theis equation. Various analytical methods have been continuously developed to account for situations such as:

- Presence of leaky aquitards, with or without storage and above or below the pumped aquifer (Hantush 1956, 1959, 1960, Hantush & Jacob 1955, Cooper 1963, Moench 1985, Neuman & Witherspoon 1969, Streltsova 1974, Boulton 1973);
- Delayed gravity drainage in unconfined aquifers (Boulton 1954a,b, 1963, Neuman 1972, 1974, 1975, Moench 1996);
- Other "irregularities" such as heterogeneous aquifers, large-diameter wells and presence of bore skin on the well walls (Papadopulos & Cooper 1967, Moench 1985, 1993, Streltsova 1988);
- Aquifer anisotropy (Papadopulos 1965, Hantush 1966a,b, Hantush & Thomas 1966, Boulton 1970, Neuman 1975, Maslia & Randolph 1986).

Various attempts have been made to develop analytical solutions for fractured aquifers, including dual-porosity approach and fractures with skin (e.g. Moench 1984, Gringarten & Witherspoon 1972, Gringarten & Ramey 1974). However, due to the inevitable simplicity of analytical solutions, all such methods are limited to regular geometric fracture patterns such as orthogonal or spherical blocks, and single vertical or horizontal fractures.

Regardless of the aquifer type and the nature of aquifer porosity, it is important to understand that often more than just one type curve or analytical method may be fitted to the observed time-drawdown curve. Selection of the "correct" one therefore depends on the thorough overall hydrogeologic knowledge about the aquifer in question. This is particularly true in case of karst aquifers because of the absence of any prescribed, rigorous and widely agreed-upon analytical method of aquifer pumping test analysis. As is the case with numeric groundwater modeling in karst environments, analysis of aquifer pumping tests is mostly performed by applying and combining methods developed for intergranular and simple fractured aquifers.

There are quite a few general textbooks and publications on analyzing aquifer pumping test results such as Ferris et al. (1962), Stallman (1971), Lohman (1972), USBR (1977), Driscoll (1986), Walton (1987), Kruseman & de Ridder (1990), and Dawson & Istok (1991). Results of aquifer tests are now routinely analysed using computer programs and manual matching of the observed data to type curves is rather antiquated. Also commonly applied is inverse modeling of aquifer test results, using either analytical or numeric methods. AQTE-SOLV (HydroSOLVE 2003) is one of the better-known and versatile computer programs for aquifer test analysis and includes over thirty different methods for various types of aquifers. Particularly useful is the United States Geological Survey published series of ten public domain (free) Microsoft Excel-based programs for aquifer test analysis (Halford & Kuniansky 2002).

Fig. 5.11 shows comparative analysis of the aquifer pumping test conducted in the Upper Floridan aquifer in southern Georgia at Test Site 14 near Pateville, USA. The test was performed at 88 m deep irrigation well with the pumping phase lasting 72 hours and the pumping rate of 90 L/s. The thickness of the Floridan aquifer in the test area is approximately 105 m. A residuum layer, approximately 20 m thick, overlies the Floridan aquifer limestone

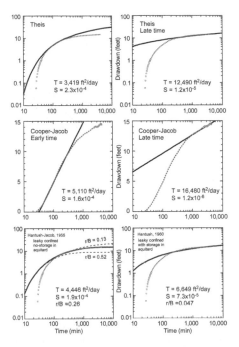

Figure 5.11. Theis, Cooper-Jacob, and leaky aquifer solutions for the time vs. drawdown data at the Upper Floridan aquifer observation well F1, Test Site 14, Pateville, Georgia, USA. See text for discussion (data courtesy of Geologic Survey Branch of the Georgia Environmental Protection Division, USA).

at the test site. Drawdown was recorded at two partially penetrating observation wells, F1 and F2, which are located 380 and 90 m from the test well respectively. Comparison of various grapho-analytical methods applied to the time-drawdown data at F1 is given for illustrative purposes only. The results of all the analyses reflect adjustments for partial penetration of the well and the hydraulic head trend observed prior to pumping. Fig. 5.11 clearly shows that the Floridan aquifer does not behave as an ideal "Theis-ian" confined aquifer and that one simple Theis type curve (or a Cooper-Jacob straight line) cannot provide satisfactory estimates of the aquifer parameters: transmissivity – T, and storativity – S. Deviation from the Theis curve indicates that some "less than an ideal" (e.g. not purely confining) mechanism is influencing the drawdown. At least several possible explanations immediately come to mind: 1) the cone of depression is reaching a more permeable portion of the aquifer which is reflected in the slower rate of the drawdown increase, 2) the overlying residuum is starting to provide water due to the hydraulic head decrease in the pumped aquifer ("leaky" aquitard solution), 3) the effects of "dual-porosity" are starting to show up (see further in this chapter), and 4) an equipotential boundary, such as a karst conduit, is starting to exert its influence on the drawdown (see next section).

Given a significant difference in the results between the methods shown, it is obvious that obtaining "true" (i.e. more representative) aquifer parameters would require additional investigations and hydrogeologic analysis, including a longer pumping test and possibly a different analytical method in other to assess the "dual porosity" nature of the aquifer. In any case, it is important to understand that the drawdown data later in the test is more

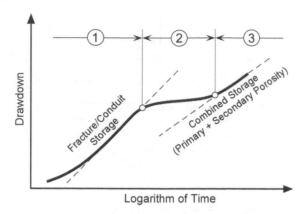

Figure 5.12.   Theoretical response of the time-drawdown curve caused by effects of dual porosity in a karst aquifer. (1) Drawdown due to the initial drainage of secondary porosity (fractures and conduits). (2) Transitional drawdown. (3) Drawdown due to stabilised drainage of all porosity types, including matrix porosity.

representative of the overall aquifer characteristics and its future response to a long-term pumping such as for water supply, irrigation or aquifer restoration purposes. In the absence of a longer observation period, it is therefore advisable to focus on "curve fitting" of the late drawdown data, rather than trying to "approximately" fit the entire data set.

Fig. 5.12 illustrates a typical response of a karst aquifer to a pumping test, revealing the nature of the aquifer porosity. Given enough pumping time and presence of all porosity types, the time-drawdown curve would show three distinct segments. The first portion of the curve, with a uniform slope, indicates a quick response from a well-connected network of secondary porosity, which may include large dissolutional openings and/or fractures. Drainage of this type of porosity, in the early stages of the test, is characterised by storage properties generally similar to that of confined aquifers. Unconfined intergranular aquifers often exhibit similar early response to pumping, due to sudden change in hydraulic pressure, with the storage coefficient significantly less than 1 percent (1×10^{-3} or less). The flattening of the curve (curve portion 2) indicates that the initial source of water in large solutional openings ("channels") and/or fractures is being supplemented by water coming from another set of porosity. This additional inflow of water starts when the fluid pressure in the channels and/or large fractures decreases enough, resulting in the hydraulic gradients from the smaller fractures and fissures towards the larger fractures and/or karst conduits. This is a transitional portion of the curve. Again, a similar response often happens in unconfined intergranular aquifers when additional water, due to gravity drainage, starts reaching the lowered water table (this is called delayed gravity response to pumping).

As the sources of water from different sets of secondary porosity features, possibly including water from the primary ("matrix") porosity attain similar level of influence, the drawdown curve exhibits another relatively uniform slope (curve portion 3). Such rock formation is often referred to as a dual-porosity formation because of the distinct hydraulic characteristics of different types of porosity present (Barenblatt et al. 1960). An exact determination of individual storage properties of different porosity types in a true karst aquifer is beyond capability of common aquifer pumping tests. Aquifer storage parameters related to

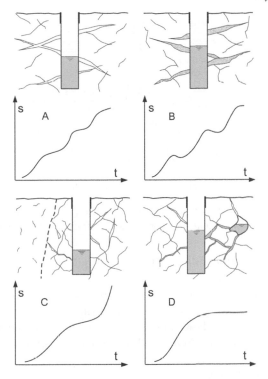

Figure 5.13.   Characteristic time-drawdown curves for some examples of well pumping tests in karst aquifers. A) Pumping from a limited number of large fracture and conduit sets which are being consecutively dewatered; B) Pumping from large fractures/conduits filled with clastic sediments which are being washed out; C) Cone of depression reaches a less permeable (less fractured) portion of the aquifer; D) Cone of depression reaches an equipotential boundary such as a large karst conduit with flowing water or surface stream (modified from Larsson 1984).

the effective matrix porosity may be approximately determined by laboratory tests, which would ideally include core samples from multiple locations and depths within the aquifer.

Depending on the duration of the test and the characteristics of the karst aquifer system itself, the time-drawdown curve may exhibit additional changes due to some factors external to the main portion of the aquifer being pumped. These factors may include influence of impermeable boundaries, less permeable portions of the aquifer, recharge, or constant head boundaries. Given the nature of karst aquifer storage properties, and almost inevitable external influences, time-drawdown curves of long-term pumping tests in karst aquifers exhibit a wide variety of shapes, often non-typical. It is therefore critical that such tests be analysed using overall geologic and hydrogeologic knowledge about the tested aquifer, rather than formally applying some predetermined "type curve formula".

Fig. 5.13 illustrates just some of the possible responses of karst and fractured aquifers to groundwater withdrawal with three successively higher pumping rates. In addition to the discussion presented within the figure caption, it is important to emphasise the importance of step-drawdown tests in karst aquifers. Such tests, routinely performed to determine well efficiency and optimum pumping rate for the main aquifer test (e.g. Rorabaugh 1953,

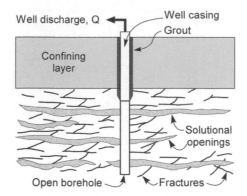

Figure 5.14. Conceptual model of groundwater flow to a well pumping in a karst formation consisting of solutional openings and a network of diffuse fractures (Greene et al. 1999).

Roscoe Moss Company 1990, Kresic 2007), may quickly reveal the nature of aquifer porosity surrounding the well.

A schematic of a formation having a network of dissolutional openings intersecting a network of "diffuse" fractures (in which the groundwater flow is slow), and being pumped by a fully-penetrating well, is shown in Fig. 5.14. The equations and boundary conditions governing the pumping of water from a well in a dual-porosity karst formation are given by Greene et al. (1999) and described below.

Flow in dissolutional openings:

$$S\frac{\partial h}{\partial t} - T\frac{1}{r}\frac{\partial}{\partial r}\left(r\frac{\partial h}{\partial r}\right) = \beta(h_f - h)$$

Initial condition in dissolutional openings: $h(r, t = 0) = H$
Boundary conditions in dissolutional openings:

$$2\pi rT \left.\frac{\partial h}{\partial r}\right|_{r\to 0} = Q$$

$$h(r \to \infty, t) = H$$

Flow in fractures: $S_f\frac{\partial h_f}{\partial t} = -\beta(h_f - h)$

Initial condition in fractures: $h_f(r, t = 0) = H$

where S is the storativity of dissolutional openings, S_f is the storativity of the network of diffuse fractures, t is the time, T is the transmissivity of dissolutional openings, h is the hydraulic head in the dissolutional openings, h_f is the hydraulic head in the network of diffuse fractures, H is the initial head, r is the radial distance from the pumping well, and β is the rate of fluid exchange between the network of fractures and the dissolutional openings.

Assuming that the aquifer is homogeneous and isotropic, its properties are estimated by solving the above equations for various choices of T, S, S_f and β, and comparing the calculated type curves with the observed data until an acceptable fit is obtained. Fig. 5.15 shows the results of such curve fitting to the data observed during testing of the Dickey well in the confined Madison limestone karst aquifer near Spearfish, South Dakota, USA

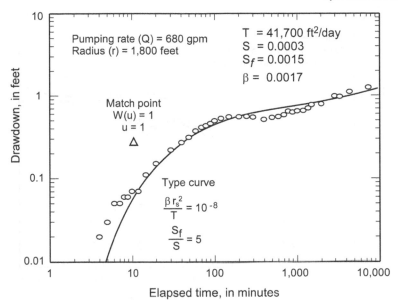

Figure 5.15. Best fit type curve match to the drawdown data observed at the Kyte well during the 6-day aquifer pumping test at the Dickey well, Madison limestone aquifer (Greene et al. 1999).

(Greene et al. 1999). The observed data are for the Kyte well, located 1800 feet from the Dickey well which was pumped for 6 days at 680 gallons per minute. Note that, if the test had lasted only one day for example, the effects of dual porosity would have not been apparent. As mentioned earlier, unconfined aquifers often exhibit similar time-drawdown curves due to delayed gravity response.

For illustration purposes, Fig. 5.16 shows the results of fitting the Kyte well data to a type curve obtained using Neuman solution for unconfined aquifers with delayed gravity response. Arguably, the fitted type curve looks even "better" than the one on Fig. 5.16. Boulton (1973) showed analytically that the same set of equations can be applied to: 1) the delayed flow to a well in an unconfined aquifer, and 2) leakance from the upper aquifer through the aquitard overlying the pumped confined aquifer. Kresic (2007) argues that possible interpretation of the aquifer test results shown in Figs. 5.16 and 5.17 thus become more ambiguous knowing that the Madison limestone aquifer at the test location is overlain by the Minnelusa aquifer which in its lower part acts as a confining layer consisting of interbedded sandstones and dolomitic limestone.

Whatever the reason for the apparent dual porosity-like response to the pumping test is (e.g. true dual porosity nature of the karst aquifer, or leakance from the overlying aquitard, or both), the selection of aquifer parameters for some future quantitative analysis should always consider the late drawdown information. Interestingly enough, as illustrated by Fig. 5.17, a simple Cooper-Jacob solution for both the early and the late drawdown data gives values of T and S very similar to the dual-porosity solution of Greene et al. (1999). The Cooper-Jacob early storativity is almost identical with the fracture/conduit storativity of Greene et al., while the Cooper-Jacob late storativity is very similar to the "diffuse fracture" storativity of Greene et al.

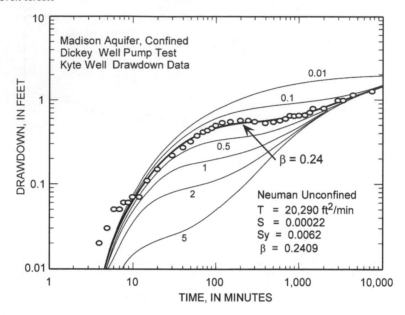

Figure 5.16.   Neuman solution type curve match to the drawdown data observed at the Kyte well, Madison limestone aquifer, during the 6-day aquifer pumping test at the Dickey well (Kresic 2007; time-drawdown data from Greene et al. 1999).

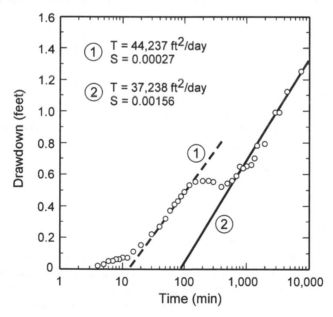

Figure 5.17.   Cooper-Jacob solution to the early and late drawdown data observed at the Kyte well, Madison limestone aquifer, during the 6-day aquifer pumping test at the Dickey well (Kresic 2007; time-drawdown data from Greene et al. 1999).

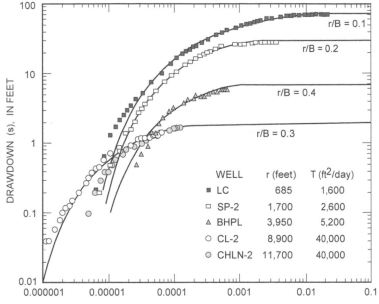

Figure 5.18. Drawdown data for five observation wells superimposed on the best fit leaky confined aquifer type curves for RC-5 aquifer test, Madison limestone aquifer, South Dakota, USA (Greene 1993).

5.4.3 Aquifer anisotropy

Karst aquifers are by definition heterogeneous and anisotropic when analysed at a realistic field scale. Determining quantitative parameters of these two very important characteristics of the karst porous media is virtually impossible using aquifer pumping test results alone. At least some information on tectonic (fracture) fabric, geometry and nature of aquifer top and bottom, and carbonate depositional environment, should be available in order to successfully interpret results of an aquifer test in karst. Various methods of determining anisotropy of aquifer transmissivity, developed for homogeneous intergranular aquifers, have been more or less successfully applied in karst using the equivalent porous medium approach. Warner (1997) applies methodology developed by Papadopulos (1965) and a related computer program by Maslia & Randolph (1986) to estimate anisotropy of the Upper Floridan aquifer transmissivity using pumping test data from the southwestern Albany area, Georgia, USA. A degree of professional judgment is exercised by excluding from the analyses, as an outlier, a well with the highest transmissivity. Although debatable, such practice of a selective interpretation of the data is often present in karst hydrogeology given its complexity and non-conformity with many methods developed by the "classic intergranular" hydrogeology.

Greene (1993) explains in detail a method developed by Hantush (1966b) using an example from a pumping test conducted in the Madison limestone aquifer in the western Rapid City area, South Dakota, USA (see Fig. 5.18 and Fig. 5.19). The method assumes homogeneous, leaky confined aquifer conditions where anisotropy is in the horizontal plane. Fig. 5.18 shows field-data plots for five observation wells located at different distances (*r*)

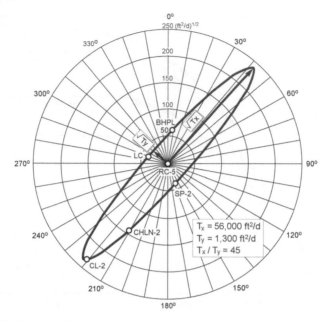

Figure 5.19.   Theoretical ellipse showing the angle and magnitude of the major and minor axes of transmissivity from the anisotropic analysis of the RC-5 aquifer test, Madison limestone aquifer, South Dakota, USA (Greene 1993).

and angles from the production well. The five plots would theoretically fall on the same curve if transmissivity was not anisotropic or leakage did not occur. The displacement of the plots from one another is the result of directional transmissivities in the Madison aquifer and because leakage through the confining bed does not equally affect drawdown in the five wells (Greene 1993).

The results of the anisotropy analysis are presented in Fig. 5.19. The major axis of transmissivity is 56,000 ft^2/d (5200 m^2/d) at an angle of 42 degrees clockwise from north. The minor axis of transmissivity is 1300 ft^2/d (120 m^2/d) at an angle of 48 degrees counterclockwise from north. It should be noted that any such analysis, based on the equivalent porous medium approach, is just an indicator of a possible anisotropy due to, for example, some preferential flow paths (conduits) in the karst aquifer.

5.4.4   Transmissivity of discrete aquifer zones

In a true karst aquifer, with heterogeneous secondary porosity (fractures and dissolutional cavities), the inflow of water to a pumped well would most likely occur from discrete intervals where the borehole intersects preferential flow paths within the surrounding rock. One such example is illustrated in Fig. 5.20 which shows results of point flow calculations based on the flowmeter data collected in the pumping well 12K147 during three different overall pumping rates (Warner 1997). The majority of the flow entering the well is from two discrete zones at about 118 and 122 feet (35 and 37 m) below ground surface. The graph also illustrates that the lower tested portion of the well does not contribute any flow to the well at the lowest pumping rate of 1080 gallons per minute (4087 L/min). Based on just this

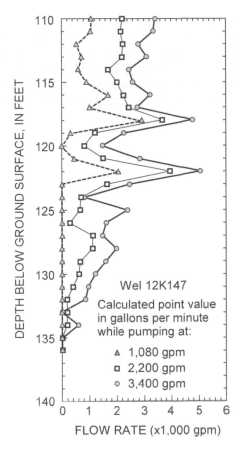

Figure 5.20. Calculated point flow rates entering a test well in the Upper Floridan aquifer near Albany, Georgia, USA during three different overall pumping rates (modified from Warner 1997).

data, one could erroneously conclude that the permeability of the 123–136-foot section is very low: when the pumping rate is doubled and then tripled, the section in question does contribute significantly to the overall pumping rate. Although water entering the well along this section may still be coming indirectly from the two preferential zones located above, it is clear that the rock mass is capable of transmitting this flow and can certainly not be qualified as "low permeable".

5.5   SUMMARY

Equivalent porous media generalisations in karst hydrogeology may be unavoidable in many cases, but should always be based on a solid overall knowledge of the specific aquifer system in question. Calculations of groundwater flow rates through various portions of a karst aquifer should be based on the hydraulic heads observed in monitoring wells only when there is a clear understanding of the underlying hydrogeology. Borehole packer tests are used to evaluate discrete intervals in the immediate borehole vicinity and should not be used alone to calculate representative flow rates in the aquifer. Combining borehole flowmeter tests and aquifer pumping tests is arguably the only method that can be used to determine aquifer transmissivity and hydraulic conductivity reasonably accurately. Caution should be exercised when interpreting results of aquifer pumping tests: whenever possible, one should account for the actual representative thickness of the aquifer being pumped (i.e. for the preferential zones providing most of the well flow) rather than for the entire well length open to the aquifer.

CHAPTER 6

Hydrochemical methods

Daniel Hunkeler & Jacques Mudry

6.1 INTRODUCTION

Hydrochemical techniques provide significant information about the functioning of karst aquifer systems and they complement hydrodynamical methods. While some features of karst aquifers can be derived from hydrographs (retention times, volumes), these methods do not yield data concerning actual transit or residence time and on location of storage (epikarst, unsaturated or saturated zone). Hydrochemical techniques, if a sufficient contrast of parameters exists, provide answers to these questions and also help to locate and quantify the acquisition of mineralisation by the water. Thus, the chemical compounds can be considered as natural tracers that provide information about the structure and dynamics of karst aquifers. In addition, hydrochemical studies are often motivated by water quality problems and aim at identifying the extent and origin of water contamination. The two main goals of hydrochemical studies, natural tracing and water quality assessment may often be combined. On one hand, a good understanding of the functioning of a karst aquifer is necessary for a thorough diagnosis of water contamination. On the other hand, anthropogenic contaminants such as NO_3^- or Cl^- can sometimes also provide information about the characteristics of karst aquifers.

6.2 PARAMETERS AND PROCESSES

6.2.1 Generalities

The chemical composition of water at a spring or at an intermediate position in the karst system depends on several factors (Fig. 6.1) including land use, recharge mechanism (diffuse vs. concentrated), climatic conditions, lithology and type of flowpaths (diffuse vs. conduit).

Different types of land use such as forests, grasslands or cultures modify the composition of infiltrating water, especially due to the use of fertilizers, but also by stimulating the decomposition of soil organic matter. A part of the water often infiltrates rapidly for example through swallow holes, bypassing the soil zone and other protecting layers such as till. Rapidly infiltrating water generally has a low level of mineralisation and may transport dissolved or particle-bound contaminants downward. The infiltrated water can reach the saturated zone by different pathways. The epikarst plays an important role in regulating the transit of water. It can store water for a considerable amount of time and distribute it

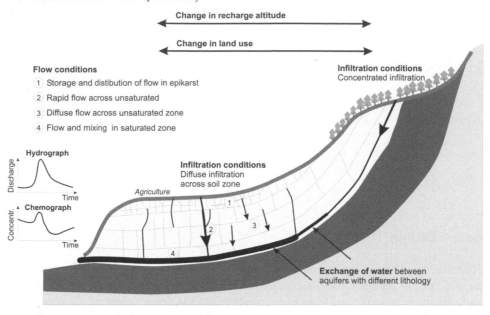

Figure 6.1.   Factors influencing the hydrochemical parameters.

between vertical conduits that may be reached after some horizontal flow within the epikarst and diffuse flow along fissures of the unsaturated zone. The time available for rock-water interactions, and hence the chemical composition of water, strongly varies depending on the flow path and storage location of the water. The flow path and residence time also influences the contaminant fate. During transport in conduits little sorption occurs due to low surface area and little biodegradation due to the rapid transfer of water. However, especially in the case of diffuse infiltration, transport often also occurs across zones of fissured rock with lower velocity and higher surface area, or contaminants may be stored in the epikarst or in LPV before reaching the spring, increasing the possibility for sorption or biodegradation. Especially during flood events, particles that may carry contaminants (Mahler & Lynch 1999, Mahler et al. 1999) and dissolved compounds from the soil zone or surface waters are often transported rapidly to the spring. The spring discharge consists of a mixture of water that has reached the saturated zone by different pathways, and may have been stored in saturated low permeability zones adjacent to the conduits. The resulting temporal variation of hydrochemical parameters is termed a chemograph. For designing sampling programs and interpretation of hydrochemical responses, it is important to know the origin of the different chemical compounds and the processes that control their concentrations. In the following, the characteristics of important hydrochemical parameters are discussed in more detail according to their main origin (Tab. 6.1).

6.2.2   Precipitation-related parameters

Rainwater is not pure H_2O. Its chemical composition can be influenced by natural factors such as sea spray in coastal areas leading to elevated Na^+, K^+, Cl^- and SO_4^{2-} concentration or air pollution (Appelo & Postma 2005). The isotopic composition of rainwater (^2H, ^{18}O,

Table 6.1. Factors controlling various hydrochemical parameters.

	Parameter	Source	Sink
Soil-related parameters	CO_2	Root respiration and degradation of organic matter	Dissolution of carbonate minerals, Degassing in conduits
	^{222}Rn	Radioactive decay of ^{226}Ra	Radioactive decay with half-life of 3.8 days
	Dissolved organic carbon (DOC)	Decomposition of litter and humus	Sorption to clay minerals, biodegradation
	NO_3^- [1]	Synthetic and organic fertilizer, nitrification of NH_4^+ from fertilizer	Denitrification under anaerobic conditions
	NH_4^+ [1]	Synthetic and organic fertilizer, decomposition of soil organic matter	Nitrification, Ion exchange
	PO_4^{3-} [1]	Synthetic and organic fertilizer	Sorption on Fe-hydroxides, precipitation with Ca^{2+}
	Cl^- [1]	Fertilizer, Rain, road salts	Conservative
	SO_4^{2-} [1]	Fertilizer	Conservative except under very reducing conditions
	K^+ [1]	Fertilizer, Dissolution of silicates	Ion-exchange
	Turbidity	Particles from soil zone, also from sediments in conduits	Filtration
Carbonate-rock related parameters	Ca^{2+}	Dissolution of carbonate minerals	Ion-exchange, precipitation of carbonate minerals
	Mg^{2+}	Dissolution of carbonate minerals	Ion-exchange
	HCO_3^-	Dissolution of carbonate minerals	Precipitation of carbonate minerals
	$\delta^{13}C$ of DIC	Dissolution of soil gas CO_2, carbonate minerals	
Parameters related to other rock types	SO_4^{2-}	Dissolution of gypsum and anhydrite in evaporates	Conservative except under very reducing conditions
	Sr^{2+}	Dissolution of celestite	Ion-exchange
	Various trace metals	Dissolution of evaporites (see text)	
Anthropogenic compounds	Metals	Fertilizers, road runoff, air pollution	Adsorption, Precipitation
	Pesticides	Agriculture	Adsorption, Transformation
	Volatile organic compounds	Industrial sites, landfills, traffic	Volatilisation, Sorption, Biodegradation

[1] partly of anthropogenic origin.

^3H in H_2O) can serve as a natural tracer to infer information about transit time and storage of groundwater as discussed in Chapter 7 of this volume.

6.2.3 Soil-related parameters

The hydrochemistry of groundwater that originates from diffuse infiltration is strongly influenced by its passage through the soil zone (Tab. 6.1). During passage through the soil, the water becomes enriched in CO_2 that originates from root respiration and degradation of soil organic matter. The soil gas CO_2 concentration is generally higher in summer than in winter due to the higher biological activity. It decreases with altitude, due to less developed soils, less vegetation and lower temperatures. It also depends on land use, with higher CO_2 levels of up to 10% in arable soil due to stimulation of organic matter decomposition. Part of the dissolved CO_2 is transformed into carbonic acid that promotes the dissolution of carbonate and other minerals. In the deeper zones of karst system, the annual amplitude of the CO_2 concentration in gaseous and aqueous phases tends to be smaller than in the soil zone and concentrations are lower due to consumption for carbonate dissolution (e.g. Savoy 2007). Rain water or surface water that infiltrates directly into the conduit system has a much lower CO_2 concentration than soil water and thus a lower capacity to dissolve carbonate minerals. In addition to CO_2, other gases present in the soil gas may dissolve, for example ^{222}Rn, an interesting natural tracer that is discussed in Chapter 7.

The soil zone is also the principal source of ***dissolved organic carbon (DOC)***, which consists of a complex mixture of organic molecules of variable size that mainly originates from decomposition of litter and humus and is usually analysed after filtration at 0.45 µm. If the samples are not filtered, the total organic carbon (TOC) is obtained, which includes particulate organic carbon (POC) in addition to DOC. The amount of DOC that is leached from soils depends on interaction between DOC release, biodegradation and physical retention by sorption or precipitation (Kalbitz et al. 2000, Neff & Asner 2001). It is also strongly influenced by hydraulic factors such as the rate of water flux and occurrence of storm events. In general, the downward flux of DOC decreases with increasing soil depth due to DOC retention in deeper soil layers and biodegradation. DOC concentrations in soil solutions tend to be higher in summer than in winter because of a higher DOC release associated with higher biological activity and lower infiltration due to evapotranspiration. However, DOC removal by microbial mineralisation may be enhanced as well. During storm events, a DOC pulse may be released, especially if preferential flow paths are present that have little DOC retention. DOC is commonly divided into a labile and refractory fraction, whereby the former can constitute up to 40% of the DOC. DOC can serve as a natural tracer indicating a rapid transfer of water from the soil zone towards springs with little time for degradation or physical retention. Furthermore, DOC is also an important parameter from the point of view of water quality. DOC can form complexes with metal cations and adsorb hydrophobic organic contaminants, facilitating their transport. In addition, the presence of DOC leads to formation of chlorinated hydrocarbons if water is disinfected with chlorine.

In addition to dissolved compounds, ***particles***, including ***microorganisms***, can be transported from the surface to the saturated zone, especially in case of concentrated recharge and conduit flow (Mahler & Lynch 1999). Transport of particles from the surface can lead to a turbidity pulse at the spring after rain events (primary turbidity), which implies fast flow paths with little filtration between surface and spring. However, turbidity can also originate from re-suspension of sediments in conduits during a flood event (secondary turbidity). Particles can be loaded with chemical and microbial contaminants.

Soil related parameters can be used as indicators for fast arrival of the water from the soil zone to the sampling point provided that they decay with increasing residence time and are not stored in deeper subsystems. This is the case for DOC that may be degraded or sorbed, ^{222}Rn that decays by radioactivity, but not necessarily for NO_3^- (see below), which can persist under aerobic conditions. The CO_2 concentration can indicate transfer of water from the soil zone and epikarst especially during summer when the CO_2 concentration is higher in the soil zone than in deeper zones where the annual variations are buffered (e.g. Savoy 2007).

6.2.4   Carbonate-rock related parameters

The chemical composition of groundwater from carbonate aquifers is usually dominated by products from dissolution of carbonate minerals including calcite, aragonite and dolomite. In addition, minor amounts of Fe- or Mn-carbonates may be present. These minerals often do not occur in pure form, but part of the cations may be substituted by others having a similar radius and the corresponding mixed minerals are denoted as solid solutions. For example, Mg^{2+} may substitute some of the Ca^{2+} in calcite and Sr^{2+} some of the Ca^{2+} in aragonite, while Sr^{2+} will not replace Ca^{2+} in calcite (Appelo & Postma 2005). A study of various carbonate aquifers in Europe suggested that trace element concentrations in groundwater are generally low. Only iodine was often found to be present in the low $\mu g/L$ range, probably originating from the decomposition of fossil organic matter (Kilchmann et al. 2004).

Dissolution of carbonate minerals in the presence of CO_2 proceeds, according to the following stoichiometric equations:

$$CO_2 + H_2O \leftrightarrow H_2CO_3 \tag{1}$$

$$H_2CO_3 \leftrightarrow HCO_3^- + H^+ \tag{2}$$

$$CaCO_3 \leftrightarrow Ca^{2+} + CO_3^{2-} \tag{3}$$

$$CO_3^{2-} + H^+ \leftrightarrow HCO_3^- \tag{4}$$

The solubility product corresponding to reaction 3 is quite small. However, because the carbonate ion (CO_3^{2-}) is "removed" by protonation (eq 4), dissolution of carbonate minerals can proceed to a significant extent, depending on the quantity of carbonic acid that is available and delivers protons (eq 2). Equations 1–4 can be summarized by:

$$CaCO_3 + CO_2 + H_2O \leftrightarrow Ca^{2+} + 2HCO_3^- \tag{5}$$

This equation again illustrate that carbonate dissolution depends on the amount of CO_2 or H_2CO_3 that is available. Two different modes of CO_2/H_2CO_3 supply can be distinguished although in nature the conditions may frequently lay in between: open system conditions, where the water is in contact with a gas phase of constant CO_2 partial pressure, and closed system conditions where no exchange between aqueous and gas phase occurs. In the former case, the amount of carbonate mineral that is dissolved at equilibrium is larger, and the final pH lower because the consumed carbonic acid is continuously replenished by gaseous CO_2 (Fig. 6.2). For a given CO_2 partial pressure, the calcite solubility decreases with increasing temperature because the solubility product for calcite dissolution (eq 3) and the solubility of CO_2 in water are lower. However, the larger CO_2 production at higher temperature due to increased biological activity usually more than compensates for this temperature effect and thus with increasing temperature during summer a net increase of calcite dissolution is usually observed.

The rate of dissolution of carbonate minerals depends on a number of factors, including the rate of detachment of ions from the mineral surface, transport processes across the

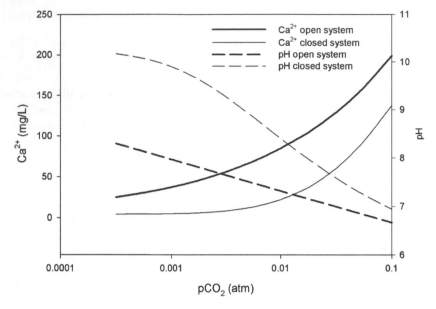

Figure 6.2.   Equilibrium concentration of calcium and pH as a function of the initial CO_2 partial pressure for dissolution of calcite under open and closed system conditions.

stagnant boundary layer adjacent to the surface and the rate of supply of carbonic acid by reaction of CO_2 with water (Dreybrodt 1998). Hence, the dissolution rate depends on flow velocity and flow conditions (laminar or turbulent), which control the thickness of the boundary layer. Therefore, it is not possible to define a typical time scale for reaching dissolution equilibrium. For calcite dissolution, it is often observed that 90% saturation is reached relatively fast, but afterwards, the dissolution rate strongly decreases, possibly due to inhibition of dissolution by other ions. The dissolution rate of dolomite is considerably smaller than that of calcite. Appelo & Postma (2005) estimate that it takes more than 100 times longer to reach 95% saturation for dolomite than for calcite under certain conditions. Due to the slow dissolution of dolomite, elevated Mg^{2+} concentrations have been used as an indicator of long groundwater residence times although it is again hardly possible to define an absolute time scale because the dissolution rate is influenced by many factors.

Due to the dominance of the CO_2-$CaCO_3$ system on the hydrochemistry, a number of parameters have been used to characterize the state of this system. These include the partial pressure of CO_2 (pCO_2), concentrations of Ca^{2+}, Mg^{2+}, HCO_3^-, the saturation index SI for carbonate minerals and the stable carbon isotopic composition of dissolved inorganic carbon (DIC). The use of ^{13}C as natural tracer is only briefly touched on in this chapter. More information on this method can be found in Emblanch et al. (2003).

6.2.5   Parameters related to other rock-types

Carbonate rocks are sometimes associated with evaporite beds. Evaporites are composed of gypsum and/or anhydrite and occasionally also halite. Water that was in contact with evaporites has usually elevated concentrations of SO_4^{2-}, a higher content of trace elements and sometimes also an elevated concentration of Na^+ and Cl^-. Typical trace elements are

Sr^{2+}, Ba^{2+} and F^- originating from dissolution of celestite, barite and fluorite, Mn, Ni, Cd and Cu from dissolution of carbonates and oxidation of sulphides, Cd from phosphates and Li and Rb from brine inclusions (Kilchmann et al. 2004). The concentration of Sr^{2+} is often particularly high and frequently correlated with SO_4^{2-} concentrations. In addition to its origin from celestite, which may be present as microcrystalline inclusions in gypsum, Sr^{2+} can also occur as solid solution in carbonate minerals.

6.2.6   Compounds of anthropogenic origin

Anthropogenic compounds often originate from zones of limited spatial extent denoted as *point sources* such as: landfills, sewage systems, effluent of waste water treatment plants that is infiltrated into the subsurface due to the absence of surface waters, leaking manure storage facilities, leaking pipes, leaking storage tanks and accidents. Except for accidents, most of these point sources release contaminants over a prolonged period of time. However, contaminants may also be released at lower concentrations over larger areas, denoted as *diffuse sources*. The most important diffuse sources are usually from the application of pesticides and fertilizers. While these compounds are usually spread over larger areas, the application occurs as discrete events in time. If manure is applied, not only chemical compounds are released but also microorganisms some of which may be pathogenic. Pathogenic microorganisms can also originate from domestic wastewater. Another diffuse source of contamination is air pollution. In the following the origin and fate of different types of contaminants is discussed.

6.2.6.1   *Nutrients*
The use of organic and synthetic fertilizers in agriculture strongly modifies the chemical composition of soil water (Böhlke 2002). Fertilizers not only contain nutrients such as N, P and K but also by-products such as Cl^-, SO_4^{2-}, metals and, in case of organic fertilizer, organic matter and microorganisms. Often only a part of the nutrients are taken up by plants with the reminder being stored in the soil or leaking downward. Leaching is particularly important for nitrate (NO_3^-), which shows little interaction with clay minerals and metal-hydroxides. Leaching of NO_3^- is higher below cultivated than grasslands or forests. Although fertilizers are applied as discrete events, leaching of NO_3^- may occur throughout the year, since the soil represents an important reservoir of nitrogen. It can be particularly high in autumn after the harvest when a vegetation cover is lacking and infiltration is often high. In contrast to NO_3^-, cations such as K^+ and cationic metals are less mobile due to interaction with negatively charged clay minerals or due to precipitation. However, also some anions can be retained such as phosphate due to strong sorption on Fe-hydroxide or precipitation with Ca^{2+}. High evapotranspiration during summer can lead to the accumulation of mobile compounds such as Cl^- and SO_4^{2-} in the soil solution, leading to a concentration peak during the first strong precipitation event in autumn.

6.2.6.2   *Metals*
Metals in soil and infiltrating water can originate from various sources such as fertilizers, air pollution, road runoff and industrial activity. Many metals are usually present in cationic form, frequently with an oxidation state II+ such as Cd^{2+}, Pb^{2+}, Cu^{2+} or Zn^{2+} (Fetter 1999, Langmuir 1997). However, some elements are present in more positive oxidation states that change depending on redox conditions such as Cr^{3+}/Cr^{6+}, As^{3+}/As^{5+} or Sb^{3+}/Sb^{5+}.

Some of the latter elements occur in water as oxy-anions with a pH-dependant charge such as $Cr(VI)O_4^{2-}$ or $HAs(V)O_4^{2-}$. Metal cations and oxy-anions show a different behaviour. Processes that decrease the mobility are precipitation and sorption to the matrix. In carbonate aquifers, precipitation with dissolved carbonate is particularly important. Many metal cations have solubility products with carbonate that are many orders of magnitude smaller than that of calcite. Sorption of metal cations is particularly effective in case of iron-hydroxides that may form by weathering of Fe-bearing carbonate rock. Transport of metal cations is enhanced by formation of complexes in particularly with dissolved organic matter, or by adsorption on organic or inorganic (Fe-hydroxides, clay) particles. Christensen et al. (2001) demonstrated that in landfill leachate, most of the metal cations are present, sorbed on particles or in form of complexes. Particle bound metals may accumulate in sediments of conduits, and may be mobilized during flood events (Vesper & White 2004a).

The mobility of the metal ions with a higher oxidation state strongly depends on the redox conditions. Under oxidizing conditions, Cr occurs as an oxy-anion with oxidation state VI+ that shows limited interaction with surfaces and is well soluble, while it tends to precipitate as hydroxide (III+) under reducing conditions. Similarly, Sb is rather more mobile under oxic than anoxic conditions. In contrast, arsenic is more mobile as As(+III) under reducing conditions while under oxidizing conditions it forms an oxy-anion (+V) that is strongly retained by iron-hydroxides.

6.2.6.3   *Organic compounds*

Organic contaminants may be released dissolved in water or, for example in case of leaking storage tanks, as an organic liquid. The spillage of organic liquids is particularly problematic as they can migrate to substantial depth forming a long term source of dissolved compounds that is difficult to remove (Loop & White 2001). The mobility of organic liquids depends on their density and viscosity (internal resistance to flow). Organic liquids are often differentiated into LNAPLs (Light non-aqueous phase liquids) and DNAPLs (Dense non-aqueous phase liquids) depending on whether they are less dense or denser than water. Examples for LNAPLs are gasoline, diesel fuel or heating oils, for DNAPLs chlorinated solvents, creosote or polychlorinated biphenyls (PCBs). If spilled or disposed of into sinkholes, organic liquids may directly enter the conduit system unless the sinkhole has a soil plug (Crawford & Ulmer 1994). Organic liquid spilled elsewhere can migrate across the soil zone reaching the epikarst, depending on its quantity and the presence of macropores. Within the epikarst lateral migration can occur. DNAPL may accumulate on the lower boundary of the epikarst, which usually consists of a relatively impermeable, irregular limestone surface. Accumulation leads to a pressure head that may eventually be sufficient for DNAPL to penetrate into fractures leading to further downward migration. LNAPLs may also accumulate in the epikarst, floating and migrating laterally to the water table of perched aquifers. For further downward migration, larger openings than for DNAPL are usually required, since LNAPLs are not in direct contact with the rock surface as long as saturated conditions persist. During downward migration, part of the organic liquid may become trapped in cavities and dry caves of the vadose zone. In the saturated zone, DNAPLs have a tendency to accumulate at the lowest points of conduits and may be remobilised together with sediments during flood events. In contrast, floating LNAPLs are trapped behind completely saturated zones of a conduit and may be remobilised at low water level when a continuous unsaturated zone is present in the conduit (Vesper et al. 2000). Many DNAPLs and some of the compounds in

LNAPLs usually have a high volatility and thus vapour clouds can form in addition to the dissolution of compounds into water.

The fate of organic compounds depends on their solubility, the tendency to volatilise from the aqueous phase, the tendency to sorb on the solid phase, and their biodegradability (Schwarzenbach et al. 2003). Organic contaminants mainly sorb onto organic matter. The tendency to sorb is a function of the organic matter content of the subsurface and the affinity of organic compounds for organic matter – usually expressed as the water-organic matter distribution coefficient K_{OC}. Sorption may also occur onto organic particles enhancing the transport of otherwise poorly soluble compounds. Biodegradation can be an effective mechanism for elimination of organic compounds such as petroleum hydrocarbons. However, some compounds such as chlorinated hydrocarbons degrade only under specific redox conditions and degradation may be incomplete (Wiedemeier et al. 1999). Only a partial transformation by biotic or abiotic processes is often observed for pesticides and the concentration of metabolites can be larger than that of the parent compound. For example, atrazine is transformed to a number of metabolites such as desethylatrazine or desisopropylatrazine. Hence, studies on chlorinated hydrocarbons and pesticides should include the analysis for degradation products in addition to parent compounds.

6.2.6.4 *Microorganisms*
Microorganisms in aquifers can be subdivided into three groups: viruses, bacteria and protozoa. Bacteria are naturally present in aquifers, catalyse a number of biogeochemical processes, and contribute to contaminant degradation (Goldscheider et al. 2006). However, they may also originate from external sources, including natural sources such as soils and anthropogenic sources such as sewage systems, the application of organic manure or grazing cattle. Faecal matter contains a very high density of bacteria. Both animal and human faeces can contain bacteria that are pathogenic to humans. In contrast, viruses susceptible to infect humans only occur in human faeces. Protozoa are unicellular eukaryotic organisms that are naturally present and feed on microorganisms. However, there are also parasitic forms with animals or humans as a host such as Cryptosporidium or Giardia. Parasitic protozoans are usually present in the environment in a resting state (cyst), several μm in diameter. Cysts can remain infective for a prolonged period of time and are difficult to inactivate by disinfection.

The transport of microorganisms is frequently approximated by particle transport although bacteria may actively interact with surfaces. In media with small pore sizes or fissures, larger microorganisms may be trapped because of their size. However, this process is likely to be less important in carbonate aquifers where fissures may be enlarged by carbonate dissolution and where conduits exist. Another retention mechanism is physical-chemical filtration. The efficiency of filtration depends on the frequency of collision of microorganisms with surfaces and the fraction of the colliding microorganisms that stick to the surface. Sticking mainly occurs by electrostatic interactions and, since most of the microorganisms are negatively charged, is more effective for positively charged surfaces. The surface charge varies depending on the mineral and pH. Calcite, usually the dominant mineral in carbonate aquifer, generally has a positive surface charge and thus is favourable for retention of microorganisms. However, since the surface area in fracture media and conduits is quite small, likely less collisions occur and thus filtration is often less effective than in porous media. In the unsaturated zone, microorganisms may also be retained at air-water interfaces in pores. Microorganisms may also attach to particles rather than immobile surfaces, enhancing their transport (Mahler et al. 2000). Hence, elevated concentrations of bacteria often occur during flood events. In addition to filtration, an important elimination mechanism of viruses and bacteria is die-off due to environmental conditions that are harsh from the perspective of microorganisms that are adapted to conditions in the intestine of warm-blooded creatures (John & Rose 2005).

6.3   SAMPLING STRATEGIES AND METHODS

6.3.1   Spatial versus temporal sampling

To a greater degree than in slowly reacting aquifers (porous, finely fissured), the sampling strategy has to be well designed. Two major sampling strategies are used: spatial sampling in a defined geographical or geological area and temporal sampling at a limited number of locations. Spatial surveys are best performed during periods with stable discharge when hydrochemical parameters tend to vary little, such as summer droughts or low waters in winter. Spatial data from karst areas can generally not be used for plotting isoconcentration contour maps, unlike for data from granular aquifers, where concentrations tend to change gradually with distance. Spatial surveys make it possible to evaluate discrete changes in hydrochemistry between different springs or different sampling points within a karst system. These variations can reflect factors such as land use, lithology, residence time or altitude of recharge.

Temporal sampling frequently targets the reaction of karst systems to rain events. The resulting data are usually plotted as chemographs (concentration vs. time curves). The shape of chemographs not only reflects intrinsic features of a karst system (e.g. degree of karstification, presence of cover layer) but also depends on the state of the system (amount of water in storage, delay since last flood event) and external factors (intensity, temporal and spatial distribution of precipitation). Therefore, if possible, the reaction of the system to several flood events should be recorded. It has also to be taken into account that the event scale variations may be overlain by seasonal variations e.g. due to seasonal variations in CO_2 or DOC production.

6.3.2   Sampling locations

The most common sampling point is the karst spring. Because of the hierarchical nature of karst drainage patterns, waters having a different origin and transmitted along different flow paths converge on the karst spring. Sampling at the spring provides information about the functioning of the entire karst system. However, the interpretation of the data can be complicated by mixing of waters from different sub catchments. If access points exist, they can constitute intermediate sampling points between effective rainfall and outlet. Caves often provide access to water from the vadose zone (e.g. stalagmite drip water) and the saturated zone. In addition, water samples may be taken in man-made structures such as boreholes or artificial galleries. Sampling can target already organised flow (e.g. in a natural cave) or zones of diffuse flow (e.g. in an artificial gallery). Sampling at intermediate locations can provide information about hydrochemical characteristics of water in different subsystems such as the epikarst and the unsaturated zone or in different sub-catchments. Such information can help in the interpretation of spring chemographs by providing information about the chemical composition of different components contributing to spring discharge.

6.3.3 Sampling frequency

Again two different strategies can be distinguished, fixed-interval sampling and event-based sampling. Fixed-interval sampling can be sufficient for systems that show a highly inertial behaviour e.g. due to a little-developed drainage network or due to particular climatic conditions such as aridity. However, many karst systems are characterized by a rapid response to rain fall event, which will often be missed with fixed-interval sampling unless the intervals are very short. Therefore, fixed-interval sampling is often supplemented by event-based sampling targeting flood events. During flood events, samples may again be taken at regular but more closely spaced intervals – as often as every hour or even less. Alternatively, the sampling frequency can be adjusted depending on how rapidly the discharge changes. Auto-samplers are particularly useful for event-based sampling. Some of them start and adjust the sampling frequency automatically as a function of external parameters such as discharge, electrical conductivity or precipitation. While discrete sampling may be sufficient for example to detect groundwater contaminants, it can be advantageous to measure some parameters continuously, especially for studies that aim at a detailed characterization of the functioning of a karst system. Furthermore, as the sampling frequency increases, the sample number and hence the analytical cost rapidly increase making continuous measurement an attractive alternative. However, currently only a limited number of parameters can be measured reliably on a continuous basis (see Section 6.4.1). Before defining the sampling frequency, it is recommended to characterize the reaction of the system to precipitation events by measuring some basic parameters (discharge, temperature, electrical conductivity) continuously. During the actual sampling programme, measurement of these parameters should be continued to facilitate data interpretation. Some additional considerations for appropriate sampling frequencies in contaminant studies are discussed in Section 6.6.

6.3.4 Sampling methods

Auto-samplers are convenient for the investigation of temporal variations of hydrochemical parameters. Care has to be taken that they always collect flowing water, even when the water level is low. Auto-samplers have also some disadvantages. In most auto-samplers, the water remains in contact with the gas phase, which modifies the concentration of dissolved gases and volatile contaminants (e.g. BTEX, chlorinated hydrocarbons). The loss of CO_2 can lead to carbonate precipitation. Furthermore, immediate preservation of the samples (Tab. 6.2) e.g. by adding acid in case of NH_4^+ or metals is difficult. Nevertheless, samples from auto-samplers provide reliable information for conservative ions (Cl^-, NO_3^-, and SO_4^{2-} in oxidizing conditions), or chemical compounds that are present at concentrations well below saturation with respect to associated minerals (Sr^{2+}, Mg^{2+}). Furthermore, there are some types of equipment available that can take gas-tight samples (e.g. ISCO). Hand sampling enables complete filling of the bottles, which limits the exposition to the gas phase, and immediate pre-treatment of the samples is possible. When sampling for organic contaminants, appropriate tubing (usually high-density polyethylene or Teflon) and sampling vessels (glass with Teflon-coated septa) have to be used to avoid loss of the compounds by sorption.

Table 6.2 summarises sample pre-treatment requirements and holding times. Samples for major cations, Fe^{2+}, NH_4^+ should be acidified to avoid precipitation of carbonate minerals (Ca^{2+}, Mg^{2+}) or to avoid oxidation of reduced compounds (Fe^{2+}, NH_4^+). Before acidification, the samples have to be filtered to avoid the dissolution of particles due to the

Table 6.2. Sample handling and analytical methods.

Parameter	Laboratory analysis[1]	Bottles[2,3]	Preservation[3]	Maximum storage[3]	Typical detect. limit (mg/L)	Cont. meas.
Soil-related parameters						
CO_2	GC	P,G	Analyze immediately			+
^{222}Rn		G		7 d		+
Dissolved organic carbon (DOC)	Combustion and analyses of CO_2	G	None	7 d	0.1*	+
NO_3^-	IC/UV-Vis S	P, G	Filtration, Add H_3PO_4 or H_2SO_4 to pH < 2, refrigerate	48 h	0.3*	
NH_4^+	IC/UV-Vis S	P,G	Filtration, Refrigerate; Filtration, Add H_2SO_4 to pH < 2, refrigerate	7 d	0.3*	
PO_4^{3-}	IC/UV-Vis S	G (A)	Filtration, refrigerate	48 h	0.3*	
Cl^-	IC/UV-Vis S/ Titration	P,G	Filtration	28 d	0.1*	
SO_4^{2-}	IC/UV-Vis S	P, G	Filtration, refrigerate	28 d	0.3*	
K^+	AAS/AES/ ICP-MS/IC	P, G (A)	Filtration, Add H_2SO_4 to pH < 2	6 months	0.3*	
Turbidity			Analyze immediately			+
Carbonate-rock related parameters						
Ca^{2+}	AAS/ICP-MS/ IC/Titration	P, G (A)	Filtration, Add H_2SO_4 to pH < 2	6 months	0.3*	
Mg^{2+}	AAS/ICP-MS/ IC/Titration	P, G (A)	Filtration, Add H_2SO_4 to pH < 2	6 months	0.3*	

	HCO_3^-	Titration	P, G	Filtration, refrigerate	24 h	10
	SI	Calculated				
	$\delta^{13}C$ of DIC	IRMS	P, G	Filtration, refrigerate		
Parameters related to other rock types	SO_4^{2-}	IC/UV-Vis S	P, G	Filtration, refrigerate	28 d	0.3*
	Sr^{2+}	AAS/ICP-MS	P, G (A)	Filtration, Add H_2SO_4 to pH < 2	6 months	0.001–0.0001
	Various trace metals	ICP-MS	P, G (A)	Filtration, Add H_2SO_4 to pH < 2	6 months	0.001–0.0001
Anthropogenic compounds	Metals	AAS/AES/ ICP-MS	P, G (A)	Filtration, Add H_2SO_4 to pH < 2	6 months	0.001–0.0001
	Pesticides	GC/HPLC	G	Refrigerate	7 d	variable
	Volatile organic compounds	GC	G	Add HCl to pH < 2, or NaOH to pH > 10 refrigerate	7 d	variable

(1) GC = Gas chromatography, IC = Ion chromatography, UV-Vis S = UV/Visible spectrometry, AAS = Atomic adsorption spectrometry, AES = Atomic emission spectrometry, ICP-MS = Inductively coupled Plasma-Mass Spectrometry, HPLC = High pressure liquid chromatography, IRMS = Isotope ratio mass spectrometry.
(2) P = Plastic (High Density Polyethylene), G = Glass, G(A) = Glass acid rinsed.
(3) According to Standard Methods for Examination of Water and Wastewater (Eaton et al. 1995).

addition of the acid. When sampling for trace metals it should be verified that filtration and acidification does not introduce contamination by treating some high purity water as a sample (blank). Similarly, when sampling for DOC it should be ensured that the filter does not release organic carbon. Samples for major anions are generally filtered before analysis since particles can interfere with the analytical system. Similarly, samples for determination of HCO_3^- by titration have to be filtered because dissolution of carbonate particles during titration would influence the results. In contrast, samples for analyses of volatile organic compounds should not be filtered since pressure changes during vacuum filtration can lead to loss of these compounds. Hydrophobic organic compounds and metals are frequently transported bound to particles and hence the particulate fraction has to be included in the analyses to detect these contaminants. Organic contaminants can be released from the particles by extraction with a solvent, metals by acid-digestion of the complete sample (Vesper & White 2003). For detection of microbial contaminants, samples have to taken using sterilized bottles.

6.4   ANALYTICAL METHODS

6.4.1   Methods for continuous measurement

In this section, only those instruments are discussed that can easily be installed without requiring special infrastructure and/or a fixed electricity supply. Hence they are suited for springs and also for intermediate sampling points. In general, measurement methods that rely on physical principles are more stable than for example electrochemical methods that often show drift. Parameters that can be reliably measured include specific conductance, temperature, dissolved O_2 and CO_2, UV-absorbance, natural fluorescence and turbidity. A number of companies offer instruments for some of these parameters (e.g. WTW Instruments, Germany; Van Essen Instruments, Netherlands; Yellow Springs, USA; In-situ, USA; ISCO, USA; Hach, USA). Turbidity is usually measured using the nephelometric principle whereby a light beam is directed onto the sample and the scattered light at an angle of 90° is measured. Dissolved oxygen concentrations are frequently measured using Clark electrodes, which rely on electrochemistry. This method is not sufficiently stable for continuous measurements over several weeks. However, recently, instruments that rely on changes in the fluorescence properties of a membrane depending on the oxygen concentrations have become available (Hach, USA), which should be suitable for continuous recording of oxygen concentrations. DOC concentrations can be measured indirectly either based on UV absorbance at 254 nm or by its natural fluorescence at excitation wavelengths around 350 nm and emission around 420 nm. The UV method has the disadvantage that turbidity also adsorbs light. Some measurement systems use a turbidity correction based on simultaneous measurement of adsorption at a different wavelength, which is however not very reliable, or require an on-line filtration of the sample before measurement. In principle, the concentration of specific ions can be measured with ion-selective electrodes. However, these systems usually show substantial drift over time and require frequently recalibration.

6.4.2   Laboratory methods

6.4.2.1   *Overview*
The most common ions are usually analysed by titration, ion chromatography (IC), UV/VIS-spectrophotometry and less frequently by atomic emission or absorption spectrophotometry

(AES or AAS). Metals can also be analysed by AES or AAS but increasingly inductively coupled plasma – mass spectrometry (ICP-MS) is used. For organic compounds, usually gas chromatography (GC) or high-pressure liquid chromatography (HPLC) is used, which will not be discussed because extraction and analytical methods vary depending on the compound that is analysed. Typical detection limits for common methods are provided in Tab. 6.2.

6.4.2.2 *Titration*
Titration relies on the reaction of a given ion to a regent of known concentration. The equivalence point corresponding to complete reaction is determined by means of a potentiometric measurement or a visual indicator. Titration is still used for alkalinity (OH^-, CO_3^{2-}, HCO_3^-), and to a lesser extent for hardness (Ca^{2+}, $Ca^{2+} + Mg^{2+}$). Automatic devices are available for example from Metrohm, Switzerland.

6.4.2.3 *Ion chromatography (IC)*
The principle of ion chromatography is to elute a sample containing a mixture of ions within a column that specifically retards the different ions making it possible to separate them. A detector, located at the end of the column provides a signal that is proportional to the content of compound in the sample. The compound concentrations are calculated by comparing the sample response with that of standards. With this method, several ions can be quantified during one analysis. The method is widely used for analyses of major anions (except HCO_3^-) and cations using electric conductivity detection. More recent developments make it possible to detect trace heavy metals using a spectrophotometric detector.

6.4.2.4 *UV/VIS spectrophotometry*
The method relies on the transformation of an ion of interest into a complex that adsorbs light at a specific wavelength. Using Beer-Lambert's relation, the adsorption can be related to the concentration of the ion. This method can be used to quantify concentrations of most of the anions (Cl^-, NO_3^-, NO_2^-, SO_4^{2-}, PO_4^{3-}, H_4SiO_4) and some cations (NH_4^+, Fe^{2+}). While the method is generally very sensitive, it has the disadvantage that for each ion a different reagent has to be used and that with some compounds interferences with other ions occur. The method is particularly useful for compounds that are often present at concentrations below the detection limit for standard HPIC such as NO_2^-, PO_4^{3-}, NH_4^+ or can not be measured by standard HPIC. In addition to laboratory instruments, portable field instruments exist (e.g. Hach, USA), which are especially useful for labile compounds such as NO_2^- or Fe^{2+} although their precision is lower than for laboratory measurements.

6.4.2.5 *AAS, AES and ICP-MS*
Atomic adsorption spectrometry (AAS) is based on the possibility to promote atoms to a higher energy level by electromagnetic radiation. Before the measurement, the sample has to be volatilised and the ions transformed to atoms in a flame or furnace. Since each element adsorbs at a different wavelength, only one element can be measured at a time. The method can be used for about 40 different elements. For karst waters, it has been frequently used for major (Ca^{2+}, Mg^{2+}, Na^+, K^+) and minor (Sr^{2+}, Fe^{2+}, Mn^{2+}) cation analysis.

When an excited atom returns to the ground state, it emits again electromagnetic radiation of a given wavelength, a phenomenon that can also be used to quantify the concentration of an element. In case of atomic emission spectrometry (AES), the atoms are promoted to a higher energy level by thermal excitation. Because, many of the elements are excited and the emitted radiation is specific for each element, the method makes it possible to measure concentrations of several elements simultaneously. However, with some elements, interferences can occur and the detection limit is rather high (because only 5% of the atoms are excited). The method is mainly used for simultaneous measurement of concentrations of alkaline and earth alkaline elements. For minor and trace elements, Inductively Coupled Plasma – Mass Spectrometry (ICP-MS) is better suited with its low detection limit and high resolution. Using this instrument, the sample is introduced as an aerosol into plasma and emerges as a cloud of atoms and some ions. Part of the cloud is diverted into a mass spectrometer that analyses numerous elements almost simultaneously, with detection limits below the μg/L range.

6.5   INTERPRETATION OF HYDROCHEMICAL DATA

6.5.1   Introduction

Since hydrochemical parameters often show strong temporal variations in karst systems, care has to be taken when samples from a single programme are evaluated. However, especially under baseflow conditions, a single sampling programme can sometimes provide insight into the origin of the water. For example, high sulphate concentration suggests that the water partly originates from evaporite formations. In addition, single programmes can provide information on anthropogenic impacts by highly soluble compounds such as nitrate (Perrin et al. 2003) but also chlorinated hydrocarbons (e.g. Williams et al. 2006). Most frequently, karst hydrochemical studies focus on temporal variations in the hydrochemical composition of waters. Time series of hydrochemical data can be used to address a number of different questions and can be evaluated using qualitative and quantitative approaches.

6.5.2   Assessing aquifer functioning using parameters related to limestone dissolution

The chemical composition of karst groundwater is usually dominated by products from dissolution of limestone, especially calcite. The measurement of temporal variations of these products provides insight into the functioning of a karst system. There is a choice of different parameters available, including Ca^{2+}, HCO_3^-, total dissolved solids (TDS), hardness and electrical conductivity. Since these parameters are usually highly correlated, it is often sufficient to measure one of them. The electrical conductivity is often the parameter of choice since it can be measured continuously with relatively little effort. The measurement of electrical conductivity is often combined with temperature measurement, which can be carried out using the same instrument. The combination of rainfall, discharge, electrical conductivity and temperature is the most fundamental data set of hydrochemical studies.

Data illustrating the reaction of these parameters to a series of flood events provide information about the overall functioning of the investigated system. The shape of the chemograph reflects the organization and the effectiveness of the drainage system. A flat chemograph indicates a low degree of karstification unable to transfer an input signal to the outlet (Fig. 6.3, above, curve A). Such a system is a good mixer, consisting of a regularly distributed network of fractures with absence of conductive channels. A large storage volume or a thick overburden can also contribute to a dampened response to storm events. On the contrary, a chemograph with sharp positive or negative peaks (Fig. 6.3, curve B)

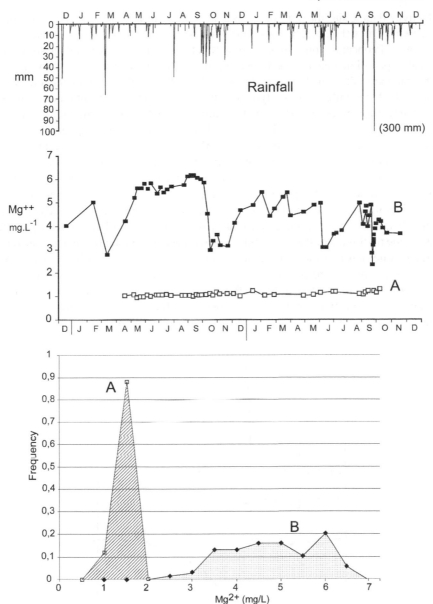

Figure 6.3.   Above: Rainfall and variation of the magnesium concentrations of two carbonate springs. Below: Frequency distribution of magnesium concentrations. Data from Lastennet (1994).

indicates an effective channel network able to transmit the infiltration component to the spring during a flood period and/or concentrated infiltration.

An alternative way to evaluate the data is the use of mineralisation frequency distribution diagrams rather than time series (Bakalowicz 1977). Such graphs illustrate how frequently concentrations fall within a certain range and can also be constructed with electrical conductivity data. Using the same data sets as above, two different distributions are obtained for the two springs (Fig. 6.3, below). Curve A has only one high frequency peak, covers a small range of concentration values and is symmetrical. This indicates that the spring drains one water type, originating from a fissured aquifer with little karstification. Curve B reaches lower frequency values and has several peaks. This pattern indicates again that the system is well karstified, and different water types reach the spring. The dominant mode corresponds to the composition of reserves stored in the aquifer while the lower frequency modes reflect flood events where reserve water is mixed with another component (e.g. epikarst).

A more detailed analysis of individual storm events can often provide additional information about the arrival of different water types at the spring and mixing between different waters, as illustrated in Fig. 6.4. During the initial phase of a flood, the concentration of limestone-dissolution related parameters can stay constant or even increase, indicating that highly mineralised water is pushed out of the system, a phenomenon usually denoted as piston effect. Only later does the concentration of limestone-dissolution related parameters decrease due to dilution of the pre-event water with water from the rain event. The recently infiltrated water has a low level of mineralisation because it does not have sufficient time to reach saturation with respect to calcite and/or because it originates from concentrated infiltration with a low initial pCO_2. Often the response of different subsections of the catchment to precipitation events is superimposed leading to complex patterns of electrical conductivity and temperature at the spring (Hess & White 1988).

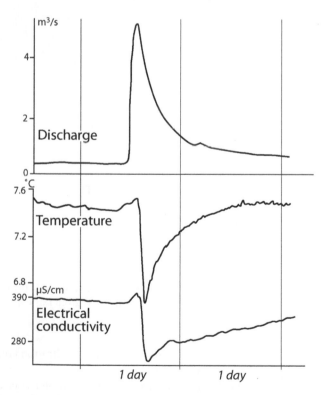

Figure 6.4.   Temporal evolution of discharge, temperature and electrical conductivity at a karst spring in France (Tissot & Tresse 1978).

The piston effect often becomes even more apparent if the data are plotted as a function of discharge rather than time (Fig. 6.5). Often a hysteresis is observed and the flood cycle can be divided into different parts. Initially, the discharge increases while the concentration (or electrical conductivity) remains constant (A). Then, the concentration strongly decreases while the discharge remains elevated, indicating the arrival of rainwater (B). Next, the concentration remains low while the discharge decreases, indicating further outflow of recently infiltrated water (C). Finally, the concentration increases again due to an increasing proportion of water that was in storage and because the water has an increasing amount of time to become more strongly mineralised (D).

The piston-flow water may originate from conduits in the saturated zone or from the epikarst or unsaturated zone. In the latter cases, the mineralisation may be higher than for

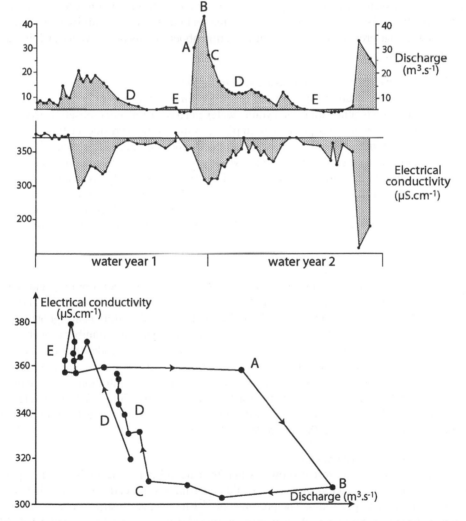

Figure 6.5. Temporal variation of discharge and electrical conductivity (above) and electrical conductivity vs. discharge diagram (below) illustrating the piston effect for the Fontaine de Vaucluse karst spring, France (Mudry 1987).

pre-event water due to the high pCO_2 close to the soil zone. The occurrence and duration of a piston effect does not only depend on the intrinsic characteristics of karst system but also on the distribution of precipitation, which influences how quickly rainwater reaches the spring, the amount of precipitation, which affects the potential for dilution, or the volumes and chemical composition of water stored in the different subsystems. The latter depends on the climatic conditions in the period preceding the precipitation event. The piston effect is usually particularly pronounced after a prolonged dry period.

6.5.3   Assessing residence time and origin of water

The measurement of major parameters related to limestone dissolution usually provides information about the relative contribution of pre-event and freshly infiltrated water to discharge at a spring. However, it provides only limited information on residence time of water, location of storage and flow paths. In this section, a range of additional parameters is discussed that provide more subtle information about the history of water arriving at a sampling point.

6.5.3.1   *Parameters related to rock-water interaction*
There is no simple relationship between residence time and overall mineralisation. Since calcite dissolution is quite rapid, even young water can have a high mineralisation especially if it was in contact with a gas-phase of high pCO_2. A more reliable residence time indicator than total mineralisation is Mg^{2+} concentration (Batiot et al. 2003, Lastennet & Mudry, 1997) provided that the catchment contains Mg-bearing minerals (e.g. dolomite). Generally Mg^{2+} is released more slowly than Ca^{2+} and hence increasing Mg concentrations or an increasing Mg/Ca ratio indicates a greater water age. However, only the relative residence time of water arriving at different moments is obtained, not the absolute residence time. Figure 6.6 illustrates the arrival of water with a young age but a high mineralisation during a flood event.

6.5.3.2   *Saturation indexes*
With increasing residence time, dissolution reactions approach equilibrium, and hence the saturation indices (SI) provide an indirect means to evaluate residence times (Hess & White 1993). Saturation indices can be calculated with little effort using geochemical codes such as PHREEQC that can be downloaded free of charge at a number of web pages (e.g. http://www.geo.vu.nl/users/posv/phreeqc/index.html). These codes take into account the effect of temperature, ion strength and the formation of complexes and ion pairs on solubility of minerals. In addition to major anion and cation concentrations, it is important to have reliable values for pH or pCO_2 for these calculations. Figure 6.7b illustrates the arrival of recently infiltrated water that did not have sufficient time to reach saturation with respect to calcite at a spring (Vesper & White 2004b).

6.5.3.3   *Soil-related parameters*
It is often of particular interest to know if water from the soil-zone reaches the spring rapidly. The occurrence of such a transfer can be evaluated using soil-related parameters that decay during storage in deeper zones. The presence of [222]Rn in water illustrates a fast transfer of water to the spring because [222]Rn is mainly produced in the soil and decays with a half-life of 3.8 days (see also Chapter 7). Dissolved Organic Carbon (DOC) or total organic carbon (TOC) can also be considered as a tracer for short transit time of water from the soil zone

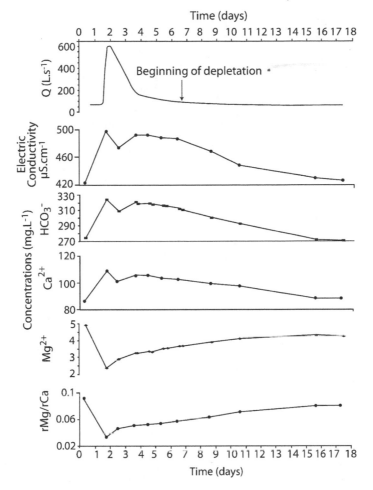

Figure 6.6. Temporal variations of discharge, electrical conductivity, bicarbonate, Ca^{2+}, Mg^{2+} and the Mg/Ca ratio at a karst spring (Lastennet & Mudry 1997).

because DOC/TOC mineralises to inorganic carbon or may adsorb in deeper zones (Batiot et al. 2003, Emblanch et al. 1998). Figure 6.7a illustrates the arrival of water with elevated TOC concentrations from the soil zone. The highest concentrations are reached during the period when the SI for calcite diminishes indicating rapid transfer of water.

6.5.3.4 *Carbonate-system related parameters (pCO$_2$ and δ^{13}C)*

In karst aquifers, storage of water may take place in low permeability volumes (LPV) and conduits of the saturated zone or as perched aquifers in the unsaturated zone or epikarst. These reservoirs are drained during baseflow but may also be activated during the initial phase of the flood pulse as illustrated by the piston effect. It can be difficult to distinguish between these two reservoirs based on the mineralisation because both of them may show substantial concentrations of limestone-related parameters. However, the measurement of pCO_2 and $\delta^{13}C_{DIC}$ can provide indication about the origin of the water and flow paths.

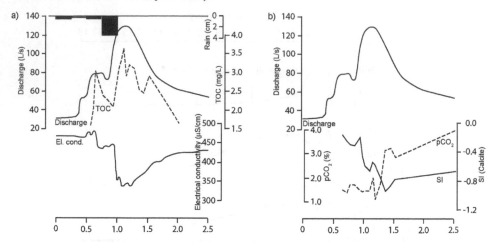

Figure 6.7.   Temporal variations of (a) TOC and (b) saturation index and pCO₂ of a karst spring at Fort Campbell, Kentucky, Tennessee, USA.

Water stored in the unsaturated zone is likely to be in contact with a gas phase containing elevated levels of CO_2. Thus carbonate dissolution evolves under open system conditions resulting in a higher final pCO_2 than in deeper zones where dissolution may take place partly under closed conditions. Furthermore, in deeper zones seasonal fluctuations of CO_2 concentrations are dampened (Hess & White 1993) and hence during summer the pCO_2 in the soil and the epikarst is expected to be higher than in deeper zones. Hence, an increase of the pCO_2 at the spring indicates arrival of water that was in storage close to the soil zone (Savoy 2007). Alternatively, pCO_2 values can provide information on the recharge mechanisms with concentrated recharge through sinkholes leading to a low pCO_2 and diffuse recharge across the soil to a higher pCO_2. For example, at a karst spring in the USA (Vesper & White 2004b), a strong increase in pCO2 during the later part of a flood peak was observed and was hypothesized to be due to arrival of water that infiltrated across CO_2-rich soils (Fig. 6.7b).

The $\delta^{13}C_{DIC}$ can vary between waters stored in the unsaturated zone and saturated zone. Under open conditions in the unsaturated zone, the $\delta^{13}C_{DIC}$ is more strongly influenced by the biomass carbon pool (around $-25‰$) compared to the carbonate pool (around $0‰$) and hence the $\delta^{13}C_{DIC}$ is slightly more negative (for details see Emblanch et al. 2003). Figure 6.8 contrasts karst springs mainly fed by rapid infiltration from the unsaturated zone (negative $\delta^{13}C_{DIC}$) with boreholes extracting long residence time water from the deepest part of the aquifer (more positive $\delta^{13}C_{DIC}$). The variable DOC concentration suggests that some springs have a longer residence time, sufficient to mineralise some of the DOC.

6.5.4   Identifying lateral inflows

Aquifers in carbonate rock may also be recharged by water originating from other lithologies. If the adjacent reservoir has a distinct chemical composition, a hydrochemical study can help to confirm such an influence. Large contrasts especially occur in case of evaporites, such as in Alpine and Mediterranean karsts, which generate water with elevated

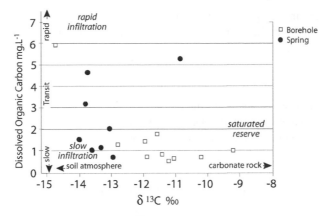

Figure 6.8. Dissolved organic carbon vs. δ^{13}C of dissolved inorganic carbon for a number of springs and boreholes in France (Mudry et al. 2002).

sulphate and sometimes also chloride concentrations. It is more difficult to identify water from crystalline rock. Sometimes, it can be enriched with tracer elements due to the presence of secondary minerals in fractures. Varying contributions of adjacent reservoirs can be illustrated using chemographs or also based on binary and ternary diagrams.

6.5.5 Demonstrating and quantifying mixing

6.5.5.1 *Flux-discharge diagrams*
The concentration variations of limestone-dissolution related parameters during a flood event can be rationalized as a mixing of two different types of water: pre-event water that was stored for a prolonged period of time in the system and recently infiltrated event water. Simple mixing of two waters becomes apparent as a straight line on a flux-discharge. The flux (e.g. g/s) is obtained by multiplying the chemical concentration by the discharge of the spring. In addition, on flux-discharge diagrams, the piston effect can also be identified.

6.5.5.2 *Hydrograph separation*
As discussed in the preceding sections, the water arriving at a karst spring can be considered to consist of a mixture of different water types. Accordingly, an attempt can be made to quantify the contribution of the different flow components based on chemical and isotope parameters that are distinctly different for the different components and show a (nearly) conservative behaviour during mixing. The approach is usually denoted as hydrograph separation and is analogous to end-member mixing analysis (EMMA), frequently used in surface water studies. Equations for two (Lakey & Krothe 1996), three and four (Lee & Krothe 2001) component mixtures have been developed. The most basic separation consists of separating the hydrograph into pre-event water (long residence time, high mineralisation) and event water (recently infiltrated, low mineralisation). The pre-event water can further be subdivided, for example, in water from the saturated zone and epikarst. Common parameters used for hydrograph separation are δ^{18}O and δ^{2}H of water, dissolved silica or alkalinity. In the following, the approach is illustrated for a two-component mixture of pre-event and event water. The concentrations associated with these two components can be assumed to be quite uniform due to smoothing of the first component in the saturated zone and of the

second component in the epikarst. If we consider that, during a flood, the concentrations of both components remain approximately constant, we can write an equation for total discharge:

$$Q_t = Q_{pe} + Q_e \qquad (6)$$

with

Q_t total discharge at the outlet at time t
Q_{pe} discharge from the saturated zone component
Q_e discharge from the unsaturated zone component

and a second equation for total mass flux:

$$Q_t \cdot C_t = Q_{sz} \cdot C_{sz} + Q_{uz} \cdot C_{uz} \qquad (7)$$

with

C_t concentration at the outlet at time t
C_{pe} concentration of the saturated zone component
C_e concentration of the unsaturated zone component

Combining equations (6) and (7), we can calculate the contribution of both components:

$$Q_{pe} = Q_t \cdot [C_t - C_e]/[C_{pe} - C_e] \qquad (8)$$

$$Q_e = Q_t - C_{pe} \qquad (9)$$

The concentration of the pre-event water can be estimated based on the concentration at the spring during depletion and the concentration of event water can be estimated based on samples from the unsaturated zone. If the latter are not available, the concentration can be estimated based on the composition of water at the moment of least mineralisation during a flood event. The hydrograph in Fig. 6.9 was separated into the two components using silica as a conservative tracer. It becomes apparent that even during the flood peak, pre-event water contributes equally or more than event water to flow. The hydrograph separation method has to be considered as a rough estimation since water from different reservoirs usually does not have a uniform composition and its composition can only be estimated approximately.

6.5.6   Global analysis using statistical methods

A thorough hydrochemical analysis leads to large sets of chemical parameters and it is not practical to plot all the parameters using the approaches described above (e.g. as chemograph). Often this is not necessary because some of the parameters behave similarly hence providing no additional information, while others are independent of each other. Principle component analysis (PCA), a descriptive statistical method, can be used to identify correlations among parameters and regroup them into new "combined"-parameters denoted as components. Often the components can be considered to represent a water type or process that affects a set of parameters in a similar way. Instead of evaluating how the concentration of individual parameters varies during a flood event, it is evaluated how strongly a water sample is influenced by a certain component. An example for application of PCA to karst studies can be found in Lastennet & Mudry (1997). In addition to descriptive statistical methods there are also so-called inferential statistical methods, such as discriminant factor

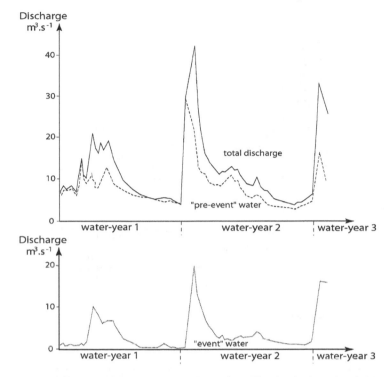

Figure 6.9. Two-component hydrograph separation using silica for the Fontaine de Vaucluse karst spring, France (Mudry 1987).

analysis (DFA). These methods help to validate hypotheses (e.g. storage location epikarst vs. saturated zone) using quantitative data.

6.6   EVALUATION OF THE ORIGIN AND FATE OF CONTAMINANTS

6.6.1   Detecting contaminants at karst springs

When sampling for contaminants, it has to be taken into account that temporal patterns of contaminant concentrations vary depending on the source characteristics and contaminant properties. It is often useful to first evaluate the reactivity of the sampling point to rain events using the hydrochemical methods described above before carrying out a contaminant study. For springs with significant temporal variations, it has to be considered whether the contaminants of interest are expected to be continuously present at the spring or only during flood events. Figure 6.10 summarizes potential chemographs for different type of contaminants as well as for some basic parameters measured to evaluate the reactivity of the sampling point.

Compounds that are present as non-aqueous phase liquids (NAPLs) in a karst system, in particular chlorinated hydrocarbons, are expected to occur at karst springs even under base flow conditions (Fig. 6.10c). Hence even a single sampling event can demonstrate the presence of these compounds. However, the concentrations are expected to be

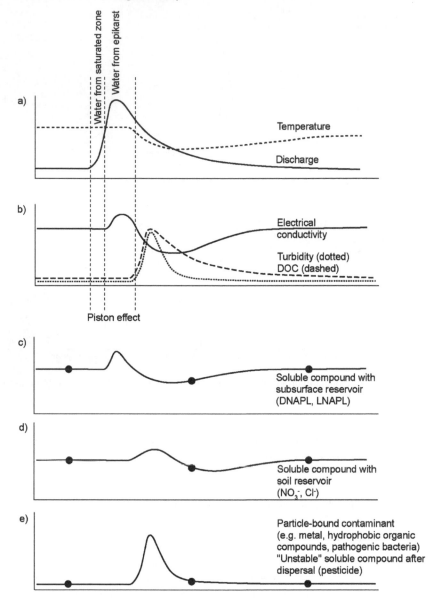

Figure 6.10.   Schematic illustration of temporal variations of different type of contaminants and basic hydrochemical parameters at karst spring. Filled circles: Hypothetical fixed-interval sampling campaigns.

well below maximal solubility due to dilution, which is usually great given that NAPL source zones have a limited spatial extent while karst springs integrate drainage from over a large area. For multi-component NAPLs such as gasoline or heating oil, only the most soluble compounds (e.g. benzene) may arrive at the spring. Depending on the travel time, petroleum hydrocarbons can be subject to biodegradation before arrival the spring.

The concentrations of organic compounds from NAPL can be higher during baseflow than floods, because less dilution occurs and because contact time between water and organic liquid is longer. However, especially after dry periods, high concentration may occur during the first flush because the contaminants have had sufficient time to dissolve (Williams et al. 2006). An increase can also occur due to the mobilization of organic liquids for example trapped in sediments of caves during flood events. For some organic compounds, such as MTBE, which has a high affinity of the aqueous phase, trace amounts may originate from atmospheric contamination.

Highly soluble compounds that are not continuously released such as NO_3^- from fertilizers or road salts may nevertheless occur continuously, due to storage in soil, epikarst or deep phreatic zones, where they remain in solution (Jones & Smart 2005, Peterson et al. 2002). Variations in the temporal concentration of nitrate can be considerable. During flood events, concentrations may increase due to flushing of the soil (Fig. 6.10d). However, it is also possible that much infiltration occurs along preferential flow paths with little NO_3^- and thus dilution occurs (Peterson et al. 2002). Furthermore, temporal variations may be related to arrival of water from different regions of the catchment with different land uses (e.g. Perrin). Compounds with limited solubility and a tendency to degrade, such as some pesticides, are especially likely to arrive during storm events succeeding their release (e.g. Panno & Kelly 2004). During later storm events, concentrations can be lower due to degradation and sorption to organic matter.

Compounds with a high tendency to associate with the solid phase, such as hydrophobic organic compounds, certain metals (Vesper & White 2003) and also microorganisms (Ryan and Meiman 1996) are predominantly transported during flood events and sampling should concentrate on these events (Fig. 6.10e). Vesper & White (2003) observed a strong correlation between turbidity and the presence of heavy metals, while Pronk et al. (2006) found that the presence of faecal bacteria is correlated with increasing DOC concentrations because both originate from the soil zone. Hence measurement of turbidity and DOC concentrations can help to choose appropriate times for sampling or to choose which samples should be submitted for analysis. Sediment-bound compounds may still arrive at the spring during flood events months or years after their application.

6.6.2 Quantifying contaminant levels

The greatly varying concentrations complicate the comparison of measured concentration with regulatory values. Depending on the legislation in place it may be necessary to determine maximal concentration to evaluate possible acute effects on ecosystems or to calculate average concentrations to assess long-term exposure. To detect maximum concentrations, an event-based sampling during flood events is necessary even for highly soluble compounds, which can increase during flood events (see above and Fig. 6.10c,d). Average concentrations can be calculated in different ways including arithmetic mean, geometric mean or flow-weighted average. For soluble compounds such as nitrate or chlorinated hydrocarbons, the arithmetic mean calculated based on fixed-interval sampling with monthly or even semi-annual frequency is often not much different from that calculated based on much larger data series that include event-based sampling (Currens 1999, Williams et al. 2006). The reason for the good agreement is that deviations from mean concentrations during flood events usually only occur for a short duration. For contaminants that predominantly occur during flood events, fixed-interval sampling leads to an underestimation or even non-detection of the contamination (Fig. 6.10c–e). The flow-weighted average is particularly appropriate to evaluate exposure to contaminants because each concentration is weighted by the volume of water in which it is present. The flow-averaged concentration is calculated by multiplying the concentration representative for each time interval with the discharge. By adding the contributions of all intervals, the mass flux is obtained, which is divided by the cumulative discharge for the same period to obtain the flow-averaged concentration. The flow-averaged concentration can be calculated based on different data sets, fixed-interval concentration and discharge measurement, fixed-interval concentrations and continuous discharge measurement and for most precise evaluation, by combining fixed as well as event-based measurements with continuous discharge measurement. The tendency of these approaches to under or overestimate the actual flow-averaged concentration is discussed in Currens (1999) and in Williams et al. (2006). The mass flux, an intermediate step in the calculation of the flux-averaged concentration, can be of interest in itself. Mass flux calculations can for example be carried out before and after a remediation to demonstrate its efficacy or to evaluate the fraction of an agrochemical applied in the catchment area that arrives at the spring.

The need for event-based sampling to detect particle-bound contaminants and the data requirements to calculate accurate flow-averaged concentrations leads to a large sample number and hence high costs. One possibility to reduce analytical cost is to combine several samples before analysis. Using this approach, contaminants that only occur during short periods can still be detected and, if the samples are combined proportionally to the discharge, flow-averaged concentration can still be calculated. For example, Panno & Kelly (2004) combined 4 samples taken at 6 h intervals in a study on nitrate and herbicide contamination.

6.6.3   Identifying contaminant sources

The identification of contaminant sources in karst terrain is difficult due to complex transport paths and because contaminants can be transported over long distances. In some cases, it may be possible to narrow down the location of the source area, by sampling different tributaries in karst caves. Another approach is to try to identify suspected contaminant sources and, using hydrogeological data and tracer tests, demonstrate a link between the suspected contaminant source and spring (Crawford & Ulmer 1994). For organic contaminants, stable isotope methods may also be used to substantiate that compounds found at a source correspond to those detected at the spring because different sources of the same compound often have a different isotopic signature (Hunkeler et al. 2004).

At industrial sites or waste disposal sites frequently organic liquids may have been spilled in the past but whose current location is unknown. Characterization of the geology can help to develop hypothesis about potential pathways of organic liquids. If organic liquids are trapped in the epikarst, a soil gas survey can help to locate zones of organic liquid accumulation. Gas analysis can also be useful in conduits although gaseous contaminants may be transported over extended distances. Caution has to taken since the presence of gaseous compounds can lead to explosions.

For contaminants from agricultural activities, vulnerability mapping (Zwahlen 2004) can provide some indications as to the location of zones where contaminants may preferentially enter aquifers and hence where land use changes should be enforced.

6.6.4   Investigating contaminant fate

In addition to detecting and quantifying contaminants, the aim of studies can also be to investigate the fate of contaminants in karst aquifers. Such studies may especially focus on processes that retain or eliminate contaminants. It is difficult to substantiate these processes based on contaminant concentrations because the input concentration is usually poorly known and since concentrations strongly vary due to dilution. The measurement of stable isotope ratios of contaminants can provide information about transformation processes. Especially for small molecules, transformation is often faster for molecules with light isotopes and hence the heavy isotopes become increasingly enriched in the residual contaminant fraction (Clark & Fritz 1997). Shifts in stable isotope ratios have been used to demonstrate denitrification in karst aquifers (Einsiedl et al. 2005, Panno et al. 2001). In exceptional cases, it may be possible to quantify contaminant attenuation by comparing the contaminant input on the whole catchment with the output at the spring. For example, Panno & Kelly (2004) estimated that 21–31% of the applied N fertilizer, but only 3.8–5.8% of the atrazine and 0.05–0.08% of alachlor reaches the spring. An alternative approach to investigate retention and degradation processes is the use of comparative tracer tests, which include a conservative tracer and one or several reactive tracers. The method has been applied to investigate the behaviour of microorganisms (Auckenthaler et al. 2002).

6.7   SUMMARY

Hydrochemical analyses not only provide information about water quality, but also insight into the functioning of karst aquifers. The simple measurement of temporal variations of electrical conductivity and temperature can provide information about the contribution of water that was in storage (pre-event water) and water from the rain event to flood discharge. Additional parameters can provide further insight into the relative residence time of water in the karst system (Mg, TOC) or some indications on storage location (^{13}C, ^{222}Rn). Usually the karst spring is the main sampling point, but sampling at intermediate locations such as caves can provide additional information about the behavior of different subcatchments. When sampling for contaminants, it has to be taken into account that different types of contaminants show different temporal variations. Soluble contaminants, such as nitrate or compounds dissolving from organic liquids, tend to be permanently present in karst spring water, sometimes even at higher concentrations during base flow conditions than during flood events, due to the absence of dilution. In contrast, contaminants that tend to combine with particles (metals, hydrophobic organic compounds) or can be considered as particles themselves (microbial pathogens) mainly arrive during flood events. These compounds are often only detected with a targeted sampling of flood events. TOC concentration and/or turbidity can serve as proxy to select appropriate sampling times for these contaminants.

CHAPTER 7

Isotopic methods

Robert Criss, Lee Davisson, Heinz Surbeck & William Winston

7.1 INTRODUCTION

Karst groundwaters can be characterized by natural or anthropogenic tracers including isotopes (this chapter), hydrochemical constituents (chapter 6), and artificial tracers such as fluorescent dyes (chapter 8). Isotopic data elucidate the origin and age of karst waters and the rocks that enclose them, and show that mixing and carbonate reactivity are major variables. Many isotopic systems are useful, but those highlighted here offer particularly important insights into the nature of karst systems. In our experience the most reliable approach that leads to predictions useful for groundwater management is high frequency time series analysis of stable isotopes, followed by a modest number of isotopic age-dating measurements for residence time verification. We review both types of isotopic systems and introduce interpretive methods that factor in dispersion and reactivity in order to delineate sources and residence times.

Hydrogen and oxygen isotopes are intrinsically associated with the water molecule, providing special opportunities for hydrogeologic studies. Both H and O have naturally-occurring stable isotopes that provide a dual system for tracking the origin of water and its subsequent movement. In cool groundwaters these isotopes behave in a conservative manner, as they tend not to exchange extensively with substrate materials at low temperatures. H and O isotopes prove that most karst waters are derived from ordinary meteoric waters. Spatial and temporal variations normally occur in these parent meteoric waters and are basically inherited but imperfectly homogenized in karst waters, facilitating water tracing while providing a means of calculating ground water residence time.

Carbon isotopes elucidate the interactions of water with atmospheric carbon dioxide, organic matter, and carbonate bedrock. Dissolved inorganic carbon (DIC) tends to be derived in roughly equal proportions from carbonate bedrock and carbon dioxide in soil air. Though not reviewed below, C and O isotope studies of travertines and speleothems can be used to study past environmental conditions and to identify changes in water sources over time (e.g. Dorale et al. 1998).

Hydrogen and carbon also include radioactive isotopes, notably tritium (T) and radiocarbon (^{14}C), that can be used to determine the age of water and in some cases provide additional tracers. Most karst waters contain elevated tritium concentrations due to the incorporation of young water contaminated by thermonuclear fallout. This effect limits the utility of simple tritium counting in water dating, but the T/He method can correct for this. Carbon-14 dating can be routinely performed on DIC by use of accelerator mass spectrometry (AMS).

Noble gas isotopes are dissolved in water in regular proportion to their content in ambient soil air, but depending on temperature. Applications include determination of water recharge temperatures, the breakthrough time of surface waters, and the delineation of radioactive contaminant sources. Continuous radon measurements have special potential to quantify fast transport processes.

7.2   HYDROGEN, CARBON AND OXYGEN ISOTOPES

7.2.1   Isotope abundance and significance to karst hydrogeology

Hydrogen, carbon and oxygen are the dominant constituents in karst waters, karst bedrock, speleothems, travertine deposits, and in dissolved bicarbonate which is normally the dominant anion in karst water. Fortunately, these elements all include stable isotopes that exhibit variations large enough to be routinely measured with mass spectrometers. Hydrogen and carbon also include rare radioactive isotopes with hydrologically-useful half lives. The masses, relative abundances, radioisotope half lives, and other information about the isotopes of these and other elements are given by Walker et al. (1989).

Elemental hydrogen is dominantly constituted of protium (^{1}H), but deuterium (D, or ^{2}H) is a stable trace isotope that is ubiquitous on Earth at $\sim$150 ppm. Tritium (T, or ^{3}H) is a rare radioisotope with a half life of 12.3 years. Tritium is naturally produced by cosmic-ray bombardment reactions in the atmosphere, normally in very low concentrations, but large augmentations have been caused by man's nuclear activities, which peaked in 1963 (Fig. 7.1).

Elemental carbon comprises two stable nuclides, ^{12}C, which dominates, along with about 1.1 atom percent of the isotope ^{13}C. Another isotope, ^{14}C or radiocarbon, has a half-life of 5730 years, and is naturally produced in very low concentrations by cosmic-ray bombardment of atmospheric nitrogen. Man's nuclear activities have also augmented its abundance, roughly doubling the natural atmospheric concentration during the peak of surface testing.

Elemental oxygen is dominantly constituted of stable ^{16}O, but contains two ubiquitous stable isotopes ^{17}O and ^{18}O with atomic abundances of about 400 and 2000 ppm, respectively. Most available data are for ^{18}O. The ^{17}O isotope has been largely ignored in hydrologic studies to date, and is neglected below, but unusual abundance patterns occur in certain atmospheric molecules and mechanisms exist whereby these can be transferred to condensed phases, providing possibilities at the scientific frontier. All radioisotopes of oxygen have half-lives of a few minutes or less (Walker et al. 1989), rendering them useless to karst studies.

7.2.2   Isotopic variations and notation

Natural processes cause the relative abundances of the isotopes to vary, or fractionate producing easily measured variations that render them useful as hydrogeologic tracers (e.g. Criss 1999). Fractionation reflects the equilibrium or non-equilibrium partitioning of an isotope among coexisting phases containing the element of interest, or between the separated parts of a single phase. Much larger variations in the concentrations of radioactive and radiogenic isotopes are produced by their decay or production.

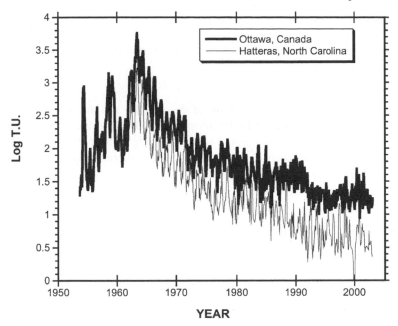

Figure 7.1. The tritium concentration (1 tritium unit T.U. = 1 tritium atom in 10^{18} hydrogen atoms) in precipitation has varied by greater than two orders of magnitude due to fallout of thermonuclear-produced tritium from surface testing (data from IAEA 2004). Note that coastal samples (Hatteras) are depleted in tritium relative to the cold continental interior (Ottawa), opposite to the distribution of deuterium.

Special notation is used to describe isotopic abundances in natural samples. The most basic parameter is the isotope ratio R, representing the atomic abundance of a trace isotope relative to that of a reference isotope of the same element. For stable isotopes, the reference nuclide is normally chosen as the dominant nuclide, either ^1H, ^{12}C or ^{16}O, so that R represents either the D/H, $^{13}C/^{12}C$, or $^{18}O/^{16}O$ ratio. The T/H ratio is similarly used in tritium studies. For all these cases, R is a small number, a decimal point followed by 1 or more zeroes. While R is a useful variable for theoretical work and formulae, and is widely used in geochronology, it is normally not a convenient way to report stable isotope concentration data on natural samples. For the latter purpose the delta-notation was devised long ago, such that:

$$\delta = 1000(R_x - R_{std})/R_{std} \qquad (1)$$

where R_x is the isotope ratio of the sample and Rstd is that of the defined reference standard. The isotope ratio of the sample is thus reported as a normalized deviation from that in the reference material; the factor of 1000 in equation 1 converts this difference to per mil (‰). For the stable H and O isotopes, the most commonly used reference material is SMOW (standard mean ocean water); for carbon isotope studies the PDB (Pee Dee belemnite) carbonate standard is normally used (O'Neil 1986). The PDB standard is of marine origin, so the $\delta^{13}C$ values of the marine carbonate rocks that host most karst aquifers are similar at about 0 ± 3‰. In contrast, biogenic fractionation in terrestrial plants produces much lower $\delta^{13}C$ values between -13 and -28‰. All δ-values in this paper conform to these definitions.

Different conventions are used to report other isotope abundances of interest. For tritium the TU, or "tritium unit", is used, where 1 TU represents a T/H ratio of 1 atom in 10^{18} hydrogen atoms. In contrast, ^{14}C abundances are normally reported as pmc values, or percent modern carbon:

$$pmc = 100R_x/R_{std} \qquad (2)$$

where R_x in this case is the $^{14}C/^{12}C$ ratio of the sample, and the standard is calibrated against the ^{14}C atom abundance of 1950 atmospheric CO_2. Any pmc values >100 indicate the incorporation of carbon originating from thermonuclear testing fallout, whereas values <100 are a proportional measure of radioactive decay, or alternatively of isotopic exchange with carbonate rock devoid of ^{14}C.

7.3   OXYGEN AND HYDROGEN ISOTOPES IN METEORIC WATERS

7.3.1   Geographic variations

Isotopic variations in meteoric waters are inherited in modified form by karst waters. The physical effects that control the geographic patterns of precipitation also govern the concentrations of D, T and ^{18}O. Most important is the fact that the vapor pressures of water and ice increase exponentially with temperature. As a general rule, a saturated air mass at the coast will progressively lose moisture and become isotopically depleted as it adiabatically rises in elevation or moves inland toward a continental interior. The geographic variations produced by these effects are complex in detail, but an illustrative comparison is that annual precipitation commonly exceeds 10 m in tropical rainforests but is less than 10 cm at the South Pole!

Patterns of the isotopic content of rainfall can be similarly understood in terms of the vapor pressures of individual isotopic species of water (e.g. Dansgaard 1964). "Heavy" species such as HDO, HTO and $H_2^{18}O$ have lower vapor pressures than ordinary $H_2^{16}O$, with the disparities increasing as temperature decreases. As a result, the D and ^{18}O concentrations of meteoric precipitation are typically higher in warm tropical areas and coastal locations than in cold regions, continental interiors, rain shadows, or at high elevations (e.g. Dansgaard 1964). Such effects may be visualized as the preferential "rain out" of heavy isotopes from advecting air masses. A global map of average $\delta^{18}O$ and δD values is available (IAEA 2004), as are more detailed maps for some continents (e.g. Sheppard 1986).

Tritium is subject to all the aforementioned fractionation effects, but includes even more important terms involving stratospheric production and downward transport (see sec. 7.6). In any given year, tritium is much more abundant at high latitudes than at low latitudes and close to coasts (Fig. 7.1), opposite to the distribution seen for D and ^{18}O.

7.3.2   Temporal variations

Concentrations of D and ^{18}O in meteoric waters at a given geographic location vary over time. In part these variations reflect seasonal changes in temperature at single locations; they are indeed greater in regions with large seasonal temperature changes than in areas where such differences are small. Typically, winter precipitation has less D and ^{18}O than summer precipitation. The source region and advection path of individual air masses also exert control. Large isotopic variations can even occur during individual storms at a given site.

The pattern of tritium abundance is greatly complicated by anthropogenic sources. Natural tritium levels in precipitation were dwarfed by stratosphere inputs from nuclear detonations (Fig. 7.1). Tritium is now, at a given location, about $100\times$ less abundant than in 1963, a far greater reduction than can be attributed to decay. Geographic variations of T suggest greater dilution of a descending, stratospheric T component in regions with high rainfall and rapid cycling of atmospheric water, an effect not applicable to D and ^{18}O. However, seasonal variations of tritium and their relation to rain amount at a given location conform more closely to those for D and ^{18}O. Thus, at a given time and place, tritium variations are governed by the same types of fractionation effects that affect stable isotopes in water.

7.3.3 Meteoric water line

A simple and very useful empirical correlation called the meteoric water line (MWL) relates the δD and $\delta^{18}O$ values of meteoric precipitation (Craig 1961):

$$\delta D = 8\delta^{18}O + 10 \tag{3}$$

The degree of correlation is high, probably >0.95, though it cannot be calculated because the data base has become so huge. In detail, different "local water lines" having slightly different slopes and y-intercepts have been proposed for individual regions, but equation 3 represents a very good global average.

Clearly, equation 3 allows the δD value of a given rain or snow sample to be estimated to good accuracy from the measured $\delta^{18}O$ value, or vice versa. One might conclude that making both isotopic measurements on a single sample would be pointless, but this would preclude the most important scientific application of equation 3! In particular, waters having non-meteoric origins or affected by secondary processes commonly lie off the meteoric water line. For example, the δD and $\delta^{18}O$ values of normal seawater are about 0 ± 5 and $0 \pm 1\%o$ respectively, so normal seawater does not lie on the meteoric water line. Hydrothermal waters, formation waters, oil field brines, evaporated waters, landfill leachates, oceanic pore fluids, etc. are all distinguishable, and excepting leachates and cool pore fluids, most lie far to the right of the meteoric water line (Sheppard 1986). In many cases analyses of sample suites define linear trends on graphs of the δD vs. $\delta^{18}O$ values that project back to a parent water that lies on the MWL. Evaporation of meteoric waters or their exchange with heated rocks commonly generate waters that deviate from the MWL.

7.4 SOURCE IDENTIFICATION OF KARST WATER

7.4.1 Meteoric percentage of karst waters

The hydrogen and oxygen isotope ratios of most cool groundwaters at shallow depths resemble the average ratios of local meteoric waters. Most karst waters similarly lie on or near the MWL and clearly are derived from local precipitation that becomes substantially homogenized in ground water reservoirs. Table 7.1 illustrates this similarity for several

Table 7.1.   Oxygen-18 content of some large springs and precipitation.

Spring	Flow, m^3/s	$\delta^{18}O$ value of spring, ‰	$\delta^{18}O$ of local precipitation	Reference
Manatee Spr, FL	5.1	−3.5	−3.74 ± 0.9[#]	Katz 2004
Comal Spring, TX	9.3*	−4.25	−4.03[#]	Blake 1992, IAEA 2004
Big Spring, MO	12.1	−6.52 ± 0.21	−7.18 ± 4.06[#]	This chapter
Maramec Spring, MO	4.2	−6.64 ± 0.69	−7.18 ± 4.06[#]	Frederickson & Criss 1999
Kirkgozler, Turkey	10	−7.25 ± 0.24	−5.6 ± 2.1	Dincer & Payne 1971
Grande, Spain	0.63	−7.54 ± 0.47	−6.81 ± 2.8	Andreo et al. 2004
Figeh, Syria	7.7	−9.02 ± 0.16	−8.55 ± 0.8	Kattan 1997
Saivu Spring, CH	0.1	−9.12	−9.45	Perrin et al. 2003
Capo Pescara, Italy	7.5	−9.8	−8.45[#]	Governa et al. 1989, IAEA 2004
Siebenquellen, Austria	0.31	−12.26	−11.06	Maloszewski et al. 2002
Big Warm Spring, NV	0.44	−16.0	−16.0[#]	Rose et al. 1997

* Historical flow; spring now nearly dry due to pumping.
[#] Precipitation record >50 km from spring.

large karst springs, where the "local" precipitation column reports the nearest available data (measured <200 km from the spring), and also shows regional progressions. For example, Manatee Spring, a first-magnitude spring in subtropical, coastal Florida, is higher in ^{18}O than Comal Spring in Texas, which has more ^{18}O than Big Spring in the Ozarks of Missouri, located in the continental interior, which is higher in ^{18}O than Big Warm Springs in Nevada, located in the semiarid, intermontane rain shadow region called the Great Basin. The $\delta^{18}O$ values of these springs reflect their increasing distance and isolation from the coast as well as their generally increasing elevation. The $\delta^{18}O$ differences between springs and local precipitation tend to be small, and can be attributed to differences between the recharge area and the precipitation sampling sites. The largest ^{18}O differences between springs and precipitation occur in mountainous areas, where recharge can originate at high elevations where $\delta^{18}O$ values of precipitation are lower than near the spring orifice (e.g. Austria, Tab. 7.1). Isotopic disparities can also arise if the fractional amount of infiltrating precipitation varies by season.

For less voluminous karst waters, isotopic differences among distinct local environments are common. Even the average $\delta^{18}O$ values of springs can differ substantially from the local meteoric average. Only 3 of 8 well-characterized perennial springs in our St. Louis, Missouri database have average $\delta^{18}O$ values (−7.21 ± 0.37; −7.13 ± 1.22; −7.16 ± 1.08‰) that are basically identical to the 10-year average $\delta^{18}O$ value for local precipitation (−7.18 ± 4.06‰; this study). The other five springs have significantly higher values of −6.98 ± 0.67, −6.27 ± 0.72, −6.26 ± 0.79, −5.09 ± 1.26, and −3.69 ± 1.55‰. Three of the latter group drain sinkhole plains having up to 50 sinkholes per km^2; these plains contain small ponds that undergo evaporation and can contribute ^{18}O-enriched waters to the springs. The most extreme example, Weldon Spring, now derives 80% of its flow from a leaky artificial lake that has undergone significant evaporative enrichment (Criss et al. 2001).

Other types of karst waters can be similarly influenced by secondary processes, providing exceptions to the generalization of Yonge et al. (1985) that the $\delta^{18}O$ values of cave drip waters approximate those of average local precipitation or its derivative recharge. For example, Ingraham et al. (1990) established that the δD and $\delta^{18}O$ values of drip waters, popcorn drip waters, and cave pools in Carlsbad Caverns are all distinct. In particular, the drip waters lie along the MWL, while the cave pools and popcorn drips have higher salinity and heavy isotope contents due to evaporation in the cave passages. Far larger isotopic enrichments occur in the unsaturated zone in soils (Barnes & Allison 1988). Such effects are important in the epikarst; for example, Bar-Matthews et al. (1996) found that cave drip waters in a semi-arid region are about 1.5‰ enriched in ^{18}O relative to local precipitation, due to evaporation.

Large springs exhibit much less isotopic variability than local precipitation (Tab. 7.1), and in a given region tend to vary less than small springs but more than deep well waters, which are practically constant. Interesting isotopic variability is evident when data are gathered in time series, as discussed in section 7.5. In short, the oft-cited isotopic similarity between karst waters and the local meteoric average is a useful generalization, but exceptions are common, and single samples of karst waters may differ greatly from average meteoric water, particularly if the karst water sample represents a small reservoir. Even long-term averages of the stable isotopic values of karst waters may differ from the local meteoric average, particularly if they have undergone evaporation at the surface or in the epikarst. Samples affected by such processes normally plot well to the right of the meteoric water line.

7.4.2 Isotopic mixing and solute correlations

Many processes affecting karst waters are revealed by correlations between the δD and $\delta^{18}O$ values, or between either of these values and the concentration of a dissolved solute. On such diagrams, simple mixing of two distinct sources produces waters of intermediate composition that lie along straight line segments connecting those end members. The proportions of the end members contributing to any given mixture are given by the "lever-rule"; thus, if mixed in equal proportion the intermediate water plots exactly midway between the two end members. For simple, two component mixing, the relation is:

$$C_{mix} = FC_A + (1 - F)C_B \qquad (4)$$

where F represents the fractional contribution of water A to the mixed water volume, $1 - F$ is that of component B, and the C's represent the concentration of a given dilute solute, or alternatively, the δD or the $\delta^{18}O$ values of the water. This equation may be extended to include more mixing components by incorporating additional terms (see Criss 1999).

Simple mixing of distinct waters will produce straight-line segments on graphs involving any pairs of the aforementioned variables, such as a plot of the dissolved bicarbonate concentration vs. the $\delta^{18}O$ values of the water. Not all parameters behave in this way, however. In the case just mentioned, a strongly curved mixing trend normally results if, instead of the bicarbonate concentration, the $\delta^{13}C$ value of the dissolved bicarbonate is plotted against the $\delta^{18}O$ value of the water (see Criss 1999).

Simple mixing is commonly invoked in hydrologic studies. Nevertheless, an interesting feature of karst systems is the extent to which their hydrochemical variations are not explained by simple mixing. Lakey & Krothe (1996) demonstrated hysteresis loops on a graph of δD vs. $\delta^{18}O$ values for a karst spring in Indiana subject to storm-induced variations in discharge and chemistry. Analogous, complex relationships commonly occur in springs when a solute concentration is plotted against δD or $\delta^{18}O$ values. Fig. 7.2 shows the

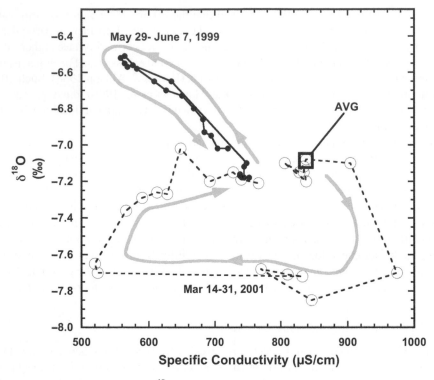

Figure 7.2. Covariations of the $\delta^{18}O$ values and electrical conductivity in Bluegrass Spring, Missouri, following two spring storms. The response patterns define remarkably distinct hysteresis patterns, but both feature rapid displacement from the long-term average spring composition (AVG) followed by slow recovery. This response is observed in many karst systems where storms induce rapid geochemical variations.

response of Bluegrass Spring, Missouri to two storms, one in late May 1999 that delivered 4.8 cm of rainfall with a $\delta^{18}O$ value of $-4.9‰$, and another in March 2001 that delivered 2.6 cm of rainfall with a very low $\delta^{18}O$ value of $-14.7‰$. The spring's response to the first event appears at first glance to show a linear mixing trend that projects to the dilute rainfall composition, but in detail, the data define a flattened hysteresis loop whereby the points move along a counterclockwise path over time, being rapidly displaced from the initial, approximately mean composition and then slowly recovering back toward it. However, the spring's response to the March 2001 storm defines a completely different path, with the spring again being displaced from its approximate average composition and then slowly returning to it, but in this case along a broad clockwise loop (Fig. 7.2). Simple mixing concepts and elementary hydrograph separation are useful tools in hydrology, but only provide coarse approximations to complex real behaviour.

7.4.3   Stable isotope tracing of karst springs

Hydrogen and oxygen isotope data can be used to trace groundwater and springs whenever their parent waters have distinctive δD or $\delta^{18}O$ values relative to other local waters.

Favorable circumstances are present when elevation differences between recharge and discharge points are large, when lateral distances are great, or when special sources such as rivers or evaporated water bodies contribute significantly to groundwater reservoirs.

The Great Basin of Nevada provides an excellent example where regional groundwater flow involving great elevation change renders isotopic tracing possible (Davisson et al. 1999). Inter-basin groundwater flow occurs in Paleozoic carbonates dissected by basin and range normal faulting. Valley floor elevations decrease nearly 1000 m over a 350 km distance, driving regional flow originating as mountain snow melt in the north to springs far to the south. The average $\delta^{18}O$ values of local meteoric precipitation vary by nearly 7‰ in the huge region underlain by this aquifer, but only the highly depleted snowmelt from the north is an important recharge source, and this has $\delta^{18}O$ values that are sharply lower than precipitation collected near the spring discharge points. This contrast has been misinterpreted as evidence for locally-derived, pluvial period recharge, when it instead signifies spring recharge from distal sources.

Normally, only small geographic variations in the δD and $\delta^{18}O$ values of meteoric precipitation occur over the limited lateral scales (<100 km) typical of karst systems, unless the range of local elevation is large (e.g. Kattan 1997). This lack of variation, combined with the homogenization and dilution of any distinctive waters in the subsurface, commonly impedes the isotopic tracing of karst springs. One exception is that meteoric waters that have undergone significant evaporation in lakes, reservoirs or sinkhole ponds are easily identified by their high δD and $\delta^{18}O$ values compared to local meteoric waters. Impounded reservoirs in karst regions commonly leak (Aley et al. 1972) and the contributions of such evaporated surface waters to karst springs can be easily detected (Criss et al. 2001).

7.4.4   Contribution of karst waters to surface streams

Profound interactions between surface waters and groundwaters in karst regions are evidenced by swallow holes, springs, resurgences, internal drainage, and other conspicuous features. Methods of hydrograph separation quantify this interrelationship and reveal its dynamic nature. At any time, the observed total flow ("tot") of a surface stream can be divided into a "base flow" ("bf", i.e. groundwater) fraction having allegedly constant character, and an "event water" ("ew") fraction representing the temporally-variable input of the most recent rainfall whose isotopic character will generally be distinctive from the mean. Under these assumptions, equation 4 becomes:

$$F = (\delta_{tot} - \delta_{ew})/(\delta_{bf} - \delta_{ew}) \qquad (5)$$

where F is the fractional base flow contribution to the total streamflow, and the δ-values refer to either the δD or $\delta^{18}O$ values of the indicated component (e.g. Sklash & Farvolden 1979). Suitable samples will characterize all of the δ-values on the right hand side, so F can be quantitatively determined for any given stream flow sample. Analogous results can be derived for chemical concentrations (Pinder & Jones 1969), but in practice the event water contribution is not a simple dilutor (see below).

Numerous geochemical studies have applied hydrograph separation to springs and rivers in karst areas (e.g. Lakey & Krothe 1996, Winston & Criss 2002). Well established, first-order conclusions are that 1) ground water generally constitutes nearly 100% of total stream flow during normal conditions; 2) event water contributions are maximized during flooding, but typically achieve their greatest relative proportion after peak flow, 3) flows of both event water and ground water dramatically increase during flooding, and 4) event water proportions are largest in the greatest floods. These findings refute traditional hydrologic assumptions that ground water flows decrease during flooding (e.g. Fetter 2001).

7.5   OXYGEN ISOTOPE RESIDENCE TIME OF KARST AQUIFERS

7.5.1   Temporal isotopic variations in springs

Although the average δD or $\delta^{18}O$ values of meteoric precipitation vary little across small watersheds, temporal variations of these isotopes can be large. In many places the δD or $\delta^{18}O$ values of meteoric precipitation vary over a roughly sinusoidal, seasonal cycle with respective amplitudes of 120 and 15‰ or more. This seasonal forcing perturbs the isotopic character of the karst watershed, causing variability that provides important information about system timescales and character (Fig. 7.3).

A key concept is that karst watersheds imperfectly integrate and homogenize meteoric input from different storms and seasons, so that the isotopic variations in groundwaters are damped relative to those of the meteoric input. Were this damping completely effective, the $\delta^{18}O$ and δD values of karst groundwater would match the local, long-term meteoric average.

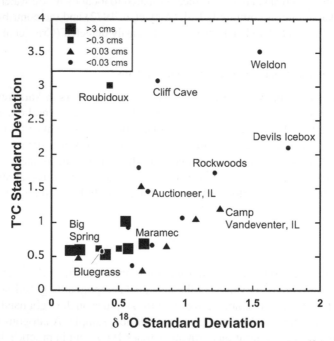

Figure 7.3.   Correlation between the variability (standard deviations) in the $\delta^{18}O$ values (‰) and temperatures (°C) in 26 springs of the Missouri Ozarks and Ozark border and 2 springs in western Illinois, USA. The largest Ozark springs such as Big Spring (12 cms, or m³/s) tend to have the lowest parameter variability and the longest residence times. More than 1000 $\delta^{18}O$ determinations are represented in the collective data set, and the temperature data are even more numerous and include measurements by USGS.

Many deep, slow-moving groundwaters in porous media approximate this condition and essentially are completely damped. In contrast, the shallow, flowing karst waters that feed ordinary springs exhibit partially-damped seasonal isotopic cycles.

Perhaps the simplest model for the δ-values of karst groundwater would be a simple running average of the δ-values of sequential rainfall increments, summed over a selected, moving interval that approximates the groundwater residence time. Shorter intervals produce less damping and more isotopic variation; at the other extreme, complete damping represents the case where the mixing interval becomes very long. Improved models weight the δ-values of each rainfall increment according to the amount of precipitation represented (see below).

7.5.2   Relative magnitudes of isotopic, physical and chemical variations

The processes that integrate and homogenize the isotopic character of groundwater systems also tend to dampen variations in other chemical and physical properties including electrical conductivity, temperature, chemical composition, and discharge. As a rule of thumb, if the variance of a given parameter is small, the variance of all these parameters will be low, and vice versa for high variance (Fig. 7.3). In addition, large karst systems tend to have less variance than small systems, and deep systems are less variable than shallow ones.

The greatest geochemical variations in karst springs accompany fluctuations in their discharge (e.g. Winston & Criss 2004, Stueber & Criss 2005). If discharge variations represent variable additions of dilute event waters to a base flow component that is constant in flow rate and composition, then plotting the δD or $\delta^{18}O$ values, or a chemical concentration, against the inverse discharge of the spring will result in a linear trend whose y-intercept indicates the event water composition (eq. 4). However, scattered relationships or loops on such graphs are typical and provide evidence for more complicated behavior. Direct graphical comparison of the covariance of two parameters is informative, such as a plot of a chemical concentration vs. electrical conductivity. Simple addition of dilute event water to base flow would produce positively-sloped, linear trends on such diagrams; the y-intercepts would indicate the event-water composition and would plot close to the origin. Real examples are more complex and display both positive and negative slopes. Two contrasting chemical groups can be defined, a "normal" group of ions whose concentrations decrease as flow increases, as expected for dilution, and a group whose concentrations sharply increase as flows increase (Stueber & Criss 2005). Members of the "normal" group include the most abundant ions in karst waters, such as those arising from carbonate bedrock dissolution, while the "abnormal" group includes ions and agrochemicals that are sorbed to soils or are present in acidic soil waters that are mobilized by event water pulses.

7.5.3   Linear reservoir model

The damped-average model of Frederickson & Criss (1999) describes temporal isotopic variations in diffuse springs (Fig. 7.4). According to this model, individual rainfall increments are sequentially supplied to a well-mixed but leaking linear reservoir, whose flow rate is proportional to the volume of water contained at any time. Isotopic variations in the spring are independently predicted from the amount and isotopic character of precipitation events sequentially collected in a rain gauge, given a single free parameter, the isotopic

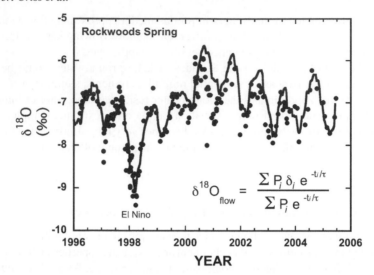

Figure 7.4.   Observed oxygen isotope variations in Rockwoods Spring, Missouri (dots), compared with the predictions of the linear reservoir isotope model (solid line and indicated equation) for an assumed residence time of 1 year. The seasonal amplitude of the δ^{18}O variations is about 2‰, ten fold less than that for meteoric precipitation in this region. Note that the large negative isotopic excursion during El Nino is almost perfectly simulated by this model. Updated after Frederickson & Criss (1999).

residence time (τ) for the spring. In particular:

$$\delta^{18}O = \sum P_i\delta_i e^{-t_i/\tau} \Big/ \sum P_i e^{-t_i/\tau} \tag{6}$$

where P_i and δ_i are the amount and δ-value of each increment of meteoric precipitation, and t_i is the time interval since that particular increment was added. According to this model the amount of any given rain increment that remains in the reservoir decreases exponentially with time after its addition.

7.5.4   Hydrologic pulse model

Numerous data sets link variations in the chemical and isotopic concentrations of springs with variations in their flow rates. Simple mixing relationships (eq. 4) or hydrograph separation techniques (eq. 5) are commonly invoked to explain these variations in terms of the addition of dilute event waters. The real relationships are more interesting.

Variations in groundwater flow following storm events conform closely to the theoretical "delta-function" model of Criss & Winston (2003) that is based on Darcy's law. According to this model, the rainfall event produces a sudden increase in hydraulic head, inducing an increase in the flow rate followed by recession:

$$Q/Qp = 2.4395 \, (b/t)^{3/2} \, \mathrm{Exp}(-b/t) \tag{7}$$

where Qp is the peak flow rate, b is the "time constant" of the response, and t is the time elapsed since the pulse in head. This expression simulates the entire storm hydrograph of

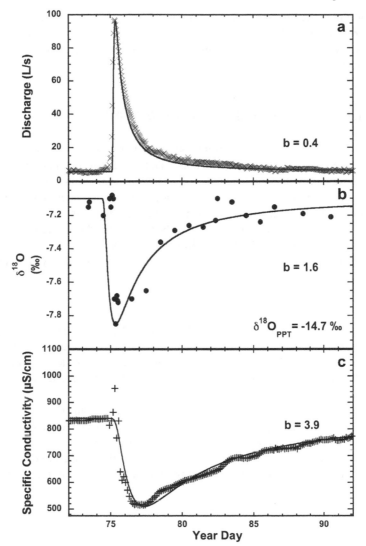

Figure 7.5. Graphs showing the variations in flow rate (x, top), $\delta^{18}O$ values (dots, middle) and electrical conductivity (+, bottom) following a storm in March 2001 that delivered 2.9 cm of rain with a $\delta^{18}O$ value of $-14.7‰$. Curves show the normalized patterns predicted by equation 7 for respective time constants of 0.4, 1.6 and 3.9 days. All curves are generated using the same initial start time to coincide with the causal head pulse. This proves that this model simulates the behaviour of several geochemical parameters and that the coupled responses are all driven by the same hydrologic perturbation (after Winston & Criss 2004).

many springs, streams and small rivers, and offers a more realistic expression for recessional flow than the standard exponential model (Criss & Winston 2003). The basic shape of the hydrograph defined by equation 7 also simulates many physical, chemical, and isotopic variations that accompany discharge variations (Winston & Criss 2004). Figs. 7.5a,b,c use this function to simulate the observed disturbance and recovery of discharge, $\delta^{18}O$ values, and electrical conductivity of Bluegrass spring following a rainfall event. Note, however, that each parameter has its own time constant, in this case, 0.4 days for the flow (Fig. 7.5a), 1.6 days for $\delta^{18}O$ (Fig. 7.5b), and 3.9 days for electrical conductivity (Fig. 7.5c). The mutual disparities of these time constants cause "hysteresis" loops when one parameter is plotted against another, as in Fig. 7.2.

7.6   RADIOISOTOPE DETERMINATION OF WATER AGE

Groundwater age-dating is an important tool in water resource investigations because accurate age determinations within a groundwater basin can elucidate rates of aquifer replenishment and provide an independent safe-yield determination (e.g. Criss & Davisson 1996). In addition, groundwater ages are commonly related to water quality, and can provide a means to distinguish natural from man-made contaminant sources. Furthermore, groundwater ages can confirm climate-related changes inferred from stable isotope data on deep aquifers (e.g. Clark et al. 1998, Sturchio et al. 2004).

Most isotopic age-dating schemes are based on the radioactive decay equation:

$$N/N_i = e^{-\lambda t} \tag{8}$$

where N is the measured radioisotopic abundance that, when normalized to the abundance at the time of recharge (N_i), is exponentially related to the product of the characteristic decay constant λ and the elapsed time t. Most uncertainty in age-dating stems from determining the initial abundance N_i, and in groundwater this is often influenced by complexities of dispersion and reactivity of the particular radioisotope.

Numerous schemes have been developed for groundwater dating (Fig. 7.6). For the most part, these methods either exploit nuclear fallout signatures to date relatively young waters, or use naturally-occurring radionuclides to date ancient waters.

Karst transport occurs along linked fracture-like networks having large but variable surface areas that promote simultaneous mixing and chemical dissolution. Both processes

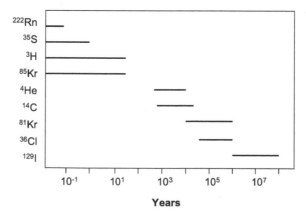

Figure 7.6.   Several radioisotopic dating schemes exploit thermonuclear fallout to date young (<50a) groundwater. No radioisotope system has been reliably developed for dating groundwater between 50 and 500 years old, but older groundwater can be characterized using long-lived cosmogenic isotopes.

affect concentrations of dissolved constituents over time as groundwater flows down gradient from recharge zones. Age dating groundwater reliably in these aquifers requires reconciliation of both the physical mixing during transport and chemical reactivity. Accordingly, sampling and analysis should focus on both temporal and spatial variations in order to capture concentration-dependent processes affecting absolute abundance of a radioisotope. In many cases, precipitation-related changes in isotopic and water quality characteristics of discharging springs or wells may provide the most accurate age indications by revealing mixing end-members (see 7.5.3, 7.5.4).

7.6.1   Tritium

Because of its radioactive half-life of only 12.3 years and its low but consistent natural atmospheric concentrations imparted to meteoric waters at the time of recharge, tritium should provide an ideal chronometer for young (≤ 40 years) groundwater. However, T concentrations in precipitation underwent huge augmentations about 40 years ago due to surface and atmospheric testing of thermonuclear weapons (Fig. 7.1). Nevertheless, the fallout of this "bomb-pulse" between 1963 and ca. 1980 was a boon to the hydrologic sciences because tritium, being part of the water molecule, provided an intrinsic water tracing property. In this sense, T fallout initiated a grand tracer experiment that enabled the observation and quantification of groundwater recharge and flow over time and space scales commensurate with the scale of the tracer input. Because of its relatively short half-life, though, observation of the rising or descending limb of the T fallout profile is required to quantify aquifer transport properties (e.g. Vogel et al. 1974). Today, much of the "bomb-pulse" is not well defined in groundwater due to tritium decay and dispersion, yet a T content exceeding typical natural levels (5–10 TU) still provides a strong indication of bomb-pulse recharge. Unfortunately, the testing has caused high uncertainty regarding the T content at the time of recharge, rendering the T content by itself useless as an absolute age-dating method.

Tritium fallout offered additional opportunities to quantify transport properties of karst groundwater systems, particularly over large time (months to years) and space (100 s of km^2) scales impractical for artificial tracers such as injected dyes. In karst regions the dispersion of bomb-pulse T was conspicuous. For example, Dincer & Payne (1971) recorded T fallout from precipitation and its admixture into shallow karst waters of western Turkey. They showed that T in karst aquifers was largely diluted relative to recent precipitation due to mixing with a large reservoir of groundwater recharged prior to weapons testing. In western Turkey, where annual precipitation is between 30 and 50 cm/yr, groundwater ages determined from a mixing model ranged from 20 to 100 years old on a 25 to 50 km length scale of transport. Furthermore, the temporal variability of T in spring-fed rivers was dramatically illustrated, ranging an order of magnitude between baseflow and runoff following the peak atmospheric fallout period. This difference decreased in subsequent years, making the younger and older flow components difficult to distinguish using ^{3}H alone.

Extensive isotopic investigations of karst groundwater conducted in the relatively higher precipitation region of Germany and Austria furthered the understanding of subsurface interconnectivity and porosity (e.g. Seiler et al. 1989, Rank et al. 1991, Maloszewski et al. 2002). δ^{18}O and T data were gathered during the entire fallout period on spring discharges and for precipitation collected in the Austrian Alps. The δ^{18}O values of annual precipitation varied widely by ~10‰, while the springs varied by less than 1‰, consistent with high groundwater dispersion rates. These data were used to simulate annual T concentrations by using a dispersive piston-flow model that invoked dual porosity consisting of fissures and chemical dissolution channels (Rank et al. 1991, Maloszewski et al. 2002).

A dual flow explanation was also forwarded for karst groundwater in Florida (Katz et al. 1999), where binary mixing was consistent with both young and old (>20 years) groundwater ages defined by using T and chlorofluorocarbon analysis. Unfortunately in this case, only single samples were collected near baseflow conditions, that precluded better definition of recharge mechanisms and age variations.

7.6.2 Tritium-helium-3 and krypton-85

High-precision noble gas mass spectrometers allow direct measurements of helium-3, the decay product of T, permitting combined T and ^{3}He data to serve as an accurate parent-daughter geochronometer (Schlosser et al. 1988). In particular, this tritiogenic helium ($^3\text{He}_{trit}$) numerically added to the measured tritium concentration in a sample yields the initial T_i content at the time of recharge. Such analytical approaches to derive initial tritium content are vastly superior to commonly used indirect estimates. In practice, however, several components comprise the measured ^{3}He and the more abundant isotope ^{4}He:

$$^3\text{He}_{meas} = {}^3\text{He}_{trit} + {}^3\text{He}_{equil} + {}^3\text{He}_{excess} + {}^3\text{He}_{rad} \qquad (9)$$

and

$$^4\text{He}_{meas} = {}^4\text{He}_{equil} + {}^4\text{He}_{excess} + {}^4\text{He}_{rad} \qquad (10)$$

where the subscript *meas* is the total dissolved abundance of an isotope analytically measured, *equil* is the amount dissolved in a non-turbulent surface water in equilibrium with the atmosphere, *excess* is the amount dissolved in water exceeding the equilibrium amount (a common phenomenon in groundwater, see Heaton & Vogel 1981), and *rad* is the amount produced from radioactive decay of other radioisotopes. For example, $^4\text{He}_{rad}$ is a product of uranium-thorium decay, and its accumulated abundance can be significant in older waters (e.g. >1000 years old). Radiogenic ^{4}He is easily observed by $^3\text{He}/^4\text{He}$ ratios <1.0 when normalized to that of air. The accumulation of $^4\text{He}_{rad}$ in groundwater has been used as a dating method itself in karst aquifers, where diffusive flux from deep crustal sources (e.g. Mazor 1991) or shallow uranium-enriched deposits (Price et al. 2003) can be sources of steady-state accumulation. The *equil, excess,* and *rad* terms are derived with the aid of measurements on the other noble gases co-dissolved in the water sample, while the $^3\text{He}_{rad}$ term is assumed (~0.2% of the total ^{3}He). Ultimately, these terms are subtracted from the *meas* term to yield a $^3\text{He}_{trit}$. Typical uncertainties in calculated age are less than ±5 years, and often are <2 years.

Although ^{85}Kr is naturally produced in the upper atmosphere by spallation and (n, γ) reactions on ^{84}Kr, most originates as a fission product from nuclear fuel reprocessing. ^{85}Kr has a half-life of 10.76 years and is ideally suited to groundwater tracing. The accumulated abundance of ^{85}Kr in air (nearly 1000 disintegrations per minute/millimole Kr, corresponding to a $^{85}\text{Kr}/\text{Kr}$ ratio of ~10^{-11}) provides at least four half-lives of detectable activity in groundwater (Rozanski 1979, Smethie et al. 1992). Examples of ^{85}Kr measurements in karst groundwater are lacking, but in porous media it provides a strong complimentary technique to other dating methods (e.g. Ekwurzel et al. 1994). Research could focus on similar comparisons to other dating methods, but in classical karst systems having mean residence times on the order of a few years.

Dispersive mixing of dissolved constituents in karst groundwater can be nearly as pronounced as those dissolved in rivers. Dispersion presents a unique challenge in tracer studies because of the inherent spreading of a tracer slug and the commonly observed skewing of the breakthrough curve. Under normal porous flow conditions, dispersion represents only a minor concern for interpreting T-^{3}He and ^{85}Kr abundances for age calculations (Schlosser et al. 1989, Solomon & Sudicky 1991, Ekwurzel et al. 1994). However, dispersion coefficients in karst groundwater can be several orders of magnitude greater than in porous systems having the same length scales, and to date no careful analysis of this effect on age-dating methods has been attempted. In addition, karst groundwaters are probably also affected by variable recharge rates and co-mingling of waters from different recharge sources along fracture or solution conduits. This will be most relevant for T-^{3}He dating, since the calculated ^{3}He$_{trit}$ content largely depends on the dissolved solutes derived from excess air content. For mixtures of dilute solutes, the isotopic abundance of the mixture requires accounting of the dissolved concentration of each mixing component as follows:

$$R_{mix} = \left(\frac{X_1 C_1 R_1 + X_2 C_2 R_2 + \cdots}{\sum X_i C_i} \right) \qquad (11)$$

where X is the volume fraction of a mixing component, C is the dissolved concentration of a solute whose isotopic abundance R is of interest. Non-linear mixing of isotopic abundances results when mixing end-members differ in concentration.

7.6.3 Radiocarbon techniques

The half life of 5730 years for radiocarbon (^{14}C) renders it highly useful for dating carbon-bearing materials in the range of 100 to 50,000 years, which greatly facilitates studies of archeological materials. Groundwater has also been dated, most commonly by determining the ^{14}C abundance in dissolved inorganic carbon (DIC). Many successful studies have been conducted using DIC ^{14}C measurements, yet debate continues about the extent to which carbonate wall rocks, devoid of ^{14}C, dilute the ^{14}C content of groundwater recharge. As a result, absolute age determinations of groundwater using ^{14}C are limited to special cases where the absence of carbonate can be demonstrated or ^{14}C correction models can be validated. For the most part, absolute ages ≤1000 years old have high uncertainty. Small, spring-fed karst aquifers typically have groundwater residence times much less than 1000 years and have not been the focus of ^{14}C studies. In contrast, groundwater in regional karst systems can be >1000 years old, and the use of ^{14}C to characterize them has led to various correction approaches to estimate the extent of exchange with ancient carbonates. Many regional karst systems occur in water-stressed regions of the world and groundwater residence time plays a central role in water resource sustainability. In the absence of ^{14}C, there is no other comparable scheme suitable for dating groundwater >1000 years.

At least half of the carbon in groundwater originates as soil zone CO_2 produced by plant root respiration or from decomposition of humic material (Pearson & Hanshaw 1970), while the remaining carbon is from dissolution of carbonate minerals in the unsaturated and saturated zones. Some exceptions have been reported (Mook 1980, Andrews et al. 1994,

Rose & Davisson 1996). The distinctive $\delta^{13}C$ values between organic and inorganic carbon (-25 to $-12‰$ for plants, -2 to $+2‰$ for marine carbonates, and -12.0 to $+2.0‰$ for soil carbonates) have been used to characterize sources of DIC and estimate their relative contributions from water-rock reactions within the groundwater environment, particularly during recharge. This isotopic correction method has led to the development of isotopic mass balance approaches (Pearson & Hanshaw 1970, Mook 1980), combined isotopic and aqueous chemical mass balance approaches (e.g. Fontes & Garnier 1979), and isotopic exchange models under open and closed conditions (e.g. Wigley 1975). These approaches are only valid where carbon sources are uniform and the groundwater flow system is simple. A prerequisite for any groundwater age-dating study is to identify recharge zones, flow paths, and mixing zones before applying mass balance approaches for ^{14}C corrections.

Isotopic exchange according to the reaction

$$H^{14}CO_3 + Ca^{12}CO_3 \leftrightarrow H^{12}CO_3 + Ca^{14}CO_3 \tag{12}$$

has concerned groundwater radiocarbon dating studies for over 30 years. Early lab experiments determined the HCO_3-$CaCO_3$ exchangeability in one-dimensional column studies using calcite and ^{13}C or ^{14}C labelled bicarbonate (Munnich et al. 1967), or investigated the isotopic fractionation among CO_2, HCO_3, and calcite (Wendt et al. 1967). Later work quantitatively confirmed that exchange was a recrystallization process with first-order reaction kinetics (Mozeto et al. 1984). Half-lives for exchange were approximately one year, and only about a third of the labelled ^{13}C used in experiments was reversibly desorbed into solution. Other workers characterized this isotopic exchange as a retardation process (e.g. Garnier 1985).

Recently, Gonfiantini & Zuppi (2003) combined the radioactive decay equation with a stable isotope exchange model to create linear relationships of field ^{14}C and $\delta^{13}C$ observations in karst groundwater. They concluded that the radioactive decay rate of ^{14}C is three times faster than the isotopic exchange rate between water and rock in these systems, suggesting that only a small age correction is necessary. For example, they consider the ^{14}C exchange half-life with aquifer rock is on the order of 10,000 years in karst systems hosted in marine carbonates. They argued that the aforementioned laboratory measurements of isotopic exchange rates are not representative of field conditions because of the high exchange capacity of the powdered calcite used in these studies. Many other studies that have used ^{14}C to characterize karst groundwater presuppose similarly slow water-rock exchange rates in their simplified age-correction models.

In contrast, Davisson et al. (1999) concluded that isotopic exchange rates are much higher in the White River flow system, a regional karst groundwater in central Nevada. Groundwater ^{14}C values are <10 percent modern carbon (pmc) throughout this >100 km long groundwater system and are independent of path length. The $\delta^{18}O$ values increase with downgradient distance and indicate mixing of local recharge with regional groundwater flow. Furthermore, hydrological mass balance of recharge and flow was characterized by well established hydrologic boundaries and spring discharge at the terminal end of the flow system, which constrained the groundwater ages to <10 ka on mass balance considerations (Eakin 1966, Davisson et al. 1999). In order to explain the observed low ^{14}C abundances and the increasing $\delta^{18}O$ values downgradient, dissolved carbonate in the groundwater required a half-life of only ~500 years. This fast exchange rate is more consistent with laboratory

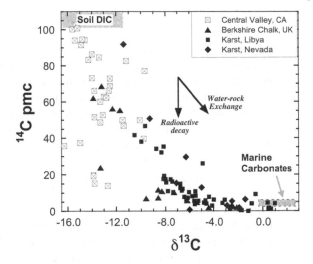

Figure 7.7. Groundwater residing in carbonate aquifers commonly exhibits simultaneous enrichment in $\delta^{13}C$ and depletion in ^{14}C, consistent with isotopic exchange with the host aquifer rock derived from ancient marine carbonates. In contrast, $\delta^{13}C$ and ^{14}C of groundwater residing in porous siliceous alluvium of the California Central Valley exhibit no such trend. Data from Edmunds et al. (1987), Davisson & Criss (1996), Gonfiantini & Zuppi (2003), and Rose & Davisson (2003).

experiments that considered surface exchange sites (Mozeto et al. 1984), implying that many calculated ages in regional karst groundwaters may be erroneous.

In summary, the use of ^{14}C measurements in karst groundwater must be treated with great caution. The propensity of DIC to react with the carbonate matrix is very high and observed ^{14}C contents in karst groundwater are best viewed as a measure of the extent of isotopic exchange with the host rock. When coupled to $\delta^{13}C$ measurements, the extent of that reaction becomes apparent and can be a valuable tool in water-rock reaction studies (Fig. 7.7).

7.6.4   Radon gas techniques

Among all noble gases, radon (^{222}Rn) has the highest water solubility (Clever 1985) and is the only one that can easily be measured continuously. This and its short half-life of 3.82 days make it an interesting natural tracer to study fast transport processes in karst systems.

Radon in soil gas and water has its origin in the radioactive decay of ^{226}Ra present in the soil. Depending on where the radium resides, a variable fraction of that produced can escape to the pore space and thus be available for transport. Only radon produced within a sub-micrometer of the grain surface can diffuse to the pore space in the short time available. Even high radium content inside heavy mineral grains cannot create high radon concentrations in soil gas. On the other hand, even relatively low radium content can produce high radon levels in soil gas and water if the radium is adsorbed on grain surfaces. This is frequently the case in well-developed soils, where dissolution and adsorption processes have led to secondary radium accumulation on the surfaces of soil grains. Iron- and manganese-oxyhydroxides play a major role in these adsorption processes for they are excellent radium scavengers. The fraction of the radon produced available for transport (emanation coefficient) varies between less than 10% for sandy or gravelly soils to over 70% for some soils over karst in the Swiss Jura Mountains. In typical Swiss Plateau fluvio-glacial soils 20% of the radon produced is available for transport.

Soils in the Jura Mountains frequently show extremely high radium concentrations as well as high emanation coefficients. This "anomaly" is probably due to fast, plant assisted limestone dissolution and radium accumulation on small, abundant, porous granules composed of clay minerals, iron- and manganese-oxyhydroxides, and organic matter (Gaiffe 1987, Surbeck 1992, Von Gunten et al. 1996). It would be worth looking for similar natural radium "anomalies" elsewhere. Soils covering karst terrains in moderately humid climates would be good candidates. Soil gas radon concentrations at a depth of >50 cm vary between about 100 Bq/m^3 and several 100 kBq/m^3, with most Swiss soils being on the order of 10 kBq/m^3. Radon in soil gas will dissolve in percolating water and be carried down to the aquifer (Surbeck & Medici 1991).

Little is known about radon production in the epikarst, the interface between soil and karst. Limited data from the Swiss Jura Mountains show a sharp decrease in radon concentration below the soil, where epikarst starts. Cave deposits seem not to be an important radon source, but few data are available (Monnin et al. 1994).

Limestone is a poor radon source. Low radium concentrations, about 10 Bq/kg in Swiss Jura Mountain samples, together with the location of radium within bulk grains rather than on their surfaces, imply both low production rates and low emanation coefficients. Radon from deep sources is similarly unlikely. No candidate for abundant, deep radon production occurs in the Jura Mountains and the short, 3.8 day half-life would greatly limit any such contributions from reaching the surface. In short, the radon concentrations in karst caves and karst waters very likely originate in soils, and transient radon increases in karst waters indicate the arrival of soil water.

Perret (1918) long ago reported that the radon concentrations in springs vary with discharge. A generation later, during the hunt for uranium ores, this knowledge seems to have been lost (Payot 1953). Sampling took place as if spring water radon contents would be constant, depending only on the host rock. Still later, in the wake of alarming news on the adverse health effects of indoor radon, studies of radon in water were renewed but were handicapped by the traditional view that radon originates in crystalline rocks and not in soil (e.g. Buchli 1987). Later, Hoehn & Von Gunten (1989) recognized the potential of radon as a tracer to study river water infiltration into gravel. Surbeck & Medici (1991) emphasized soil radon as an important source for radon in karst water, but Monnin et al. (1994) still considered deposits in karst channels and fractures to be possible radon sources.

Continuous measurements, with a temporal resolution of one hour (Surbeck 1993, Surbeck & Eisenlohr 1993), confirmed that the radon produced in soil overlying a karst system is the most important source for radon peaks observed during storms (Eisenlohr & Surbeck 1995). Increases of radon concentrations clearly indicate the arrival of soil water.

Complementing radon measurements with CO_2 measurements will provide additional information about the origin of karst water (Savoy & Surbeck 2003). Both CO_2 and radon are produced in soils but have different physical and chemical behaviors. An example for an underground karst river in the Jura Mountains is shown in Fig. 7.8. Soil at this site is several meters thick and well developed, with little residual carbonate. There is clear evidence for an important epikarst layer.

Fig. 7.8 shows the response of Rn and CO_2 concentrations to three successive rain events after a long dry period. The first two events put pressure on the system, pushing water stored in the epikarst down to the saturated zone. Water stored in the epikarst from previous storm events has lost its radon by radioactive decay, but CO_2 is still present. Thus,

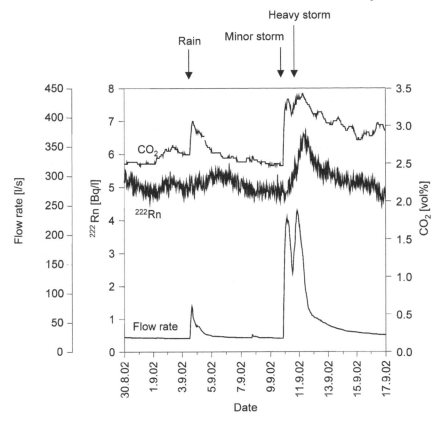

Figure 7.8. Temporal variations of flow, Rn and carbon dioxide in an underground river in the Jura Mountains. The first event mobilized stored epikarst water that had lost radon by decay but still retained a high CO_2 content. Soil water was flushed by the third event, as indicated by a sharp increase in the radon concentration. Radon and CO_2 concentrations both decrease subsequently, due to Rn decays and CO_2 reaction with limestone. Estimated statistical $1 - \sigma$ uncertainties are: flow rate: 2%, Rn: 4%, CO_2: 2%.

CO_2 concentration increases without an increase in the radon concentration, indicating the arrival of water from the epikarst. However, sufficient water had been added by the third event to flush the soil, which lead to an increase in the radon concentration; a clear sign for the arrival of water that had recently been in contact with the soil. Both radon and CO_2 concentrations decreased after the storm, because of radon decay and reaction of the CO_2 with the limestone. The "half-life" of CO_2 in this karst system is about 11 days.

In summary, using electric conductivity as an additional parameter, one can discriminate between different origins of water components in the underground river as follows:

a) Water with low radon, low CO_2, and high conductivity has been in contact with the limestone for more than a month.

b) Water with low radon, high CO_2, and medium to high conductivity is from the epikarst.

c) Water with high radon, high CO_2, and medium to high conductivity is soil water.

d) Water with low radon, low CO_2, and low conductivity represents rainwater that has found a fast transport path to the subsurface.

Table 7.2.   Isotope systems useful for karst hydrology. IRMS: Isotope Ratio Mass Spectrometry, AMS: Accelerator MS, NGMS: Noble Gas MS.

Isotope	Required sample vol. cm^3	Analytical technique	Cost $US	Hydrologic parameters	Strengths	Weakness
D/H	0.1	IRMS	10	Water source(s), mixing	Inexpensive; Conservative tracer	Requires expertise to interpret
T	1000	Scintillation Counting	100	Water Age	Quantitative	Variable source-term
$T/^3He$	1000	IRMS	1000	Water Age	Parent-daughter technique: high precision	Analysis not yet commercial
$^{13}C/^{12}C$	100	IRMS	20	DIC Source(s)	Inexpensive	Water-rock interaction; Requires expertise to interpret
^{14}C	1000	AMS	400	DIC Age	Quantitative	Water-rock interaction
$^{18}O/^{16}O$	1 to 10	IRMS	10	Water source(s), mixing	Inexpensive; Conservative tracer	Requires expertise to interpret
Noble Gas	10	NGMS	500	Recharge Time	Field validated	Analysis not commercial

7.7   SUMMARY

Isotopic data provide a powerful probe for characterizing the origin and age of karst waters (Tab. 7.2), and show that subsurface mixing and carbonate reactivity are major variables. The general statements and specific examples presented in this chapter illustrate the practical value of isotope methods for delineating catchments, determining contaminant sources, and improving groundwater management. Analytical techniques and sample requirements vary among labs and specifics can be found in references cited throughout this text.

Hydrogen and oxygen isotopes provide a dual, conservative tracer system for tracking water, and have a great advantage over other tracers in that the relative volumes of water represented by any component are automatically taken into account. A singular finding of H and O isotope studies is that most karst waters are derived from local meteoric waters, with the spatial and temporal variations in these parent meteoric waters being inherited but imperfectly homogenized in the karst reservoir. The spatial variations can be used for groundwater tracing if they are significant across the basin, as they are in intermontane aquifers. The degree to which temporal variations are damped in groundwater reservoirs provides unique information on groundwater residence times, but accurate models are available only for well-mixed "diffuse" flow systems. Fallout from nuclear testing has compromised the use of simple tritium counting to determine groundwater ages, although elevated tritium in groundwater is a sure sign that a "bomb-pulse" component is present. The T/He method can correct for source variations but is analytically challenging.

Carbon isotopes elucidate the interactions of water with atmospheric carbon dioxide, organic matter, and carbonate bedrock, and prove that dissolved inorganic carbon (DIC) is typically derived in roughly equal proportions from carbonate bedrock and soil carbon dioxide. Radiocarbon provides unique geochronologic opportunities in the crucial age range of about 1000 to 50,000 years, and can be used to determine the average age of DIC through the use of accelerator mass spectrometry. Relating the DIC age to the groundwater age requires hydrologic and geochemical assumptions. Nevertheless, making accurate groundwater age determinations is a worthy goal as they can elucidate the rate of aquifer recharge and providing an independent estimate of safe-yield.

Noble gas isotopes are dissolved in water in regular proportion to their content in ambient soil air, but depending on temperature. Applications include determination of water recharge temperatures, the breakthrough time of surface waters, and the delineation of radioactive contaminant sources. Continuous radon measurements have special potential to quantify fast transport processes.

CHAPTER 8

Tracer techniques

Ralf Benischke, Nico Goldscheider & Christopher Smart

8.1 INTRODUCTION

Investigation of the origin, movement and destination of groundwater is a fundamental goal of hydrogeology where groundwater trajectories and catchments are not immediately apparent. Tracing is a powerful tool for such investigations, particularly in karst areas. In hydrogeology, a tracer is any type of substance in the water or property of the water that can be used to obtain information on the groundwater flow and transport of matter.

Initially, convenient, but inappropriate tracers such as chaff and sawdust were used, until the invention of uranine in 1871 provided the first powerful, safe, water-soluble tracer dye. The first quantitative tracer test was carried out in 1877, when uranine, salt and shale oil were injected into the sink of the Danube River in the karst of the Swabian Alb, Germany, and reappeared two days later at the Aach spring, over 12 km away (Knop 1878). Although most tracer tests involve deliberate addition of tracers, sometimes traces occur accidentally. For example, a fire at a chemical warehouse in Basel, Switzerland, in 1986, released 2.6 tons of rhodamine pigments into the River Rhine. The tracer was detected 14 days later over 700 km downstream (Schmitz 1989, Suijlen et al. 1994). In 1909, an early radioactive trace was made using 38 kg of uranium oxide ore to link the River Reka in the Slovenian karst with the Timavo springs at the Adriatic coast (Vortmann & Timeus 1910). Increasing concern over the environment and health has virtually eliminated radioactive tracing today.

Today, several professional and scientific organisations work on water tracing, such as the Association of Tracer Hydrology (ATH) and the International Committee on Tracers (ICT) of the International Association of Hydrological Sciences (IAHS). Several IAHS publications (Gaspar 1987a,b, Peters et al. 1993, Leibundgut et al. 1995, Dassargues 2000) deal with water tracing. Also the IAH has published several edited volumes on karst including tracing (LaMoreaux et al. 1989, 1993, Paloc & Back 1992). Käss (1998, 2004) provides excellent texts on tracing in hydrogeology, including karst (Hötzl 1998).

Natural or environmental tracers are those that occur without intervention by the investigator, including natural substances (e.g. ionic composition, stable isotopes of water), anthropogenic contaminants (e.g. chlorinated organic solvents, road salt, nuclear bomb tritium) and physical properties (e.g. temperature). Artificial tracers are those injected or induced by the investigator, for example fluorescent dyes, soluble salts or even flood impulses. Natural hydrochemical and isotopic tracers are discussed in chapters 6 and 7 respectively, while this chapter focuses on artificial tracing.

A basic tracer test consists of placing a tracer into a stream and then monitoring for the tracer downstream. In karst, tracers are often injected into sinking streams, and then springs are monitored to identify underground connections, define spring catchments, localise groundwater divides and define the geometry, morphology and hydraulics of the conduit. More sophisticated tracer analysis allows derivation of hydrogeological data for the parameterisation and calibration of groundwater flow and transport models. Tracer tests are also used to simulate the transport, fate and attenuation of different types of contaminants both in the unsaturated and saturated zone.

A multi-tracer test is where several distinctive tracers are injected at different sites during the same experiment, allowing characterisation of several underground connections under particular conditions. In a comparative tracer test, several different substances are simultaneously injected at the same site, allowing comparison of the fate and transport of different tracers in order to gain knowledge either about the tracers or the geochemical system. Serial tracer tests are used to study changes in aquifer links over time. The use of intermediate points in the karst system as sampling or injection sites, such as dripping points in caves, underground streams or observation wells, makes it possible to better characterise water movement and contaminant transport in different parts of the system.

A tracer test requires preliminary investigations to determine possible groundwater linkages, including a search for published and unpublished information and field research. Legal restrictions and requirements and socio-economic aspects need to be considered as well. Ethically, a tracer test should have the lowest environmental and aesthetic impact.

8.2   TYPES OF ARTIFICIAL TRACERS

8.2.1   Overview

There are many types of tracers reviewed by Käss (1998). This chapter focuses on the most important tracers, summarised in Tab. 8.1. Water tracers for the movement of groundwater are ideally conservative in that they should be stable in the environment and not interact with the aquifer. The "ideal conservative tracer" is a substance that is unreactive, is absent from but readily soluble in water, easy to detect quantitatively, non-toxic, invisible, inexpensive, and easy to handle (Käss 1998). Fluorescent dyes are the most practical and widely used tracers because many are reasonably conservative, safe, inexpensive and highly detectable. Oxygen-18 and Deuterium might be considered ideal tracers, as they are part of the water molecule. However, these isotopes are expensive to obtain and analyse and so are rarely used as artificial tracers. Reactive tracers can be used to simulate the transport of contaminants and are acted on by processes such as adsorption, oxidation or filtration.

Two types of artificial water tracers are used: water-soluble substances and particles. Fluorescent dyes and salts are the most widely used water-soluble tracers. Radioactive tracers have extremely low background concentrations and detection limits, but have high analytical cost, demanding safety requirements and stringent legal restrictions (Behrens 1998); they are thus not considered here. Neutron activated stable isotopes are less constrained, but require irradiation for analysis. Particulate tracers simulate the transport of pathogens.

8.2.2   Fluorescent dyes

Many natural and synthetic organic substances that contain aromatic functional groups (carbon ring structures) are fluorescent; they absorb light at certain wavelengths (absorption,

Table 8.1. Properties of the most important groundwater tracers. The detection limits for fluorescent dyes represent an order of magnitude and are valid for clean waters and a modern spectral fluorimeter. The limits for salts strongly depend on the analytical method. The toxicological evaluation is based on Behrens et al. (2001).

	No.	Tracer	Detection limit (µg/L)	Natural background	Toxicology	Analytical interference with	Other specific problems
Fluorescent dyes	1	Uranine	10^{-3}	Absent	Safe	2, 6	Strong sorption at low pH
	2	Eosin	10^{-2}	Absent	Safe	1, 4	Very sensitive to light
	3	Sulforhodamine B	10^{-2}	Absent	Ecotox. unsafe	4, 5	
	4	Amidorhodamine G	10^{-2}	Absent	Safe	2, 3, 5	
	5	Rhodamine WT	10^{-2}	Absent	Genotoxic	3, 4	
	6	Pyranine	10^{-2}	Absent	Safe	1, 2	Not reliable (degradation)
	7	Naphthionate	10^{-1}	Absent	Safe	8, DOC	
	8	Tinopal	10^{-1}	Absent	Safe	7, DOC	Strong sorption
Salts	9	Sodium	*Dependent*	High	Safe	–	
	10	Potassium	*on method:*	Moderate	Safe	–	
	11	Lithium	*0.1 µg/L*	Very low	Safe with restr.	–	
	12	Strontium	*to 1 mg/L*	Moderate	Safe with restr.	–	Strong sorption
	13	Chloride		High	Safe with restr.	–	
	14	Bromide		Low	Safe with restr.	–	
	15	Iodide		Very low	(Not evaluated)	–	Chemically unstable
Particles	16	Dyed spores	*Detection*	Absent	Safe	Natural particles	Not quantitative
	17	Microspheres	*of single*	Absent	Safe	Natural particles	Time-consuming analysis
	18	Specific bacteria	*particles*	Absent	(Not evaluated)	(Other bacteria)	Time-consuming analysis
	19	Bacteriophages		Absent	(Not evaluated)	–	Time-consuming analysis

excitation or extinction) and re-emit light at higher wavelengths (fluorescence or emission). Some fluorescent dyes are excellent groundwater tracers and come close to the definition of an ideal tracer; they are commonly absent in natural waters, have a low detection limit, are highly water-soluble, non-toxic, relatively inexpensive and easy to handle. Most fluorescent dyes can be detected instrumentally at concentrations up to 1000 times less than the threshold for visual detection, allowing quantitative tracing at subvisible levels. Some fluorescent dyes breakdown in sunlight, or through microbial decay. While constraining test and sample duration, this also limits their environmental persistence.

The same dye manufactured by different companies or sold in different countries may have different names. In the following, only the most widely used names and synonyms, and (if available) the Chemical Abstracts Service Registry Numbers (CAS RN) are given. Dyes are also commonly referred to by their D&C and Colour Index numbers. Internet sites such as chemfinder.com may be consulted for further information and material safety data sheets.

Uranine (sodium fluorescein, CAS RN 518-47-8) is an inexpensive, safe dye with exceptionally low detection limits: $\sim$0.005 µg/L or even 0.001 µg/L under ideal conditions. The limit of quantitation is an order of magnitude higher. Uranine is visibly green above $\sim$10 µg/L, and red above $\sim$1 g/L. It is highly soluble (600 g/L at 20°C) and not harmful to humans and the environment. The uranine molecule forms a cation at pH < 2, a neutral molecule at pH < 5, an univalent anion at pH < 7, and a bivalent anion at pH > 7. The latter has the strongest fluorescence. The two anionic forms are conservative, while the neutral molecule and the cation are prone to sorption. Uranine is less suited to acid waters than to more basic karst. Uranine does not adsorb to clay minerals commonly present in karst aquifers.

Sunlight and strong oxidants, like chlorine bleach, destroy uranine, so it is not suitable for daytime use in surface waters or in chlorinated waters. Water samples should be stored cool and dark, and analysed as soon as possible in case of microbial degradation (Sayer 1991). Widespread use of uranine as a water tracer and industrial colorant for over 100 years (e.g. to label vehicle antifreeze) means that it is present in many groundwaters, especially those receiving road runoff or landfill leachate. Before carrying out a tracer test with uranine (or any other tracer), the background concentration should thus be measured.

Eosin (CAS RN 17372-87-1) is an orange coloured tracer dye. Like uranine it is highly soluble (about 300 g/L), not harmful, conservative but very sensitive to light. Its detection limit is significantly higher than for uranine, $\sim$0.05 µg/L; larger quantities are thus required. Overlapping uranine and eosin fluorescent spectra require special analytical attention (see section 8.3.5). Eosin is preferred in acid waters, when uranine cannot be used.

Rhodamines are a large group of chemically similar red fluorescent dyes, some of which are favourable tracers, while others are toxic and should not be used. The nomenclature is confusing, as there are often many synonyms for one dye. For example: Amidorhodamine G, Sulforhodamine G and Amidorhodamine BG are synonyms for the same dye. Rhodamine B (CAS RN 81-88-9), Rhodamine WT (CAS RN 37299-86-8), Rhodamine 6G (CAS RN 989-38-8) and Rhodamine 3G (CAS RN 3262-60-6) are not recommended for water tracing because of their toxicity and/or strong sorption properties (Käss 1998, Behrens et al. 2001).

Sulforhodamine B (CAS RN 3520-42-1) and Amidorhodamine G (CAS RN 5873-16-5) are less prone to adsorption. The former may be harmful to aquatic ecosystems at concentrations over 160 µg/L for an exposure period >48 h (Behrens et al. 2001). Rhodamine dyes absorb and emit light at higher wavelengths so can be readily distinguished from uranine and other tracers during analyses. They are also less sensitive to light and to pH than uranine.

Their detection limits are about 0.03 µg/L, and the limits of visibility are ~30 µg/L. Fluorescence of some rhodamine dyes depends quite strongly on temperature, so it is necessary either to bring samples or correct readings to a standard temperature (e.g. Smart & Laidlaw 1977).

Pyranine (CAS RN 6358-69-6) has similar fluorescence properties to uranine and is non-toxic (Benischke & Schmerlaib 1986, Behrens et al. 2001). Although the electrical charge and fluorescent properties of the pyranine molecule are highly pH-dependent, it never forms a cation and consequently shows low sorption properties. Pyranine can thus be used in acid waters, although care is necessary to ensure analysis at a standard pH. Although several successful tracer tests with pyranine are reported in the literature (Käss 1998, Reichert 1991), it cannot be considered a reliable tracer, as it often shows inexplicably low recovery (Bäumle et al. 2001, Goldscheider et al. 2001a, 2003).

Naphthionate (sodium-naphthionate, CAS RN 130-13-2) is a safe, non-adsorbent, ultraviolet-blue fluorescent dye that is invisible at concentrations <1 g/L. Naphthionate has a detection limit of ~0.1 µg/L, so that large injection masses are required. Although it has no interference with red and green dyes during analysis (Wernli 1986), it fluoresces at the same wavelengths as dissolved organic carbon (DOC), which limits the use of this tracer. Microbial decay of naphthionate has been observed (Goldscheider et al. 2001b), so samples should be stored cool and analysed shortly after sampling.

Tinopal CBS-X (CAS RN 27344-41-8) is another ultraviolet-blue dye frequently used as a tracer, but it is strongly adsorbed on clay so that recovery is often low with significant retardation. Tinopal is not recommended for acid groundwaters although in well-developed conduit systems, it can be used over long distances (Käss 1998).

8.2.3   Salts

A number of salts have also been used as tracers (Käss 1998). Salts dissolve in water into anions and cations, thereby increasing the specific electrical conductivity (EC). Most anions are conservative tracers because they show low sorption properties. In contrast, cations are prone to ion exchange (reactive tracers). The sorption strength of cations increases in the order of: $Li^+ < Na^+ < K^+ < Mg^{2+} < Ca^{2+} < Sr^{2+}$. Lithium salts are thus preferred as tracers.

Most salts have higher detection limit, background concentrations and variability in natural waters than fluorescent dyes. Higher injection masses are thus required, possibly resulting in concentrations that may be harmful near the injection point. Most salts do not cause visible colouring, are stable in light and resist microbial degradation. Salts show little analytical interference with fluorescent dyes and can be used in combined tracer tests.

Sodium chloride (NaCl) or common salt is very inexpensive, widely available, highly soluble and relatively safe. It dissolves into the sodium cation (Na^+) and the chloride anion (Cl^-), both of which can be analysed for as tracers. However, salt traces on relatively short distances are most conveniently monitored using electrical conductivity. As the analytical detection limits and natural background concentrations are relatively high, large, possibly harmful injection quantities are required.

Lithium is generally used as lithium chloride. It has low sorption properties and is the most mobile cation. The analytical detection limit is low (0.1 µg/L) and the natural background concentrations in karst groundwater are also low, often <1 µg/L (Käss 1998). Lithium salts are toxicologically "safe with restriction" (Behrens et al. 2001). Lithium has been successfully used for tracer tests in karst and chalk aquifers (Behrens et al. 1992, Witthüser et al. 2003).

Potassium shows a slightly higher tendency to adsorption than sodium but occurs at lower background levels in natural groundwater. Lower injection amounts are thus required. The most commonly used potassium salt is KCl, which is highly soluble and toxicologically safe.

Strontium is safe, although a limit of 15 mg/L is recommended for drinking water (Behrens et al. 2001). Strontium chloride hexahydrate ($SrCl_2 \cdot 6H_2O$) is most often used for tracer tests. Traces of strontium are present in all type of groundwater, $\sim 40\,\mu g/L$ in carbonates and >15 mg/L in gypsum (Käss 1998). Strontium adsorbs strongly and is most suitable for use in well-developed karst conduit systems, although it can also be used as an analogue for the transport of reactive heavy metals.

Bromide has little tendency to sorption and precipitation, and is chemically and microbiologically stable. The natural background concentrations in groundwater are 100 to 1000 times lower than Chloride, typically <0.1 mg/L (Matthess 1994, Käss 1998). On the other hand, bromide analysis is relatively time-consuming and the detection limits are relatively high using ion sensitive electrodes ($\sim 50\,\mu g/L$) and ion chromatography ($\sim 15\,\mu g/L$). Bromide is toxicologically "safe with restrictions". In organic-rich waters or oxidising water purification systems it can be transformed into cancerous bromates and organo-bromides (Behrens et al. 2001).

Iodide occurs at even lower background concentrations in natural groundwater than bromide, often $<10\,\mu g/L$ (Matthess 1994). In contrast to chloride and bromide, it is chemically and microbiologically reactive. Iodide is thus not recommended for long transit times, and for tracer tests in groundwater with high dissolved organic carbon concentrations.

8.2.4   Particulate tracers

Natural particles in water comprise mineral and organic matter, and microorganisms (Atteia & Kozel 1997). Particles $<1\,\mu m$ are classified as colloids. Contaminants, such as heavy metals and radioisotopes are often transported adsorbed to particles and colloids. Viruses and small bacteria are $<1\,\mu m$, large bacteria and most protozoans are $>1\,\mu m$ and are also commonly clustered in particle aggregations known as flocs. Particle tracers provide analogues for microbial pathogens, common contaminants of karst groundwater (Auckenthaler et al. 2002, Auckenthaler & Huggenberger 2003). Their limited analytical interference with other tracers makes them useful in multi-tracer tests.

Clubmoss spores of *Lycopodium clavatum* have an average diameter of 33 μm. They can be relatively easily identified under the microscope, particularly if they are marked using fluorescent staining. Spores are usually collected in plankton nets and then counted under a fluorescence microscope (Käss 1998). This sampling and analysis are not well controlled, precluding development of a quantitative breakthrough curve. Early spore tracing in the Austrian Alps suggested a radial-divergent drainage pattern (Zötl 1961) since shown to be incorrect, probably because of contamination arising from handling the spores (Bauer 1989, Herlicska et al. 1995).

Fluorescent microspheres are available with different diameters (0.05–90 μm), physical-chemical characteristics (e.g. density, electrical surface charge), and optical properties (excitation/emission wavelengths), allowing various multi-tracing options. Uncharged yellow-green (458/540 nm) polystyrene spheres with a diameter of 1 μm have replaced clubmoss spores for tracing. Microspheres $<1\,\mu m$ can be used to simulate colloid transport.

Microspheres can be quantitatively analysed from water samples, allowing definition of a breakthrough curve. Comparative tracer tests with microspheres and water-soluble tracers in karst, fissured and porous aquifers often show that uncharged and negatively charged microspheres travel faster, possibly as particles are less likely to diffuse into the aquifer matrix (Göppert et al. 2005). Neutral and negatively charged particles have lower recovery rates than conservative water-soluble tracers, while positively charged microspheres often disappear completely, presumably due to adsorption (Behrens et al. 1992, Bäumle et al. 2001, Kennedy et al. 2001).

Bacteriophages (phages) are viruses that attack a particular host bacterium. Phages range in size from 0.02 to 0.35 μm, making them ideal for simulation of the transport of viruses. Specific phage types that naturally occur in the marine environment can be used as artificial tracers in ground- and freshwater (Rossi et al. 1998, Rossi & Käss 1998). These phages are readily produced, do not interact with other microorganisms, are safe and stable and show favourable transport properties. Phages are analysed in water samples. It is possible to detect one single phage and to distinguish phage types. They are suitable for tracer tests in large karst aquifer systems, where very high dilution is anticipated (Harvey 1997), although the analytical sensitivity also demands careful handling.

Bacteria may be suitable particulate tracers, though analysis may be difficult. The bacteria type most often used in water tracing is *Serratia marcescens*, which can be used as a model substance to study the transport of pathogenic bacteria. However, *S. marcescens* is an opportunistic human pathogen (Kurz et al. 2003) and requires care. Bacteria can also be used with other non-toxic tracers (Harvey 1997, Hötzl et al. 1991).

8.3 PREPARATION AND OPERATION OF TRACER TESTS

8.3.1 Preliminary investigations and legal aspects

Tracer tests can provide excellent information on groundwater movement and contaminant transport, but they may take many months to execute, and can fail if poorly executed. Using too little tracer, sampling the wrong springs or mistiming sampling can lead to non-detection, an undesirable, ambiguous result. Likely groundwater pathway and flow velocity can often be derived from geological, hydrological, chemical and speleological information. An inventory of sinking streams, caves and springs, along with possible flow routes indicated by geology, geomorphology and hydrology is essential in selecting appropriate injection and sampling points, design of a sampling strategy and interpretation of results.

Legal regulations for groundwater tracing tests vary with country and jurisdiction. In some cases, official permission must be requested in advance from the water authorities. In other countries (e.g. Switzerland), it is sufficient to inform the authorities and provide the outcome for the national database (Schudel et al. 2003). Proof of environmental and human safety is often of paramount importance in obtaining permission for a tracer test. Behrens et al. (2001) provide a valuable resource base on tracer safety. Regardless of legal stipulations, concerned communities, individuals and authorities are best informed in advance of a tracer test.

8.3.2 Selection of the tracer type and injection quantity

In general, tracers should be "toxicologically safe" or "safe with restrictions" (Behrens et al. 2001), have low detection limits, and show little analytical interference. The tracers should

Table 8.2.   Factors k and B for the tracer mass equation (Käss 1998).

Tracer	k	Framework conditions	B
Water soluble tracers: mass [kg]		Surface streams	0.1–0.9
Uranine	1	Karst aquifers (conduits)	
Eosin	5.5	Pure sand/gravel aquifers	
Sulforhodamine B	4	Highly fissured aquifers	
Amidorhodamine G	2	Impure sand/gravel aquifers	2–4
Pyranine	5.5	Injection into groundwater	
Naphthionate	15	through unsaturated zone	
Tinopal	3	Karst aquifers (matrix),	
NaCl	20000	Poorly fissured aquifers	
LiCl	1000	River bank filtration	
KCl	10000	Turbid sampling water or	
Particle tracers: number of particles		tracer background level > 0	
Microspheres	1E+12	Injection through thick or	5–10
Bacteriophages	1E+13	loamy unsaturated zone	
Bacteria (*Serratia marcescens*)	1E+13	High clay/silt contents	

ideally be absent from the groundwater or show low background concentrations. It may be possible to use tracers with relatively high but stable background, e.g. chloride during steady flow conditions. Photosensitive fluorescent dyes (e.g. eosin) are not recommended in surface waters. When visible colouration must be avoided, naphthionate, tinopal, salts, and particle tracers can be used.

The tracer used also depends on the objectives of the experiment. For tracer tests intended to characterise the movement of groundwater or the transport of stable and mobile contaminants, conservative tracers such as uranine or eosin should be selected. In karst conduits, several additional fluorescent dyes (e.g. naphthionate, eosin), some cations (e.g. lithium, sodium) and most anions (e.g. chloride, bromide) are reasonably conservative. Tracers are less likely to behave conservatively when used in porous media, organic-rich environments like soils, or in long duration tests. Non-conservative tracers may be used to establish underground linkages, or to simulate the transport of specific contaminants. For multitracer tests, the more detectable and conservative tracers (e.g. uranine) should be assigned to the longer expected trajectory, the highest dilution or the most important injection point.

The optimal tracer quantity will permit clear detection at the sampling point without excess expense, unacceptable colouration and environmental loading and additional work in the laboratory. If the hydraulic and transport parameters are known, it is possible to calculate injection mass based on the advection-dispersion equation resulting in a target concentration at the sampling point (e.g. Field 2003). More commonly, the required parameters are not known, and the tracer quantity must be estimated from an empirical relationship. For example, Käss (1998) suggests:

$$M = L \cdot k \cdot B$$

where M = required tracer quantity i.e. mass [kg] for soluble tracers and number of particles for particulate tracers; L = distance to the most important sampling site [km]; k = coefficient for the tracers; and B = factor for the hydrogeological conditions (Tab. 8.2).

The target tracer concentration is not a variable in this relatively liberal approximation, so that there is some risk of visible colouration. Worthington & Smart (2003) suggest a more

Figure 8.1. Left: Injection of 1 kg of uranine from a plastic container into a karst shaft in the Hochifen-Gottesacker area, Austro-German Alps. A helicopter delivered a tank with 800 L of flushing water (Goldscheider 2005). Right: Injection of 155 kg of naphthionate, dissolved in a tank of 2000 L, into an observation well in the confined karst aquifer of the mineral springs of Stuttgart, Germany (Goldscheider et al. 2003).

adaptable formula for karst tracing including system discharge and target concentration:

$$M = 1.9 \cdot 10^{-5}(LQC)^{0.95}$$

where L is distance [km], Q is discharge [L/s], and C is target peak concentration [μg/L].

8.3.3 Selection of the injection points and injection techniques

The tracer injection point depends on the objective of the tracer test. Clarification of the conduit network in karst requires injection directly into sinking or cave streams. Hydraulic characterisation demands a discrete spike injection. Large eddies or sediments should be avoided as they can modify the input signal by retaining tracer, and influence the breakthrough. If stream injection points do not exist, tracer can be flushed into solutionally enlarged fissures, dolines and karst shafts (Fig. 8.1). Flushing should start some time before the tracer injection to establish steady infiltration, and continued for some time in order to drive the tracer to the stream. Where the karst is mantled with sediment, there may be no adequate injection point. However, the cover can be excavated to bedrock and tracer flushed into an opening proven to accept injected water. The tracer can also be injected at the land surface or into a ditch and flushed (by a stream, or rainfall, e.g. Flury & Wai 2003) through the soil and unsaturated zone towards the groundwater. The low recovery rate of such tracer tests may require greater tracer injections (see B in Tab. 8.2). The diverse flow route, tracer retention in sediment and prolonged residence time means that such experiments do not allow evaluation of hydraulic characteristics.

Wells may also serve as injection points with tracer injected through a hose (Fig. 8.1). Karst wells may sometimes provide direct access to the conduit network. More commonly the tracer may take some time to travel through the rock matrix to reach a conduit. Successful injection and interpretation of such tests requires knowledge of the hydrology of the well, e.g. depth to water table, position of the well screen(s), hydrostratigraphy, and hydraulic response of the well (Flynn et al. 2005). Natural gradient tests require injection with minimal change in well volume, but induced gradients may be required to drive tracer into adjacent conduits.

Tracing of a contaminant source requires the tracer to be introduced with the contaminant. For example, tracer can be introduced into septic tanks, sewage shafts, leaking wastewater channels or dolines and caves filled with waste.

The greater the sensitivity of tracer analysis, the greater the efforts that are required to prevent autocontamination (Smart & Karunaratne 2001) in which the water samples become accidentally contaminated with the injected tracer. The preparation and injection of tracers is best undertaken in different places, by different staff and with different materials than those used for sampling and analyses. Handling is safest when tracers are dissolved or dispersed in water well beforehand, although the resulting mass may prove difficult to handle. Powdered dyes that do not dissolve readily in cold water may be progressively moistened and worked up as a paste to prevent insoluble lumps developing. Some tracers dissolve more readily in alcohol, warm water or alkaline media, or a small quantity of detergent may aid dispersion providing that this does not compromise their performance. Injection may require protective clothing and careful clean-up and disposal.

Tracer injections can either be done instantaneously or continuously. An instantaneous injection means introducing a known quantity of tracer into the water during a minimal interval. Typically the tracer is poured from a wide-mouthed container into the stream. Such injections should result in a tracer breakthrough curve with a clear maximum. Instantaneous injections are suitable to simulate the impact of accidental contaminations on the groundwater, and allow the breakthrough curve to be analysed for hydraulic properties of the flow route. Additional equipment and tracer are required for a continuous injection where a known tracer concentration is released at a known rate into the water for a sustained period. At the sampling sites, this will result in an increase of the tracer concentration to a plateau. Continuous injections can be used to simulate chronic contamination.

8.3.4   Selection of the sampling sites and sampling techniques

Groundwater sampling sites for tracer tests in karst aquifer systems comprise springs, cave streams and pumping or observation wells. Not all springs may be accessible or identified, so stream networks can be used to integrate outflows from many springs or selected reaches of stream channel, although samples may be poorly homogenised, and tracer concentrations reduced, or even undetectable. Known springs in the bed of rivers or lakes can be sampled by drawing water through a hose anchored at the opening and extending to dry land. Drips and seeps in caves or tunnels below an injection area can be used as unsaturated zone sampling sites (e.g. Veselic et al. 2001). However, samples from wells or cave seeps are unlikely to reflect the predominant flow route, thus limiting interpretation.

The sampling strategy depends on the objectives of the tracer test, the hydrogeological environment and the resources available (Fig. 8.2). Sampling sites (where tracer is anticipated) and control sites (where it is not, e.g. upstream of tracer injection) need to be

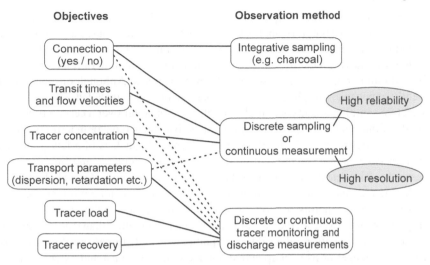

Figure 8.2. The selection of appropriate sampling methods depends on the objectives of the tracer test. Integrative sampling is sufficient to prove underground connections while the calculation of the recovery rate requires detailed discrete sampling or continuous measurement, and concurrent discharge measurements.

identified and their background fluorescence pattern established. Sampling sites are used prior to injection to established spatial background. Continued sampling at the control sites provides an indication of probable temporal variation in background during the tracer test. Simple underground pathways require relatively few sites to be sampled. More extensive multi-tracer tests require that all springs or spring groups in the region should be sampled. Proximity of springs does not necessarily indicate similarity of connection, so information from group sampling may be ambiguous. Flow paths in karst systems may be unexpectedly long, may go in unexpected directions, and may cross under topographic boundaries. Investigation of the hydraulics of a known link (e.g. travel time variations for different tracers or with discharge), requires less extensive sampling, although control is still desirable.

There are three fundamentally different sampling methods: discrete sampling, integrative sampling and continuous measurement (Fig. 8.2). Discrete (punctual or grab) sampling means collecting water samples at distinct times either manually, or using an automatic water sampler. The result allows construction of a time-concentration breakthrough curve with resolution depending on the sampling interval. In general, the shorter the travel time, the greater will be the rate of change in concentration and the necessary sampling frequency. Sampling intervals should therefore be short at the beginning, and then become progressively longer. Discrete samples can be stored in case of retrospective doubts or problems. Documented sample custody and handling history may be required for litigation. Automatic water samplers typically have 24 bottles set to be filled at preset intervals; for example, allowing daily servicing for hourly samples.

Online measurements are obtained from field fluorimeters, ion selective electrodes or conductivity cells inserted into streams and linked to data loggers. Typically, high frequency measurements (seconds) are averaged over longer time intervals (minutes), providing a real time record of tracer concentration, allowing adaptive sampling, better noise reduction and higher temporal resolution than is practical with discrete sampling. Such systems are less prone to contamination, but samples are not retained. The data are seldom as accurate as can be obtained from laboratory analysis under controlled conditions, especially for tracers influenced by ambient conditions such as pH, temperature and background fluorescence. Online measurement is most useful for precise temporal resolution of tracer breakthrough at a limited number of sites where tracer is known to appear. Although on-line equipment is currently expensive, sampling logistics are simple once it is installed.

Integrative sampling means accumulation of tracer or water over a certain time. A common mode of integrative sampling uses granular activated charcoal or macroreticular resins to adsorb dyes (Close et al. 2002, Käss 1998). Permeable packets of a few grams can be deployed in streams over intervals of hours to days, to accumulate any fluorescent tracer that passes by. The detectors are then collected, rinsed in stream water and eluted (usually in an alkaline alcohol solution) to allow analysis. Adsorption and elution are sensitive to factors such as water chemistry, flow-through rate and timing, so that conditions must be carefully controlled. Even so, the data are generally ordinal rather than true concentrations, so detector results are often referred to as "qualitative" and are best used for establishing underground linkages or as backup and supplement to less robust and extensive quantitative sampling methods. Detectors are extremely cost-effective in providing time-integrated monitoring at a large number of sites, or where vandalism may be a problem. In clean waters or short exposure intervals, they may have a better detection threshold than water samples. Multiplexing where a number of sequential water samples are automatically pumped into single sample bottles is a discrete form of cumulative sampling that may be necessary where flows are small, e.g. cave seepages.

In many cases, the three sampling techniques can be combined (Fig. 8.3). Continuous measurement occurs at a few important sampling sites, where the instruments can be protected against vandalism, e.g. at water supply springs. Discrete sampling is necessary for tracers that cannot be measured in the field, e.g. particulate tracers. Integrative detectors are typically used to determine underground connections, as deployed as insurance in case of sampler failure, or for detailed resolution of a number of springs being monitored collectively, or at sites that are difficult to reach. They may not always require elution and analysis in such roles.

To avoid photodegradation of dyes (Käss 1998) sample locations should be close to springs. Brown-tinted glass bottles with Teflon cap liners are preferable, with cool and dark sample transport and storage. When this is done, fluorescent dye concentrations often remain stable for weeks, while rapid microbial degradation of naphthionate has been observed in uncooled samples from specific water types (Goldscheider et al. 2001b). Nguyet & Goldscheider (2006) used 13 mL plastic test tubes for sampling during a tracer test in a remote region in order to minimise the weight. Charcoal bag samples are stable when they are dried and stored in the dark. Water samples that are to be analysed for salts are less sensitive than dyes. Sampling for microbial tracers usually requires sterile sampling, cool and dark storage, and rapid analysis (e.g. Harvey 1997, Hötzl et al. 1991, Rossi & Käss 1998).

8.3.5   Laboratory analyses

Both grab sampling (water samples) and integrative samples (charcoal bags, resins) require laboratory analysis. Salts can be analysed using ion chromatography (IC), ion-specific electrodes, spectrophotometry, atomic absorption spectroscopy (AAS), atomic emission

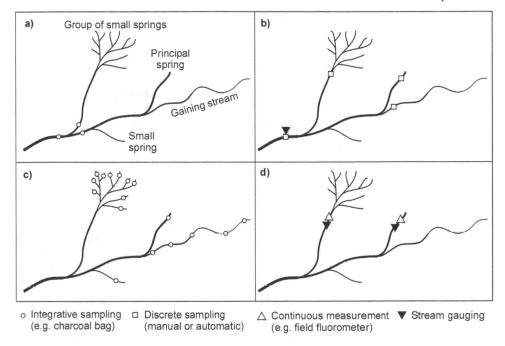

o Integrative sampling   □ Discrete sampling   △ Continuous measurement   ▼ Stream gauging
(e.g. charcoal bag)   (manual or automatic)   (e.g. field fluorometer)

Figure 8.3.   Schematic illustration of a set of karst springs with tracer sample schemes devised for different research questions. a) Reconnaissance investigations are used to demonstrate underground linkage with minimal cost. b) A precise travel time investigation requires breakthrough curves defined from discrete or continuous sampling. Overall tracer recovery requires simultaneous determination of discharge and breakthrough at a downstream location. c) Spring differentiation requires detailed spatial sampling most efficiently done using charcoal detectors. d) Conduit characterisation requires tracer breakthrough and discharge for each spring or spring group, and also allows local tracer recovery to be determined.

spectroscopy (AES) or inductively coupled plasma mass spectrometry (ICP-MS) (Dean 1995). Bacteria are usually analysed using cultivation techniques, direct count or other methods (Hurst et al. 2002, Madigan et al. 2000). The simplest analysis of fluorescent dyes is through filter fluorimetry where dye-specific colour filters control excitation and emission wavelengths. The fluorescence intensity includes background and contributions from similar dyes, so filter fluorimeters data require careful evaluation and correction.

Analysis for multiple fluorescent dyes or in the presence of significant natural organic background is best done using a scanning fluorescence spectrometer (Fig. 8.4). The excitation monochromator focuses a ray of monochromatic light onto a sample cuvette. When the water sample contains fluorescent dye, it will partly absorb the light and generate fluorescence light at longer wavelengths. The fluorescence light is detected perpendicular to the ray of excitation light with the wavelength determined by the emission monochromator. Synchronous scanning fluorimeters vary the excitation and emission wavelengths simultaneously with a constant wavelength difference to generate a fluorescence spectrum indicating fluorescence intensity with respect to excitation or emission wavelength (Behrens 1970).

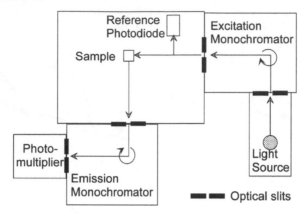

Figure 8.4.   Schematic illustration of a spectral fluorimeter. The fluorescence (emission) light is adjusted to have higher wavelengths than the excitation (extinction, absorption) light. Slit settings are widened to adjust gain (sensitivity) and noise suppression, and narrowed to increase spectral resolution. The reference photodiode allows compensation for changes in intensity of the light source.

Table 8.3.   Fluorescence characteristics of some tracer dyes (modified after Käss 1998).

Tracer	pH value	Absorption [nm]	Fluorescence [nm]	$\Delta\lambda$ [nm]	Fl. intenisty (uranine = 100)
Naphthionate	Neutral	320	430	110	8
Tinopal	Neutral	346	435	89	60
Pyranine	Acid	405	445	40	6
Pyranine	Basic	455	512	57	18
Uranine	Basic	491	512	21	100
Eosin	Acid	516	538	22	18
Amidorhodamine G	Neutral	530	551	21	14
Sulforhodamine B	Neutral	564	583	19	30

Different fluorescent dyes show distinctive spectra, i.e. absorption and fluorescence maxima, the wavelength difference between them ($\Delta\lambda$, Stokes' shift), and relative fluorescence efficiency (emission/absorption intensity) (Tab. 8.3). The presence of a dye in a sample is indicated by the appearance of the respective peak in the sample spectrum. The measured fluorescence intensity aggregates the contribution of all fluorescent compounds in a sample at the particular monochromator settings. Dyes can be distinguished in a water sample if their fluorescence spectra are clearly separated and known, e.g. naphthionate, uranine and Sulforhodamine B. However, overlapping spectra mean that peak height is not a direct indication of concentration, and very high concentrations of one tracer may mask low concentrations of another. In addition, background fluorescence (typically from natural organic matter or contaminants) is included in the spectrum. In these cases, the spectrum needs to be deconvolved, manually on a graph, or using specialised spectroscopy programs (Fig. 8.5). Such analyses require care and experience to avoid artefacts (Alexander 2005).

Fluorescence intensity is proportional to dye concentration up to tens of ppm allowing simple linear calibration of fluorimeters. More concentrated samples are less fluorescent

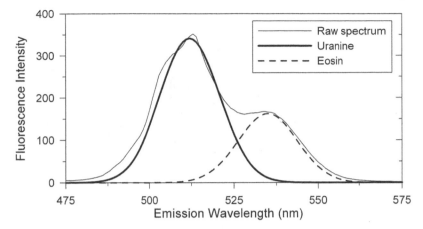

Figure 8.5.   Synchronous scan spectrum ($\Delta\lambda = 20$ nm) for a visibly green mixture of uranine and eosin with negligible background fluorescence. The scan has been decomposed into peaks attributed to the two dyes by fitting normal curves centred at 512 and 535 nm using the program PeakFit (SSI). The height of the peaks indicates the concentrations of the two component dyes.

and require quantitative dilution. Samples containing dyes influenced by pH such as uranine and pyranine may require buffers.

Fluorescent dyes adsorbed on charcoal or resin have to be desorbed. For charcoal this can be done with alkaline solvent solution (elutant), e.g. a 1:1 mixture of 2-propanol and 40% NaOH solution for uranine (Käss 1998). A small quantity of the activated carbon (e.g. 5 g) and a small volume of the elutant (e.g. 20 mL) are agitated for several hours (e.g. 4 h), filtered and the tracer concentration is then measured in the strongly alkaline eluent. The fluorescent spectrum of the eluent depends on the composition of the stream water, the duration of exposure and the duration of elution (Smart & Simpson 2002). Unexposed activated charcoal may also exhibit slight fluorescence.

Water samples are analysed for bacteriophages using a small volume added to a culture of the corresponding host bacterium. Any phage present will kill the host bacterium creating a hole in the bacterial culture, thus giving an indication of concentration of phages in the water sample. Even a single phage can be detected with this method, but large concentrations need dilution to allow accurate counting (Rossi & Käss 1998).

Fluorescent particles are most conveniently analysed and differentiated using a flow cytometer (laser-based fluorescent particle detector, Niehren & Kinzelbach 1998, Kennedy et al. 2001). This expensive instrument has difficulty distinguishing microspheres from natural fluorescent particles. This may be overcome using a robust particle counter coupled with a fluorescent detector (Goldscheider et al. 2006). Counting of the fluorescent particles on filter paper under the fluorescence microscope is more time-consuming, but allows better sensitivity and differentiation (Göppert et al. 2005, Käss 1998).

8.3.6   Instruments for field measurement

Small portable filter fluorimeters and emission spectral fluorimeters are now available for grab sample analysis (e.g. Turner Designs Ocean Optics 2006). Fluorescent dyes can be measured continuously in the field with submersible (Fig. 8.6) or fibre-optic fluorimeters where the excitation and emission wavelengths are fixed using light-emitting diodes and

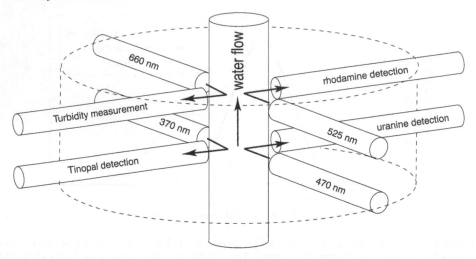

Figure 8.6.   Through-flow fluorimeter with 4 channels for the measurement of 3 fluorescent dyes and turbidity. The 370 nm channel can also be used for the monitoring of organic carbon. The instrument also measures temperature; a conductivity probe can also be included (Schnegg 2002).

optical filters (Benischke & Leitner 1992, Smart et al. 1998, Schnegg 2002). The performance of field fluorimeters is comparable to laboratory instruments, although the natural variations of turbidity, organic carbon and pH in the water may require monitoring to allow correction of the apparent fluorescence to standard conditions. The optical separation of background and different fluorescent dyes with similar optical properties (e.g. pyranine, uranine and eosin) requires careful cross calibration and appropriate correction (Behrens 1988, Smart 2005).

Conductivity meters can be used to detect salt tracers if the injected quantity is sufficiently high and if there is little background variation during the sampling period. This is most likely for relatively rapid and short tracer tests, e.g. in dilution gauging of streams (Leibundgut 1998). Ion-specific electrodes attached to a pH meter allow monitoring of specific ions, e.g. for sodium, potassium, chloride, and bromide. However, all show limited sensitivity and are liable to interference from other compounds and may be too fragile for field deployment.

8.4   EVALUATION AND INTERPRETATION

8.4.1   Data requirements, data quality and error analysis

The applicability of tracer test data depends on sampling and analytical procedures. Integrative sampling (e.g. charcoal bags) will give ordinal data resolved to the exposure period. Laboratory analysis of water samples provides the best concentration data, but with time resolution determined by the sampling interval. Online measurement provides the best time resolution, although it may give less accurate analysis. Concurrent discharge data are required when the tracer recovery is to be calculated or flow dependence assessed.

Errors in water tracing can be classified as major, systematic or random, and all three can arise from injection, sampling, handling, analysis and data processing (Smart 2005). Major mistakes include sampling at the wrong site or after the tracer has passed, or direct contamination of the samples with tracer. Systematic errors result in coherent deviation in the data from a true value that can significantly influence quantitative interpretation. For example, uncorrected high background levels can lead to overestimation of tracer concentration. Systematic error is generally tackled through a scheme of collection, processing and analysis of blanks and standards. Random error is the uncertainty remaining when major and systematic error has been removed. It can be reduced by replicating samples and analyses, and working with averages of multiple samples and measurements. Charcoal detector tracing is most vulnerable to major error, whereas more quantitative tracing is sensitive to all three types of error.

In practice, error in dye tracing is dominated by issues of background and contamination, loss of tracer (e.g. sorption) and aliasing (insufficiently frequent sampling). Rather than any formal analysis of error accepted practices currently address these issues. Background is dealt with below. Contamination can be considerably reduced using good handling protocol, proven through analysis of blanks. Tracer loss can be computed through analysis of recovery, if concurrent discharge data are available. However, batch tests may be needed to determine whether loss is from sorption, degradation or diversion of the tracer. Aliasing can be avoided by continuous analysis or by sampling frequently enough to define a serially coherent breakthrough curve. Knowledge about the uncertainty of measurements is particularly important if data are near the detection limit, particularly in the tail of a breakthrough curve where errors may significantly influence trace parameters.

8.4.2   Tracer background

Background can be defined as that part of measured tracer concentration not arising from the tracer injected. Background arises from extraneous tracer (contamination), other substances measured as tracer, or instrumental errors.

The primary source of contamination with the tracer is through mishandling. Fluorescent dyes like uranine and rhodamines are used as industrial colorants of automobile antifreeze, while blue fluorescent whitening agents are common in household detergents and ubiquitous in raw sewage (e.g. Boving et al. 2004). As a result they may be found in the environment. Fluorescence arising from organic contaminants or natural dissolved organic carbon (DOC) overlaps tracer spectra, especially in the blue end of the spectrum. This is a problem with filter fluorimeters that are unable to distinguish tracer from other material fluorescing in the same wave band. Tracer ions such as chloride, bromide, iodide, sodium, potassium and lithium, are present in almost all natural ground waters and can be substantially elevated in wastewater discharge, landfill leachate and winter road runoff. Environmental sources of background can act randomly or be influenced by runoff conditions, and so imitate a tracer breakthrough curve.

Instrument characteristics lead to analytical background. For example, fluorescence analyses generally have a background arising from stray light and thermal noise in the photodetector. Most instruments incorporate technical solutions to the problem of background, or offer the possibility of blanking out what the user considers to be meaningless background.

Environmental background is determined through pre-monitoring, i.e. collection of a series of samples prior to injection. It may be difficult to correct for background when the tracer is present, unless there is some means of separation of the two as is possible in spectrofluorimetric correction of organic background using curve fitting. The average of pre-monitoring concentrations is typically assumed to be sustained during tracer breakthrough, and is subtracted from the total observed concentration. If a suitable untraced site analogous to the traced site is available, then co-monitoring (sampling during the tracer breakthrough) can indicate if background is likely to have changed. Analogue monitoring (observation of variables known to be correlated to background) may also allow approximate correction for background fluctuations, or at least demonstrate that changes are unlikely to have occurred. Correction of dynamic background in filter fluorimetry is possible by monitoring natural fluorescence in several channels and using the relationship between these signals and background to estimate background during tracer breakthrough (Smart 2005).

8.4.3   Tracer transport in groundwater

The quantitative interpretation of tracer tests is based on the principles of solute transport (e.g. Schulz 1998, Fetter 1999). The transport of conservative tracers is dominated by advection and dispersion. Additional geochemical processes govern reactive tracers, including reversible and irreversible adsorption, precipitation, oxidation, reduction, volatilisation, degradation and transformation. Non-aqueous phase and particulate tracers are affected by sedimentation, flotation and different types of filtration. Biological processes influence microbiological tracers, including reproduction, inactivation and die-off.

Advection describes the displacement of substances with the fluid (average linear velocity). In any aquifer variability of flow paths and flow velocities results in spatial and temporal spreading of the tracer cloud, referred to as mechanical dispersion. Tracer is also spread by molecular diffusion, which occurs in response to concentration gradients in the fluid, but is generally negligible compared to dispersion in discrete conduits. During turbulent flow in karst conduits mechanical dispersion is most important in the flow direction, giving the longitudinal advection-dispersion equation (Bear 1979):

$$\frac{\partial C}{\partial t} = D_L \frac{\partial^2 C}{\partial x^2} - v \frac{\partial C}{\partial x}$$

where C = concentration, t = time, D_L = longitudinal dispersion coefficient, v = effective flow velocity, and x = longitudinal distance. This equation can be solved for a range of initial and boundary conditions. Kreft & Zuber (1978) proposed an analytical solution for porous media:

$$C(x,t) = \frac{M}{Qt_0 \sqrt{4\pi P_D \left(\frac{t}{t_0}\right)^3}} \exp\left(-\frac{\left(1-\frac{t}{t_0}\right)^2}{4P_D \frac{t}{t_0}}\right)$$

where M = tracer mass, Q = discharge, t_0 = mean transit time P_D = dispersion parameter, a normalised form of the dispersion coefficient $(D_L/(v \cdot L))$ allowing for increasing dispersion with flow paths length (L) and flow velocity (v).

Adsorption results in a retardation of the tracer cloud, i.e. transport velocity of the adsorbed tracer is less than for the fluid or a conservative tracer:

$$v_R = \frac{v_x}{R_D}$$

where v_R = retarded flow velocity, v_x = conservative flow velocity, R_D = retardation factor.

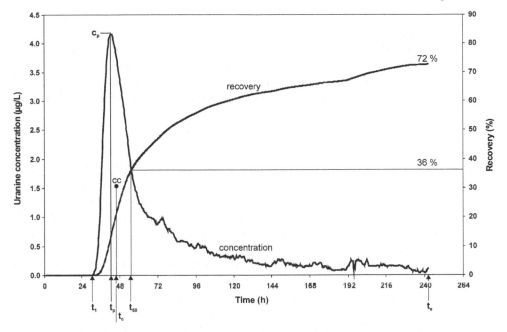

Figure 8.7.   Tracer breakthrough curve resulting from an instantaneous injection. The main features are: time of first detection (t_1), time and concentration of the peak (t_p, c_p), time of the concentration centroid cc (t_c), time where half of the recovered tracer has passed (t_{50}), end of the observation period (t_e). The recovery curve shows cumulatively the fraction of injected tracer that arrived at the sampling point. In this case, the total recovery at the end of the observation period is 72%, so t_{50} is the time with a 36% recovery.

Biodegradation is a complex process, but may be approximated by:

$$c_t = c_0 \cdot e^{-\lambda \cdot t}$$

a decline analogous to radioactive decay, where c_t = concentration of the degradable tracer at time t, c_0 = concentration of the conservative tracer, λ = constant of decay.

8.4.4   Breakthrough curves

The definitive product of quantitative tracing is a time-concentration tracer breakthrough curve (Fig. 8.7) generated by online monitoring or analysis of water samples. It represents the behaviour of the injected tracer in the flow route. The magnitude of the breakthrough is determined by the tracer mass and volumetric flow (i.e. dilution). This former effect is readily removed by dividing all concentrations by the injected mass or (in the case of particulates) injected numbers. Breakthrough curves from entirely different systems are best compared using a dimensionless form generated by dividing time by a reference time (typically mean transit time or time to peak concentration) and concentration by the peak concentration.

The primary features of a breakthrough curve are the premonitory background (if present), the rising limb, the peak, and the recession. Asymmetry is caused by dispersion, storage and other transport processes. The time delay of the breakthrough curve of a conservative tracer is determined by length of the flow path and flow velocity. Its breadth arises from dispersion and increases with time. In a karst aquifer, the breakthrough curve reflects the structure of the flow paths and may also be influenced by temporal variation of recharge. Retardation shifts the curve towards higher transit times, while degradation reduces all concentration values, mainly on the falling limb, as the degradation is time-dependent.

8.4.5   Travel time and transport velocity

A number of transport statistics can be derived from the breakthrough curve. The first arrival time depends on the analytical detection limit and background stability; it is thus more correct to speak of first detection. In contrast, the time and concentration of the peak are readily determined from a continuous data record. If discrete sampling fails to define a peak, then a peak may be estimated using adjacent values. The time of last detection is difficult to determine, as it strongly depends on the analytical detection limit and on the duration and resolution of the monitoring. An exponential fit can be used to extrapolate tracer tails.

The mean transit time cannot be directly read from the curve. In most cases, the mean transit time lies between the time of maximum concentration and the time at which 50% of the tracer has passed. The concentration centroid of the breakthrough curve provides a good approximation of the mean transit time. The mean transit time can be computed statistically (e.g. Smart 1988b) or by solution of the advection-dispersion equation for the measured data. Where the breakthrough curve is ill-sampled or contains errors, the time of peak provides the most robust estimate of travel time.

Travel times can be converted to respective velocities if a relevant travel distance is available. In karst systems, the travel path may have been explored by speleologists in caves or inferred from geology. Where it is unknown, the straight-line distance between the points of injection and recovery can be used to obtain a linear velocity. The flow velocities corresponding to the different times are the maximum flow velocity (first detection), the dominant or modal flow velocity (peak), the effective flow velocity (mean transit time) and the mean velocity (half of the recovered tracer mass) (Schulz 1998). Flow velocities in karst aquifers span a wide range from a few metres up to several hundred metres per hour (Ford & Williams 1989, Gospodaric & Habic 1976, Morfis & Zojer 1986, Behrens et al. 1992).

8.4.6   Mass recovery

The fraction of the injected tracer mass passing a sampling point provides valuable information on the system hydrology and the tracer performance. Mass recovery requires high-resolution discharge and concentration data at each site. If discharge is constant, the recovered tracer mass is the integral of the breakthrough curve times the discharge. For variable discharge, the recovered tracer mass M_R is calculated by integration of the tracer load (concentration-discharge product) over the time:

$$M_R = \int\limits_{t=0}^{\infty} (Q \cdot c) dt$$

The cumulative tracer recovery (mass or %) is often presented in the same diagram as the breakthrough curve to show that the recovery curve approaches an asymptotic final value, and to allow definition of the time at which half of the recovered tracer has passed (Fig. 8.7).

The mass recovery is often significantly less (and rarely more) than 100%, due to tracer storage (e.g. in the karst matrix) or delivery to non-monitored springs (e.g. under water or in other catchments). The tracer may also have been lost to adsorption or decomposition. Sometimes, sampling or measurement errors are responsible for poor recovery. Significant losses may imply dye storage within the aquifer, thus compromising further tracing.

8.4.7   Characterisation of conduit networks

Tracer results can be negative, poorly defined, well defined or fully configured. Negative results where the tracer is not detected are ambiguous. Poorly defined results occur when the tracer is detected only intermittently, or at levels close to background. Such results indicate a possible linkage perhaps with intermittent dilution (e.g. sampling incompletely mixed flow from a local tributary), but may also arise from technical problems. Such data cannot be rigorously analysed. Well-defined tracer results rise well above background levels and show a characteristic breakthrough curve allowing more rigorous analysis. Fully configured tracer results constitute a breakthrough curve with concurrent discharge data and high overall recovery.

The primary application of tracer tests is in defining underground linkages and this is often obtained with charcoal detectors or occasional grab sampling. Greater resolution of karst conduit networks is possible with fully configured tests where the injection tracer mass and discharge can be compared to the recovered mass and discharge (Fig. 8.8). Conservative tracers are preferred for this type of interpretation.

When the discharge at the injection point is similar to that at the sampling point, and when the recovery rate is close to 100% (case A), there is a direct connection between the two points, without flow divergence or convergence. Conceptually and comparatively, it is useful to determine the geometry of an equivalent conduit connecting the two points, assuming that it is straight, water-filled and has a constant cross-sectional area. The conduit volume (V) can be estimated by integrating the discharge from the injection time to the mean transit time or centroid time:

$$V = \int_{t=0}^{t_c} Qdt$$

The equivalent conduit cross-sectional area can be computed by dividing the volume by the travel distance, and the radius of the conduit can be computed from simple geometry.

In most cases, however, the discharge rate at the sampling site will be different than at the injection point, and the recovery rate will be smaller than 100%, indicating flow divergence and/or flow convergence between the two points (cases B–F).

The form of the tracer breakthrough curve may allow additional resolution of underground conditions. A single peak suggests a single conduit that can be characterised in general using appropriate interpretative models. Multiple peaked curves may indicate multiple flow routes (case I). Tracer passing through a conduit may be directed into storage. The storage may be an off-line (case G) or an in-line void (case H). Storage in such sites is often driven by varying flow, and may cause surprising disappearance or apparently spontaneous appearance of tracer. Sustained tails on a breakthrough curve can result from storage but may also result from unresolved multiple peaks (Werner 1998a), or matrix or dead-zone

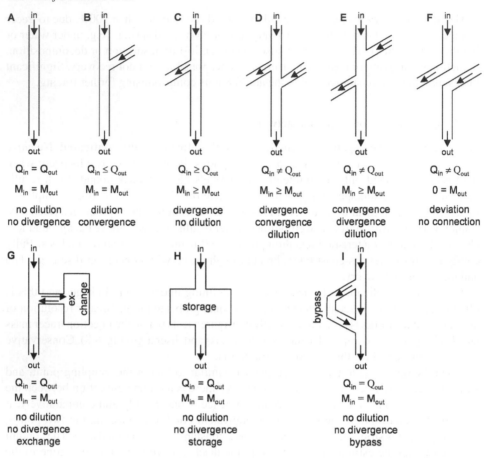

Figure 8.8.   Karst network configurations based upon the measurement of input (in) and output (out) discharge and tracer recovery under steady flow conditions. Q: discharge, M: tracer mass. A to F represent different configurations where flow converges or diverges and tracer transport will occur straight on to the outlet or will be redirected. G, H and I represent special cases, which can be deduced from the shape of the breakthrough curve (peak, tailing): G has off-line storage, H has in-line storage, I is a system with bypass. The complexity of real karst systems makes it difficult to resolve to clear configurations (modified after Ashton 1966, Atkinson et al. 1973, Brown & Ford 1971, Brown & Wigley 1969, Smart 1988a).

storage (Hauns et al. 2001). Modelling of such storage is difficult. More general dead zone models have been developed to describe the effect of temporary storage in simple uniform channels (e.g. Davis et al. 2000, Field and Pinsky 2000).

It is sometimes possible to resolve relatively complex underground flow systems, if the data from several tracer tests are combined with geological, speleological and hydrological observations (e.g. Smart 1988a), and with hydrochemical and isotopic (i.e. natural tracer) data (Pronk et al. 2006).

Replication of tracer tests under different flow conditions may reveal changes in underground routing and velocity that are of considerable importance in water resource management. The relationship between parameters like conduit volume and discharge reveals the extent of storage changes and the dynamic and static storage in a system (Smart 1988b). The travel-time discharge relationship reveals whether a system is open channel or closed conduit flow (Stanton & Smart 1981), or even where overflow routing develops (Smart 1997). Multiple peaked breakthrough curves at higher discharge may indicate more rapid overflow routing (Stanton & Smart 1981). When multiple springs are under investigation, the relative yield of water and tracer and breakthrough curves can be evaluated to identify spring families that show close or more distant relationships. Changes in these relationships with discharge are indicative of underground routing and hydraulic change (Smart 1988a).

Tracing during varying flow leads to complications of interpretation. For example, travel time is generally faster on rising than falling stage. Water and tracer can also be redirected into storage, or delayed or divided without necessarily following two physical routes.

8.4.8   Quantification of transport parameters using analytical models

Chapter 10 of this book deals with modelling. This section only provides a brief overview of how analytical models can help to better interpret tracer test results and determine transport parameters. Karst aquifers are difficult to model because as aquifers they are heterogeneous and anisotropic, and as conduit networks they have unsteady open (vadose) and closed (phreatic) channels of varying scale. There are two main types of models: global and distributed models. Distributed models subdivide the aquifer into homogeneous sub-units with water and tracer transfer between these sub-units. Global or lumped parameter models transform an input signal into an output signal using an analytical equation that substitutes for the physical properties of the aquifer. It is also possible to use a known output signal in order to determine physical parameters of the aquifer (inverse modelling). This is usually done by means of a best-fit approach, i.e. the physical parameters are systematically varied until the modelled curve fits to the measured data series. These parameters may then be applied to forward prediction of aquifer response.

Common modelling tools for the interpretation of tracer tests are based on fitting an analytical transport equation (e.g. equation 2 on page 164) to a measured breakthrough curve. For advective-dispersive transport, two fitting parameters are required, the first representing advection (e.g. mean transit time or effective flow velocity) and the second representing dispersion. Several computer codes are available for this type of interpretation, such as TRACI (Werner 1998a,b) or CXTFIT (Toride et al. 1999). The QTRACER2 program can also be used for the evaluation of tracer breakthrough curves (Field 2002).

Multiple-peak breakthrough curves produced from single tracer injections under steady flow indicate the presence of different flow paths in the aquifer. The Multi-Dispersion-Model assumes that these flow path (conduits) split directly at the injection point (swallow hole) and reunify at the sampling point (spring). Each flow paths is characterised by an individual advective flow velocity and dispersion. At the sampling point, the individual breakthrough curves superimpose to form a multi-peak curve (Maloszewski et al. 1992, Werner 1998a). Inversely, the decomposition of a multi-peak curve makes it possible to determine the individual flow velocities and dispersion of the individual flow paths (Fig. 8.9).

Karst aquifer systems are always much more complex than a model can describe. Parameters obtained from inverse modelling are useful for analysis and comparison of tracer breakthrough curves, but are greatly generalised and not representative of reality.

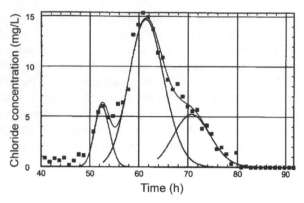

Figure 8.9.   In 1877, a tracer test with NaCl proved the underground connection between the sink of the Danube and the Aach spring in the Swabian Alb, Germany (Knop 1878). More than 100 years later, Werner (1998a) modelled the breakthrough curve using a Multi-Dispersion-Model. The breakthrough curve can be decomposed into three curves with three different effective flow velocities and dispersion coefficients. However, it is not clear what constitutes an interpretable peak and what is artefact. The initial high values suggest a level of uncertainty in the data that does not support the overall fit proposed.

8.5   SUMMARY

Groundwater tracing is a method of investigating underground water and contaminant transport by labelling water with identifiable tracer substances or physical properties. It is particularly applicable to karst areas where the underground flow dominates but is not readily comprehended. The technical and environmental requirements of a tracer mean that fluorescent dyes are most widely used, along with less tractable particulate tracers.

Tracer tests are used primarily to define underground connections, spring catchment areas and contaminant origin or destination in karst areas. Such tests require rudimentary monitoring to establish the presence or absence of a tracer. More sophisticated groundwater tracing is used to determine parameters for aquifer hydraulic and geometry or contaminant transport. Simple sampling and analysis may be sufficient for the former style of investigation, but the latter require high frequency or continuous sampling and quantitative tracer analysis. Quantitative time concentration data allow travel time and dispersion statistics to be calculated. Supplementary discharge data permits mass recovery and unsteady flow effects to be studied. A number of models are available for tracer breakthrough and aquifer analysis. They provide valuable guidance on possible configuration of the aquifer, although they remain much simplified.

CHAPTER 9

Geophysical methods

Timothy D. Bechtel, Frank P. Bosch & Marcus Gurk

9.1 INTRODUCTION

Exploration geophysics is the science of "seeing" into the earth without digging or drilling. It is the Earth Science equivalent of medical radiology, and in an analogous fashion, employs a signal that passes through (or emanates from within) the earth to record variations in some physical property of the subsurface. This is done by measuring the signal at the ground surface, sending signal from boreholes to the surface, and/or sending signals between boreholes. In order to view different types of targets within the earth, geophysicists select from an array of signals/techniques encompassing seismic, gravimetric, and electrical/electromagnetic methods (see Table 9.1), with the choice of technique based on the physical properties of the features to be imaged.

The product of geophysics is a profile of some physical property of the earth vs. horizontal distance or vertical depth (see e.g. Fig. 9.1). Many profiles may be collected to produce two-dimensional contour maps (Fig. 9.2) or cross-sectional "images", or even 3D block models. These indirect "pictures" can be interpreted to provide a geologic model of the subsurface.

Geophysics is often applied to hydrogeology because of the relationship between subsurface physical properties, and those related to the nature and occurrence of groundwater (see Table 9.1). There is no geophysical "magic bullet" that can detect all karst hydrogeologic features. For instance, electromagnetic (EM) methods may quite effectively detect a water-bearing solution channel (an electrical conductor) in limestone (a resistor), but might not see the same cavity at all if it is filled with air (also a resistor). Based on similar reasoning, and the authors' experience, the relative sensitivities of various geophysical methods to different types of subsurface karst features are characterised in Table 9.2.

9.2 GENERAL CONSIDERATIONS IN GEOPHYSICS

9.2.1 Why use geophysics?

Nowhere is the utility of geophysics more apparent than in karst terranes where borings several meters apart may yield very different depth-to-rock, porosity, or water quality. In karst formations, groundwater often occurs in a network of fractures and solution cavities, which may be irregularly spaced, and may have distinctly preferred orientations. Therefore, karst porosity is often laterally and vertically non-uniform, and anisotropic. Although

Table 9.1. Summary of common karst hydrogeophysical methods.

Geophysical method group	Geophysical method	Acronym or nickname	Related terms	Measurement type	Physical properties measured	Related hydrogeologic parameters
Seismic	Refraction	None	SEIS sometimes used for seismic methods in general	Travel times of seismic/acoustic waves	Density and/or elastic constants	Formation porosity, fracture density
	Reflection	None				
	Statistical analysis of surface waves or multispectral analysis of surface waves	SASW or MASW				
Gravimetric	Gravity	GRAV	Microgravity means high-accuracy gravity	Acceleration of gravity	Density	Formation porosity
Electrical/ Electromagnetic	Direct current resistivity	DC RES	EI means electrical imaging which is the same as electrical resistance tomography or ERT; two-dimensional and three-dimensional EI are often called 2DR and 3DR	Resistance to flow of induced electrical currents – EM measurements are either far-field (remote transmitter) or on-site or near-field (local transmitter)	Electrical conductivity/ resistivity	Specific conductivity of pore fluid, formation porosity, and formation mineralogy
	Frequency domain, time domain, very low frequency and radio frequency electromagnetics, radiomagnetotellurics	FDEM & TDEM VLF-EM, RF-EM & RMT	VLF-EM and RF-EM are often shortened simply to VLF or RF			
	Magnetics	MAG	MAG is simply DC or steady-state electromagnetics	Magnetic field intensity and/or direction	Intrinsic magnetism or magnetic susceptibility	Formation mineralogy and gross porosity
	Spontaneous potential or natural potential	SP or NP	Streaming potential is a type of spontaneous potential that is often abbreviated SP as well	Natural electrical potentials	Fluid movement through porous media or differences in fluid ionic content	Porosity, hydraulic head, specific conductivity of pore fluid, and formation mineralogy
	Ground penetrating radar	GPR	SIR means subsurface interface radar and sometimes replaces the term GPR	Travel time of electromagnetic waves	Dielectric constant and electrical conductivity	Formation moisture content and moisture chemistry

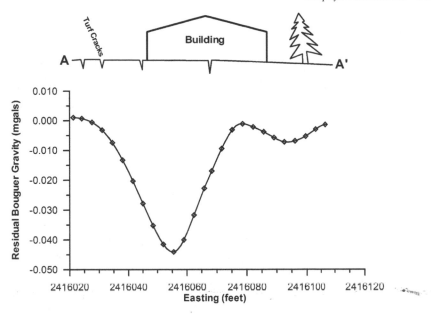

Figure 9.1.   Geophysical (microgravity) profile over a suspected sinkhole in Eastern Pennsylvania, USA.

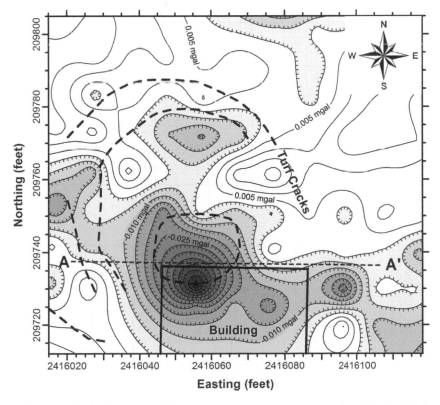

Figure 9.2.   Geophysical (microgravity) mapping of the same suspected sinkhole as Fig. 9.1.

Table 9.2. Karst applications of surface geophysical methods. See Table 9.3 for borehole geophysical methods.

Probability of Detection: 1=high, 5=low

Feature	Cross Section	Special Condition at Top-of-Rock	Feature Infill	Seismic			Gravity	Electrical/Electromagnetic						
				Refraction	Reflection	Surface Wave		DC RES	FDEM	TDEM	VLF-EM	MAG	SP	GPR
high porosity zone (fracture or solution cavity)[1]	equant	none	air	5	2	2	2	4	5	5	5	5	5	1
		none	water	5	2	2	3	3	4	4	5	5	2	1
		none	soil	5	3	3	3	2	3	3	5	5	3	2
		wet soils or infiltration	air	4	2	2	2	4	5	5	5	5	1	4
		wet soils or infiltration	water	4	2	2	3	3	4	4	5	5	1	4
		wet soils or infiltration	soil	4	3	3	3	2	3	3	5	5	1	4
		soil void	air	4	2	1	2	2	3	3	5	5	5	1
		soil void	water	4	2	2	2	2	3	3	5	5	2	1
		soil void	soil	4	2	2	2	2	3	3	5	5	3	1
		rock depression[2]	air	2	1	2	2	2	3	3	5	5	5	1
		rock depression[2]	water	2	2	2	2	1	2	2	4	5	2	1
		rock depression[2]	soil	2	2	2	1	1	2	2	2	5	3	2
	near-vertical, tabular	none	air	5	2	2	2	3	3	4	3	5	5	1
		none	water	5	1	2	1	2	2	3	2	5	2	1
		none	soil	5	2	2	2	1	3	2	3	5	3	2
		wet soils or infiltration	air	4	1	2	2	3	3	4	4	5	1	4
		wet soils or infiltration	water	4	1	2	2	2	3	3	3	5	1	4
		wet soils or infiltration	soil	4	2	2	2	1	1	2	3	5	1	4
		soil void	air	4	1	1	1	1	2	2	4	5	5	1
		soil void	water	4	1	2	1	1	2	2	2	5	2	1
		soil void	soil	4	1	1	1	1	2	2	3	5	3	1
		rock depression[2]	air	2	1	2	1	1	2	2	4	5	5	1
		rock depression[2]	water	2	1	1	1	1	1	2	2	5	2	1
		rock depression[2]	soil	2	1	2	1	1	1	2	3	5	3	2
Special Considerations				requires special data collection and processing to be useful in karst terranes	generally good for depths > 50' and S-waves are preferred	resolution decreases rapidly with survey depth	highly dependent upon depth vs. dimensions of feature	dependent upon chemistry of water, or soil moisture or mineralogy	dependent upon chemistry of water, or upon soil moisture or mineralogy	dependent upon chemistry of water, or upon soil moisture or mineralogy	highly dependent upon profile orientation vs. feature orientation	not typically useful in karst hydrogeophysics	subsurface water movement creates strong anomalies	useful only where overburden is not clay rich or otherwise electrically conductive

[1] Common karst feature of interest-table can also be applied to features with similar physical properties.

[2] Methods that are sensitive to rock depressions are typically suitable for measurement of overburden thickness as well.

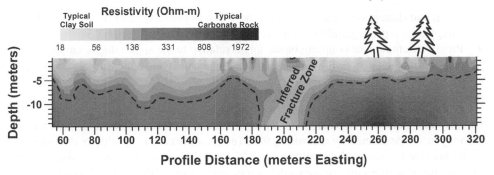

Figure 9.3.   Electrical image across a suspected limestone fracture zone in central Maryland, USA.

hydrogeologists are often tempted (or requested) to interpolate between borings, such interpolations can be misleading – or dead wrong – in karst terranes, and may be impossible even to attempt when adjacent borings display wildly differing characteristics. This results in both the motivation for the use of geophysics, and the necessary characteristics of any geophysical methods employed.

Since a geophysical measurement at a given point cannot provide the direct and detailed subsurface data of a boring at that same point, geophysics cannot replace direct testing. Instead, geophysics provides a cost-effective means of constraining hydrogeologic models. It is best used to select critical locations for borings, or to test specific predictions of a hydrogeologic model. Thus, a combination of geophysics and direct testing is best for optimising the cost, accuracy, and timeliness of a hydrogeologic study. For example, since porosity is often highly localised in karst areas, geophysics can be used to locate potential high-porosity zones, which can then be characterised in detail using an optimally placed boring. This eliminates the a-priori need for a large number of borings to confidently ensure penetration, as well as delineation, of a suspected porous zone such as, for example, the fracture zone depicted in Fig. 9.3.

Finally, geophysics can often be done where vegetation or topography might prevent easy access with a drilling rig since much geophysical equipment can be hand-carried.

In support of hydrogeologic investigations in karst terranes, the most effective uses for geophysics include:
- Detection of aquifers or high porosity zones (e.g. fractures or solution channels)
- Delineation of aquifer orientations and dimensions
- Delineation of infiltration zones or groundwater flow pathways, and
- Determination of overburden thickness.

9.2.2   Prerequisites for subsurface detection

The ability to detect a given subsurface feature depends on whether the feature represents local variation in a subsurface parameter that is measurable using a geophysical method. This means that for whatever type of geophysical measurements collected, there are certain characteristics the target or feature must possess, including:
- A sufficient contrast in some physical property (e.g. density for gravity surveys) between the target and surroundings;

- Sufficient dimensions relative to both the depth of the target and the volume of earth sampled by a given geophysical measurement; and
- Proper orientation or coupling of the target relative to the signal that is employed – to ensure that the target will interact with the signal and create a measurable anomaly.

9.2.3   The noise problem

Geophysical detection depends very strongly on the presence of noise or interference. Geophysics is performed by measuring the strength, orientation, or velocity of some form of energy that has sampled the subsurface. This desirable energy is signal. Noise is energy that has been inadvertently or unavoidably measured along with the signal. For instance, a seismic survey might be performed to measure overburden thickness. In this case, acoustic waves would be bounced off of the top-of-rock and their arrival at an array of geophones timed. However, if there is a railroad nearby, the geophones might also measure vibrations that have nothing to do with the depth to rock. In some cases, the noise can be recognized and removed from the data, but it is best to locate a target using a geophysical method that is not sensitive to local sources of unwanted energy or noise.

9.2.4   Resolution versus depth of investigation

In all of geophysics, there is a trade-off between depth of investigation and resolution. This is partially due simply to geometric spreading of signals – i.e. if you shine a flashlight on a street sign, and walk away, it will become harder to see the sign because the beam of the flashlight at the sign becomes wider – with less illumination focussed just on the target. However, there is also a frequency-dependent effect. Whether frequency refers to a variation with time (e.g. the ground motion represented by seismic wiggle traces) or variations with distance (e.g. measured changes in gravity along a profile), the farther one is from the source of a signal, the more the high frequency components (i.e. the "spiky" parts) of the signal are lost. It is best to remember the adage "the Earth eats high frequencies". This means that whatever geophysical method is used, the higher the frequency of the signal, the shallower the effective depth of the survey. This leads to a recurring (and vexing) trade-off between resolution and survey depth: Lower frequency signals can penetrate deeper into the earth, but sample a larger volume of the subsurface, and therefore can only resolve relatively large targets. Higher frequency signals sample smaller volumes and can be generated with smaller transmitters, and so have better resolution – but they cannot penetrate as deeply into the earth, because the Earth eats high frequencies!

9.2.5   Technique selection

There are numerous geophysical methods, and the importance of proper technique selection cannot be overemphasized (see US Army Corps of Engineers, 1995). A technique must be chosen that employs a signal that is sensitive to the physical property contrast between the target and its surroundings, and that extends to sufficient depth to reach the target. However, equally importantly, the selected technique must be sensitive to a signal that is unique to the target, or the desired anomaly from the target may be overwhelmed by unwanted noise. The best geophysical instrument employed by a competent operator will be wasted if the instrument is not sufficiently sensitive to the target, cannot penetrate to the required depth, or is too sensitive to local noise.

9.2.6   Location control

Proper location control is essential for all geophysical surveys. The required level of horizontal accuracy or precision varies with technique and target. For low resolution geophysical techniques (e.g. VLF-EM) or large targets, horizontal accuracy no better than several meters may be required. For high-resolution methods (e.g. GPR) or small targets, sub-meter accuracy is often required. The required level of vertical accuracy varies similarly, with one important exception; microgravity surveys will always require vertical location control with about 1.5 cm accuracy in order to properly correct for elevation effects (discussed in section 9.4). For many years, location control has been maintained by laying out stations along a profile or on a two-dimensional grid using a transit or compass and tape measure. Increasingly, location control is provided by a global positioning system (GPS) integrated with a geophysical data logger. These have the advantage of providing geo-referenced geophysical measurements that can be easily imported into a geographic information system (GIS). For very high positioning accuracy, real time kinematic (RTK) GPS, or a robotic automated total station (ATS) may be employed.

9.2.7   Non-uniqueness

The main limitation of geophysics is non-uniqueness. Based on the known physics of the geophysical measurement method, a geophysicist can produce a model of the subsurface that closely reproduces the measured field data. However, another geophysicist (or the same geophysicist after lunch) could devise a very different model of the subsurface that also closely reproduces the measured data. That is, the majority of geophysical anomalies do not lead to a unique subsurface model. This problem can be addressed by one or more of the following:
- Constraining possible models to be geologically realistic (based on available data),
- Testing competing models through carefully placed borings or other direct testing,
- Testing competing models by performing additional geophysical measurements using another technique that is sensitive to a different type of signal.

9.3   SEISMIC METHODS

9.3.1   Background

Seismic techniques involve measuring the travel time of seismic (acoustic) energy from "shots" (e.g. an explosion or weight drop), through the subsurface, to arrays of listening devices or geophones (Mooney, 1984). In the subsurface, seismic energy travels in waves that spread out as initially hemispherical wavefronts. The energy arriving at a geophone is described as having travelled a ray path perpendicular to the wave front. As depicted in Fig. 9.4, subsurface seismic energy is refracted (bent) and/or reflected at interfaces between materials with differing seismic velocities. The refraction and reflection of seismic energy through velocity contrasts follows exactly the same laws that govern the bending and reflection of light through prisms (see Telford et al., 1990).

The seismic velocity (V) of a material is proportional to its stiffness (defined by elastic moduli), and inversely proportional to density. That is, for a material with density ρ, the seismic velocities for compressional or primary (P) waves and slower moving shear

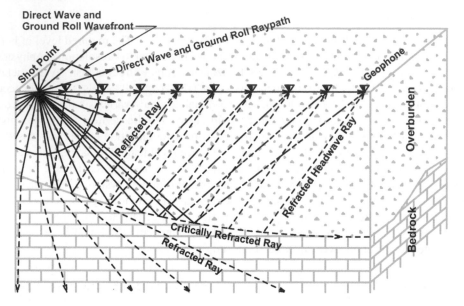

Figure 9.4.   Selected seismic ray paths from a shot point to an array of receivers.

or secondary (S) waves are (respectively): $V_p = \sqrt{(\lambda + 2\mu)/\rho}$, and $V_s = \sqrt{\mu/\rho}$, where λ and μ are Lamé's constants (which characterise the elastic stiffness of the material under normal and shear stresses). Note that seismic velocity formally decreases with increasing density (ρ). Although this seems counter-intuitive, it turns out that with increasing density, the elastic constants increase faster. Particularly in rocks, the elastic moduli are strongly dependent upon the porosity of a material, and increase faster than density with decreasing porosity such that empirically, seismic velocity increases with decreasing porosity. Therefore, higher density and lower porosity materials typically have higher seismic velocities. Dry, unconsolidated soils and sediments typically have seismic P wave velocities between 300 and 1400 m/s. Saturated soils generally have P-wave velocities in the range of 1400 to about 2300 m/s. In karst terranes where the bedrock weathers by dissolution rather than decomposition, there is usually a distinct velocity contrast at the top of rock, with typical carbonate rock P-wave velocities exceeding 2500 m/s (see e.g. Redpath, 1977).

As a seismic ray strikes a density contrast, a portion of the energy is refracted into the underlying layer, and the remainder is reflected at the angle of incidence. The reflection and refraction of seismic energy at each subsurface density contrast, and the generation of surface waves (or ground roll), and the sound (i.e. the air-coupled wave or air blast) all combine to produce a long and complicated sequence of ground motion at geophones near a shot point. The ground motion produced by a shot is typically recorded as a wiggle trace or wave train for each geophone. Arrival times for certain events on these wiggle traces are analysed to determine the velocity structure of the subsurface beneath the geophone array. The velocity structure may be depicted as boundary lines between layers with different inferred velocities, as velocity contours, or as processed and stacked wiggle trace data.

Seismic refraction and reflection methods can employ either of the P or S waves which travel through the subsurface (and are collectively referred to as body waves). P waves in a given material are always faster than S waves, with *possible* V_s ranging from 0 to 70%

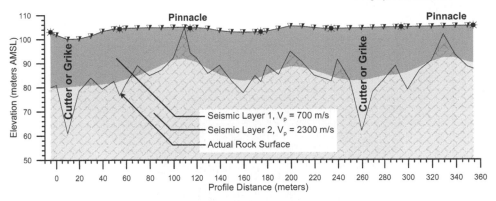

Figure 9.5. Example of a two-layer seismic refraction model compared to the actual bedrock surface (as revealed by borings and later construction activities).

of V_p, and *common* V_s in earth materials ranging from 20 to 70% of V_p. S waves present more data recording and processing difficulties due to their slower speed (and consequent arrival farther back in a complicated wave train), as well as a much greater propensity for attenuation than P waves. However, S waves are particularly sensitive to water-saturated, porous hydrogeologic targets such as solution cavities or fracture zones. This is because water can transmit P waves, but not S waves, so there is a dramatic change in the ratio of V_p / V_s in these features.

Surface wave methods rely upon high amplitude waves coupled to the ground surface (and hence called surface waves or ground roll). These waves involve mostly shearing motion, and are therefore well suited to fluid-filled or open porous targets.

9.3.2 Seismic refraction

Seismic refraction methods measure the travel time of the component of seismic shot energy that travels down to a distinct density contrast, is refracted along the contrast, and returns to the surface as a head wave along a wave front similar to the bow wake of a ship (see Fig. 9.4). The rays which travel along the top of rock are called refracted waves, and for geophones at a certain distance from the shot point, always represent the first arrival of seismic energy (Lankston, 1990). Seismic refraction is generally applicable only where the seismic velocities of layers increase with depth. Therefore, where higher velocity (e.g. clay) layers may overlie lower velocity (e.g. sand or gravel) layers, seismic refraction may yield incorrect results. In addition, since seismic refraction requires geophone arrays with lengths of approximately 4 to 5 times the depth to the density contrast of interest, seismic refraction is limited by practicality to mapping the tops of layers at depths typically less than 30 m.

Seismic refraction works best for relatively flat, dipping layers. Therefore, in karst terranes, where rock is commonly pinnacled, successful refraction surveys require highly redundant data sets (with many shot points per receiver) to be able to resolve the irregular rock profile. However, even the best refraction data often produce a smoothed version of the top-of-rock (see Fig. 9.5). Refraction interpretations that produce only a profile of the top-of-rock may provide detection and delineation of hydrogeologic features only if they express some relief at the top of rock. However, highly redundant seismic refraction data

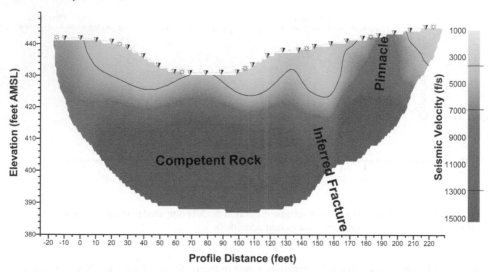

Figure 9.6.   Example of a seismic tomography image from South-Central Pennsylvania, USA.

sets can also be interpreted using modern mathematical inversion methods that treat the subsurface as a stack of discrete elements – each with a model seismic velocity. This type of inversion can provide detection of laterally restricted solution cavities, fractures, or other potential hydrogeologic features below the top-of-rock (Fig. 9.6). Even with these methods, continuous low velocity zones still can lead to errors in interpretation.

9.3.3   Seismic reflection

Seismic reflection uses field equipment similar to seismic refraction, but field and data processing procedures are employed to maximise the energy reflected by subsurface density contrasts along near-vertical ray paths (see Fig. 9.4). Reflected seismic energy is never a first arrival, and therefore must be identified in a generally complex set of overlapping seismic arrivals – generally by collecting and filtering highly redundant or multi-fold data (e.g. Steeples & Miller, 1990). Even a simplified discussion of the processing required to identify, enhance, and properly present reflected arrivals cannot be attempted here. Because of this specialised processing, costs for a given lineal footage of seismic reflection profile are usually greater than for a coincident seismic refraction profile. However, seismic reflection can be performed in the presence of low velocity zones (velocity inversions), generally has lateral resolution superior to seismic refraction, and can delineate very deep density contrasts with much less shot energy and shorter profile lengths than would be required to achieve a comparable depth using seismic refraction. In karst hydrogeology, seismic reflection can be used to detect and delineate water-bearing faults or fractures. The presence of solution cavities can often be determined, but delineation of their shape and dimensions is not typically possible.

The main limitations to seismic reflection are its higher cost compared to refraction (for sites where either technique could be applied), and its physical limitation to depths generally greater than approximately 15 m. At shallower depths, reflections from subsurface density contrasts arrive at geophones at nearly the same time as the much higher amplitude ground roll and air blast. Reflections from greater depths arrive at geophones after the ground roll and air blast have passed, and are thus more easily identified. Very shallow (<5 m), extremely high resolution seismic reflection surveys (using high frequency signal sources and receivers and ultra-high-fold data) have been successful (Baker et al., 1999), but are not yet in wide use for karst hydrogeophysics.

9.3.4   Surface wave analysis

Seismic surface waves are produced only near the ground surface. They have high ampli-tudes, which diminish with depth, and their velocity is strongly dependent upon the S or shear wave velocity of near surface materials. The effective depth to which a surface wave motion extends is dependent upon its frequency (or wavelength). Low frequency (long wavelength) surface waves have motions that extend to greater depths than high frequency waves. A typical shear wave train contains motions at a range of frequencies. Therefore, if the shear wave velocity of the subsurface varies with depth, the different frequency compo-nents of the shear wave train travel at different velocities, or become dispersed. This effect can be used to determine the shear wave velocity profile beneath an array of geophones (Park et al., 1999). A dispersion curve depicting the surface wave velocity as a function of frequency can be mathematically inverted to provide a sounding curve of shear wave veloc-ity as a function of depth. If many such sounding curves are developed for sequential arrays of closely spaced geophones, it is possible to produce a cross-sectional contour drawing, or shear wave velocity image that typically resembles the seismic tomography image in Fig. 9.6. These images are particularly sensitive to cavities and fracture zones where the air or fluid-filled porosity represents a low S velocity (low V_s) zone.

Surface wave techniques include spectral analysis of surface waves (SASW), multispec-tral analysis of surface waves (MASW), and refraction micro-tremor (ReMi). All are in relatively common usage, but are listed here in order of increasing modernity and accuracy.

Surface wave techniques have an advantage over refraction and reflection in that surface waves are generally high amplitude and therefore easy to record with a high signal to noise ratio. In addition, their interpretation is not hampered by any type of subsurface velocity structure (as refraction is hampered by velocity inversions), and no distinct velocity contrasts are required (as they are for seismic reflection and simple refraction analysis).

9.4   GRAVITY METHODS

As described by Newton (1686), any two bits of matter in the universe have a mutually attractive force proportional to the product of their masses divided by the square of the distance between them. For a given unit of mass at the surface of the Earth, the force of gravity pulls it downward with a gravitational acceleration of 9.81 m/s^2 or 981 000 milliGal (where 1 Gal = 1 cm/s^2). But this value is truly accurate only at the Potsdam Observatory (Kertz, 1969) to which, by historical convention, all other gravity measurements are refer-enced. Since the Earth is not a perfectly homogeneous sphere, the gravitational acceleration is not constant at various locations. Instead, the magnitude depends on the following five superimposed factors:

- The latitude of the station
- The elevation of the station

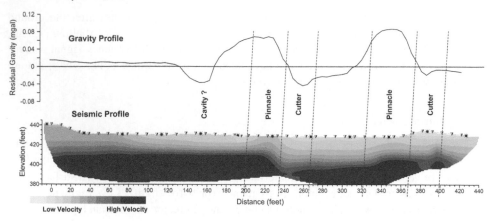

Figure 9.7.   Residual gravity profile with coincident seismic data from Central Virginia, USA. The gravity variations appear largely related to rock depth, with shallow rock beneath gravity highs and vice versa. However, there is a low gravity anomaly near profile distance 160 feet that is not expressed at the top of rock, and could therefore represent a subsurface cavity rather than deep bedrock.

- The density of materials immediately beneath a station
- The attraction of massive landforms near or obliquely above the station (i.e. the mass of a nearby mountain which actually produces a gravitational attraction)
- The positions of the sun and the moon
- Minute changes in the calibration of the gravity meter (called instrument drift)

Microgravity data must be corrected for these effects if they are to clearly reflect the presence of subsurface karst features. These corrected data displayed on profiles and/or maps are referred to as *Bouguer* gravity anomalies, and retain not only gravity information on the local mass or density distribution beneath a survey station but also the regional gravity distribution (Mayer de Stadelhofen, 1991). In order to remove regional effects and focus on local (e.g. karst) features, *residual* gravity is calculated by subtracting a smooth regional trend from the Bouguer gravity.

Bouguer or residual gravity highs are caused by subsurface mass excesses such as locally shallow bedrock pinnacles or float blocks in the soil profile; or zones of particularly massive bedrock. Gravity lows are caused by mass deficiencies such as locally deep bedrock cutters or clay seams where less dense soil displaces more dense bedrock. Voids within bedrock also produce gravity lows. Since microgravity alone cannot discriminate between a local bedrock deep and a void, it should never be interpreted without complementary geophysical (e.g. seismic) or boring data. Figure 9.7 presents coincident gravity and seismic profiles that together resolve a potential non-uniqueness problem of interpretation.

The detectability of karst features using gravity measurements depends upon the density contrast between the feature and surrounding rock, and the dimensions, and depth of the feature. In general, the lighter (less dense), larger, thicker, and shallower the karst feature is, the easier it will be to detect and delineate using a gravity survey. Because the Earth eats high frequencies, for observed gravity anomalies, the narrower or "spikier" they are, the shallower their source must be. This phenomenon can be used to estimate depths to gravity anomaly sources through spectral analysis of the gravity data (Bechtel et al., 1987).

In the same way that nearby landforms create a gravitational effect that must be removed with terrain corrections, subsurface density anomalies that lie off to the side of a survey profile will influence measurements along the profile. Therefore, in karst terranes with their lateral inhomogeneity, it is usually advisable to record data on a grid, or at least along two or more parallel profiles to be sure of where apparent anomaly sources actually lie.

Of all geophysical methods, gravity surveys may be least susceptible to interference or noise. Gravity surveys can be performed in urban settings, and near underground or overhead utilities where seismic or electrical/electromagnetic methods would be impossible. The main obstruction is low frequency ground vibration from heavy construction equipment or earthquakes (even distant ones) that may cause the meter to sway during a measurement. Even this source of error can be overcome by averaging readings over relatively long periods, or by performing measurements when the vibrations are not active.

9.5   ELECTRICAL AND ELECTROMAGNETIC METHODS

9.5.1   Background

The category of electrical and electromagnetic methods actually encompasses a wide variety of techniques, some with very different operational principles, but all bound by the fact that they are fundamentally sensitive to electrical properties of the subsurface, which are in turn strongly influenced by the porosity of the soil or rock, the degree of saturation, and chemistry of saturating fluid. Direct current (DC) electrical and alternating current (AC) electromagnetic methods, for instance, estimate the electrical conductivity (or its inverse – resistivity) of the ground. Most minerals (with the exception of graphite, clays, and some ores) have very low electrical conductivity (or high resistivity). So, the bulk resistivity (ρ_{bulk}) of earth materials is dominated by moisture content, and the resistivity (thus chemistry) of that moisture or fluid (ρ_{fluid}). Therefore, the earth conductivities measured are primarily sensitive to the key hydrogeological parameter porosity Φ ($0 \le \Phi < 1$) as given by the empirical *Archie's Law*: $\rho_{bulk} = \rho_{fluid} \, A \, \Phi^{-m}$. A and m are constants related to the geometry and inter-connectedness of pores (with typical values in rock of 1 and 2 respectively; see Keller, 1982). Spontaneous potential (SP) methods involve measuring voltages due to naturally occurring earth electrical currents. One origin of SP is the so-called flow potential or streaming potential, which arises when water flows through a capillary – as it does in karst solution channels. Ground penetrating radar or GPR is a high frequency electromagnetic method that samples the subsurface with an electromagnetic pulse that is transmitted into the earth, and reflects back to the surface at interfaces between materials with differing dielectric constants. In this respect, GPR is quite similar to seismic reflection, but uses an electromagnetic rather than an acoustic signal.

9.5.2   DC electrical methods

DC electrical methods (often called resistivity methods) employ artificial or induced electrical fields. Results are typically presented as resistivity (ρ) in Ohm-meters (Ωm). A single resistivity measurement is usually obtained by driving a known electrical current (I in Ampere) between two electrodes. In a homogeneous half-space, this current flows along field lines as depicted in Fig. 9.8A. The depth to which current flows is dependent upon

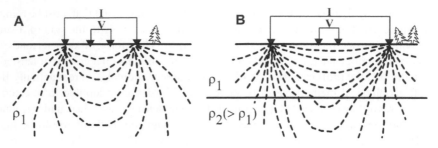

Figure 9.8.   DC Current Flow in a Homogeneous Half-Space (A), and a Layered Earth (B).

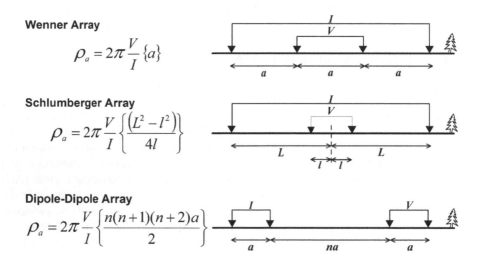

Figure 9.9.   Geometries and geometric factors (in brackets) for common DC resistivity arrays.

the distance between electrodes. Two other electrodes and a high impedance voltmeter are used to measure a voltage or potential difference (V in Volt) somewhere in the electric field.

Using Ohm's Law ($R = V/I$), the resistance (R) to current flow is simply the measured voltage divided by the applied current. This R is a property of the particular electric circuit, but can be converted to the bulk or apparent resistivity ρ_a (a physical property) of the subsurface material based on a geometric factor (k) that corrects R for the geometry of the electrodes. The generalised expression for any four-electrode array is: $\rho_a = 2\pi \frac{V}{I} k$. Although k can be readily derived for any electrode array, geophysicists commonly use one of several standard arrays of four co-linear electrodes for which the expressions for k greatly simplify (see Fig. 9.9).

Note that for a homogeneous half space, the calculated apparent resistivity (ρ_a) is the true resistivity (ρ) of the subsurface material (e.g. Fig. 9.8A). However, for a layered or laterally inhomogeneous earth (Fig. 9.8B), ρ_a represents an average of the varying true resistivities sampled by the electric field (Telford et al., 1990, Mooney, 1980).

Resistivity sounding (often called vertical electrical sounding, VES) is performed by increasing the total length of an electrode array (generally Schlumberger, see Fig. 9.9) centred at a fixed point. The effective depth of a resistivity measurement is dependent upon geology, but is commonly on the order of 1/3 the total electrode array length. Repeated measurements with increasing array length yields a curve of ρ_a versus array length (a sounding curve), which can be mathematically inverted to provide an estimate of true resistivity (ρ) versus depth (z). Sounding is suited to determining the depths to relatively flat, horizontal electrical boundaries such as clay aquitards, the top of rock, or the water table in a laterally uniform medium, and is therefore infrequently of value in karst hydrogeology.

Resistivity profiling involves moving a fixed array along a survey transect. This produces a profile of ρ_a versus distance (x), at a constant effective measurement depth. Resistivity profiling data is good for locating lateral inhomogeneities such as clay seams or fractures, and estimating their sense (i.e. resistors versus conductors), but is generally not amenable to mathematical inversion to obtain true subsurface resistivities.

Electrical Imaging (EI) combines sounding and profiling in one survey by using 30 to 90 (typically) electrodes connected by cables to a computer-controlled automated switching mechanism (Loke, 1999). This system takes readings by selecting four electrodes at a time, driving I, and measuring V, then moving on to the next four electrodes. The pattern and sequence of electrodes to be used can be pre-programmed and autonomously executed, and some systems can simultaneously read voltages at several pairs of potential electrodes while driving current between one pair of current electrodes. Combined sounding and profiling provides a two-dimensional (cross-sectional) data set of ρ_a versus distance (x) and depth (z) that can be inverted to produce a contourable data set of true resistivities (ρ). These contours are an electrical image (e.g. Fig. 9.3). EI is sometimes called electrical resistance tomography or ERT. Resistivity methods sample greater depths by increasing the electrode separation. This results in decreasing resolution with depth. Therefore, applied to karst hydrogeology, the method is restricted to near surface ($<10\,$m) karst structure mapping as well as the delineation of overburden thickness.

Although depicted in only one plane in Fig. 9.8, the electric field for a co-linear electrode array at the ground surface actually samples an oblate hemispherical area. Therefore, resistivity measurements are somewhat side looking, and a target off of the plane of a two-dimensional EI will falsely appear as an anomaly on the image. Therefore, as with gravity, multiple parallel EI's are often advisable in karst areas to ensure that the true location of targets is determined. The latest EI surveys are done as three-dimensional electrical images consisting not just of individual parallel two-dimensional images, but involving electrode arrays with varying spacings and azimuths, and even non-collinear arrays. These surveys yield three-dimensional blocks of true resistivity (ρ) versus x, y, and z that can be viewed either as fence diagrams, depth slices, or isosurfaces (e.g. Bechtel et al. 2005).

9.5.3   AC electromagnetic methods

Electromagnetic (EM) methods measure the same subsurface physical property as resistivity methods – apparent resistivity or ρ_a. However, instead of directly driving DC electrical currents, the measurements are accomplished using the phenomenon of electromagnetic induction: a primary alternating (AC) EM field induces a secondary EM field (of the same frequency) in the subsurface without galvanic contact. Both fields consist of coupled electric and magnetic components. The secondary field has an amplitude difference, as well as a peak-to-peak phase shift relative to the primary field that depends on the electrical properties of the subsurface. Geophysicists can measure certain components of the primary and secondary EM fields, and deduce the apparent resistivity of the ground (McNeill, 1980).

Various types of equipment exist – some with ground contact, some without. The choice depends on which EM field components are measured: electric and magnetic field components together, or magnetic field components exclusively. As depicted in Fig. 9.10, some systems need both a transmitter and receiver on site (active or near-field EM methods), while others need only a receiver on site (passive or far-field EM methods).

The penetration depth of an EM field depends on the frequency (f in Hertz or Hz) of the signal, and the resistivity (ρ in Ωm) of the ground, and can be estimated by the so called skin depth or d_S as: $d_S \approx 503\sqrt{\rho/f}$ (in meters). Skin depth d_S is the depth at which the amplitude of the EM field is attenuated to a value of $1/e$ (or roughly 40 percent) of its value at the surface. Note that since penetration depth increases with decreasing frequency, using different frequencies can provide information from different depths.

For near-field methods, the effective investigation depth is controlled by d_S (transmitter frequency/earth resistivity), and the transmitter-receiver separation and orientation. For far-field methods, the effective investigation depth is dictated by d_S.

AC EM methods (this section) are better suited for locating conductive targets in resistive surroundings, while DC resistivity methods (section 9.5.2) are best for detecting resistive targets in conductive surroundings. EM methods hold a slight advantage in lateral resolution in that the plan-view zone of influence for a given measurement with a given effective depth is smaller for EM than for a DC resistivity measurement with the same effective depth. On the other hand, EM instruments are susceptible to interference from the myriad electromagnetic fields and signals that populate the developed world. In addition, the proper orientation of targets (i.e. coupling) relative to the orientation of the transmitter/primary field is critically important.

9.5.3.1   *Remote transmitter (far-field) systems*

Remote transmitter systems use EM fields broadcast from radio stations all around the world. Since only a receiver is needed in the field, they are called passive methods.

Very Low Frequency-Electromagnetics (VLF-EM) was originally employed for detection of water-bearing fractures using transmissions in the 15–30 kHz range (formerly used for worldwide submarine communications, see Telford et al. 1990, McNeill & Labson 1991). Figure 9.10 illustrates the EM-field induction from a remote radio transmitter over a conductive plate embedded in resistive host rock – a typical karst structure. VLF-EM measures the amplitude ratio of the vertical and horizontal local magnetic field components (H_z/H_y) using orthogonal induction coils. In addition to the amplitude difference between the primary and secondary magnetic fields, there is often a phase shift (i.e. their peaks do not coincide in time). The inphase component is defined as the amplitude of the secondary field at the point where the primary field amplitude peaks. The quadrature component is the amplitude of the secondary field at a phase shift of 90° from the primary field peak. Anomalies in the inphase and/or quadrature ratios are produced by subsurface resistivity contrasts.

The ***Radiofrequency-Electromagnetic (RF-EM)*** method is an enhancement of VLF-EM. With regard to near surface hydrogeophysical investigations, some prototypes now use frequencies up to 1 MHz in the Long Wave radio band (see e.g. Bosch & Müller, 2001). In karst studies, conducting parallel profiles enables delineation of elongated, vertical, electrically conductive structures such as faults or fractures – without actual ground contact. This provides rapid reconnaissance of large areas. Figure 9.11 shows example contour and surface maps of RF-EM quadrature data.

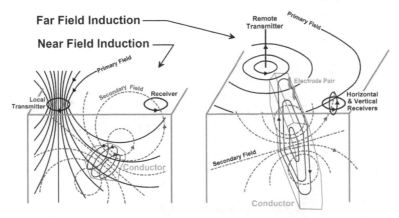

Figure 9.10. Schematic Depiction of EM Induction Methods. Current in the transmitter creates a primary field, which drives current loops in buried conductors, which create a secondary field.

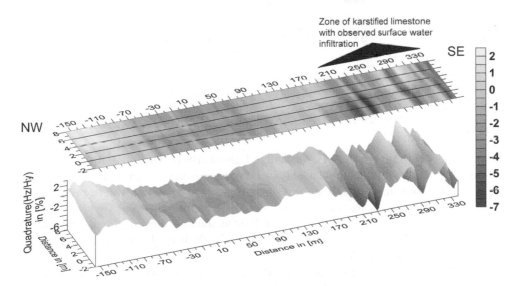

Figure 9.11. RF-EM data showing anomalies related to active karst lineaments, Canton Jura, Switzerland. The rapid car-borne system had a transmitter frequency of 234 kHz, and a 0.25 readings/second sampling rate. Profile traces are indicated by black lines. Linear anomalies between 210 and 310 m indicate fractured limestone, where infiltration of rainwater was in fact observed.

The ***Radiomagnetotelluric (RMT)*** method is an enhancement of the Very Low Frequency-Resistivity (VLF-R) method with an extended frequency range (see e.g. Newman et al., 2003). It measures the horizontal magnetic field component H_y using an induction coil, and the secondary horizontal electric field component E_x using two electrodes placed into the ground (see electrode pair in Fig. 9.10). From the ratio E_x/H_y, one can calculate the apparent resistivity (ρ_a) of the earth, as well as the phase difference (Ψ_a), between H_y and E_x (Cagniard, 1953). With this method, resistivity soundings can be achieved by using different signal frequencies and a constant electrode separation of only

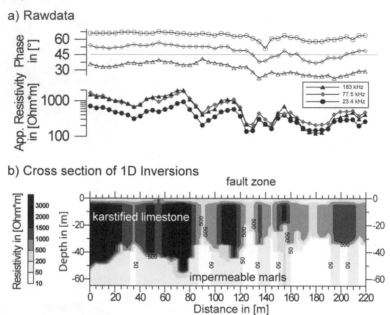

Figure 9.12.   RMT Image of Karstic Limestone in the Swiss Jura. Panel (a) shows sharp zones of low electrical resistivity – indicating possible high permeability fractures or faults. The cross section in panel (b) was obtained from one-dimensional inversions of the data, and shows the thickness of the limestone over the underlying marls as well as an apparent separation of the limestone into several blocks.

a few meters (Fischer et al., 1983, McNeill & Labson, 1991). Therefore, RMT soundings are more suitable for imaging a laterally inhomogeneous karstic subsurface than are vertical electric soundings. RMT data at a single frequency, ρ_a and Ψ_a can be presented as profiles or maps that are particularly sensitive to high permeability zones in karst terranes. Furthermore, the sounding curves of ρ_a and Ψ_a from multiple differing frequencies can be inverted to models of true resistivity versus depth with the help of magnetotelluric forward and inversion codes (e.g. Fig. 9.12).

RMT provides effective imaging of karst features and overburden thickness, and is typically suitable for targets at depths up to several tens of meters.

9.5.3.2   *Local transmitter (near-field) systems*

Near-field EM methods use an on-site transmitter as well as receiver. The measurements are accomplished by sending an AC current through a transmitter coil or wire on or near the earth. The transmitter magnetic field induces eddy currents in the earth, along with an accompanying secondary magnetic field. The receiver coil or wire measures the total field consisting of the known primary (transmitter) and secondary (earth) field components. Dual loop systems with coplanar coils (as depicted in Fig. 9.10) are most commonly employed.

As with far-field systems, the measurements from a dual loop EM system are the inphase and quadrature components (see section 9.5.3.1) of the ratio of the secondary to the primary magnetic field or H_s/H_p. These ratios can be plotted as profiles or maps to delineate electrically conductive anomalies. The inphase component is particularly sensitive to the presence of metal, and is therefore used extensively in environmental geophysics to detect and delineate landfill pits or trenches. In karst geophysical surveys, the inphase component is usually used to discriminate between anomalies that may simply be caused by buried utilities, structures or refuse, and anomalies due to actual karst features.

As seen in section 9.5.3.1 the skin depth relationship gives the ultimate upper limit to the effective depth of a far-field EM measurement. But, under near-field conditions, the effective depth for a given measurement is generally smaller than the skin depth and is controlled by the coil orientation and the inter-coil separation. For two-coil EM systems, the sensitivity versus depth for coils oriented horizontally (called a vertical dipole orientation due to the orientation of the primary magnetic field) is near zero at the ground surface, peaks at approximately 0.4 times the inter-coil spacing, and reaches its effective limit at approximately 1.5 times the inter-coil spacing. For vertical coils (horizontal dipoles), the peak sensitivity is at the ground surface, and falls-off monotonically, with an effective depth equal to approximately the inter-coil spacing. Consequently, data from different depths is obtained by varying the frequency/skin depth (called parametric sounding), or the inter-coil spacing or coil orientation (called geometric sounding). Therefore, both lateral profiling/mapping and vertical sounding of the conductivity distribution within the subsurface is possible. Note however that frequencies and coil separations must be adapted to each other for a successful measurement. Generally, profiles of true resistivity or conductivity versus depth are obtained by inverse modelling or comparison with master curves (Keller, 1971, Patra & Mallik, 1980, Wilt & Stark, 1982).

Under the condition of "low induction number", where the product of earth conductivity and magnetic permeability, and applied signal frequency are very low, direct measurements of the apparent subsurface conductivity are possible (McNeill, 1980, Knödel et al., 1997). The low induction number condition is generally satisfied if the inter-coil spacing is much bigger than the skin depth. In this case, the inphase component (real part) is negligible, and the out-of-phase (imaginary part) component of the measured field ratio (H_s/H_p) is linearly proportional to the subsurface electric conductivity. Systems with fixed frequencies and inter-coil spacings that satisfy this condition are widely available, and are commonly used for near surface conductivity mapping. These near-field EM systems provide rapid electrical conductivity mapping and sounding without ground contact in various depth ranges and resolutions. They are suitable to detecting conductive features in resistive host soil or rock such as water- or clay-filled solution cavities, or wet soil zones. Their application is of particular advantage in areas where direct or galvanic ground contact is difficult to achieve. Fig. 9.13 presents an example reconnaissance (one day) EM terrain conductivity survey.

9.5.3.3 *Time Domain Electromagnetics (TDEM)*

In contrast to the frequency domain electromagnetic (FDEM) methods described above, which are often used for resistivity profiling, TDEM methods are primarily used for resistivity sounding. TDEM provides a wide range of effective sounding depths from approximately 6 to 900 meters. As with DC electrical imaging, combining multiple soundings along a profile yields a two-dimensional resistivity cross section, while the combination of several such cross sections can provide three-dimensional imaging of subsurface resistivity variations.

TDEM measurements are carried out by sending an electrical current through a transmitter coil. When the transmitter current is shut off, induction creates a decaying primary magnetic field, which in turn induces secondary electric currents (that are essentially a shallow "image" of the former transmitter current) with their accompanying magnetic fields.

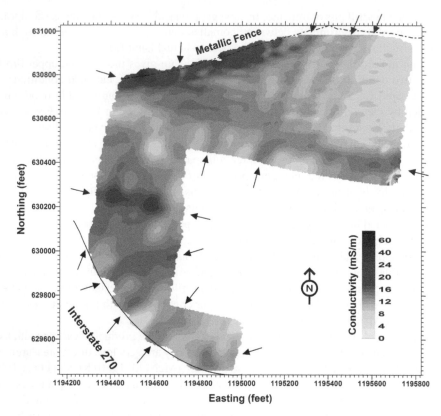

Figure 9.13.   Dual Loop EM terrain conductivity map, Western Maryland, USA. Numerous hydraulically active lineaments (fractures or grikes) are marked with arrows.

The decay of the induced field over time (which depends on the electric properties of the ground) is monitored with a separate receiver coil. Therefore, values from progressively later times provide information from correspondingly greater depths. An advantage of this procedure compared to FDEM methods is that the secondary field is measured in the absence of the primary field, which leads to an increased accuracy of the measurement (Nabighian & Macnae, 1991, Yungul, 1996, Knödel et al., 1997).

The receiver coil voltage decay time curves can be converted to curves of ρ_a versus time, which provides an approximation of ρ_a versus depth (a sounding curve). Portraying several soundings on one profile leads to a two-dimensional pseudo-section of apparent resistivity or conductivity versus time (equivalent to depth) and distance along the profile (similar to Fig. 9.12). Inverse modeling can provide models of true resistivity or conductivity versus depth and distance by fitting the measured decay curves.

In karst hydrogeology, TDEM is useful for determining the depth and thickness of overburden, depth of the water table, depth and degree of saltwater (or ionically-contaminated water) intrusion, and detection of fracture zones (e.g. Farrell et al., 2003). TDEM provides good lateral as well as vertical resolution, and a wide range of effective survey depths. The method is sometimes limited by the requirement for impractically large (e.g. $300 \times 300\,\text{m}$ or larger) transmitter coils to achieve deep measurements. In addition, the method is susceptible to interference from metal structures and power lines.

9.5.4   Spontaneous potential

The self potential (SP) method involves measuring the electrical potential field caused by naturally occurring DC electrical currents in the earth. The SP method is also called spontaneous potential or natural potential (NP). Natural electrical currents occur nearly everywhere in the earth, and may be due to myriad phenomena, including:

- Oxidation-reduction reactions around metals or metal ores spanning the water table
- Movement of ionic fluids across stratigraphic or lithologic contacts
- Underground temperature gradients (e.g. near mine fires or geothermal zones)
- Underground chemical gradients (e.g. near sulphide bodies)
- Movement of subsurface water (i.e. electrokinetic effects)

The predominant SP phenomenon in karst terranes is the last one, created by streaming potentials. These arise wherever a pressure gradient causes fluid to flow through the capillaries of a permeable medium (Revil & Pezard, 1999, Revil & Schwaeger, 1999, Ishido & Pritchett, 1999). This phenomenon is called electrofiltration, and evokes a charge separation in the capillary. As a result, one can observe a positive increase in the electrical potential in the direction of fluid flow, with the magnitude of the potential linearly related to the fluid flow velocity. Hence, SP surveys measure the actual movement of water, horizontally and vertically (in the case of infiltration), and can be used to locate preferred pathways for water flow into or through an aquifer (Wanfang et al., 1998, Fagerlund & Heinson, 2003, Aubert & Atangana, 1996). Another application of SP is cave detection (Vichabian & Morgan, 2002, Quarto & Shiavone, 1996, Nyquist & Corry, 2002, Gurk & Bosch, 2002).

SP equipment consists of two non-polarizing electrodes and a high impedance voltmeter. The measurements must be performed using non-polarizing, porous-pot ceramic electrodes filled with solutions of $Cu/CuSO_4$, $Ag/AgCl$ or similar, to avoid erroneously measuring highly localised corrosion potentials created by insertion of ordinary metal electrodes into damp earth (Perrier et al., 1997).

SP is the measured voltage drop between two electrodes. Often, one electrode is kept fixed in an electrically non-disturbed place as a reference electrode, while the other mobile electrode samples the SP along a profile or in a grid. By convention, the reference electrode is connected to the ground plug of the voltmeter. Due to geomagnetically induced currents (McKay, 2003) and/or a temperature dependency of the electrodes (Ernston & Scherer, 1986), the measured potential tends to change or drift slowly over time. Therefore, a selected station on the profile or grid must be designated as a base station, and re-occupied frequently. The change in these base readings can then be pro-rated across all other measurements to remove drift effects.

The drift-corrected SP can then be interpreted to characterise groundwater flow. Based on theoretical, laboratory and field studies of electrofiltration, recharge areas tend to become negatively charged, while discharge areas become positively charged. For lateral flow, this means that SP generally increases in the direction of flow. Zones of concentrated downward infiltration (e.g. dolines or shafts) often produce negative SP anomalies (Fig. 9.14).

In contrast to the negative SP typically produced by infiltration, artesian springs demonstrate positive anomalies. Note that while the magnitude of the SP anomaly is related to the flow velocity, the electrokinetic effect is ultimately dependent upon water flow through capillaries. Therefore, high flows through open solution channels may not produce as strong an anomaly as more moderate flow through a rubble- or soil-filled cavity.

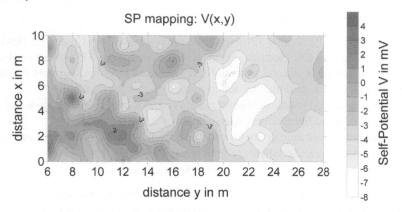

Figure 9.14. SP Mapping from a Karstic Zone in the Swiss Jura Mountains. Infiltration areas (dolines) appear as negative SP anomalies with light shading.

Generally, SP surveys in karst areas only provide a qualitative interpretation of SP data along profiles and plan-view maps – without any estimation of the depth or dimensions of a subsurface feature. SP surveys in karst areas arc usually not subject to interference from other natural sources of current (e.g. hydrothermal, or redox-related), but care should be exercised in developed areas where stray currents from non-karst sources (such as electric railways, grounding grids, cathodic protection systems, and corrosion of metal structures) may be present. Occasionally, there is a strong topographic influence on SP data (often related to the topographic influence on subsurface water movement). Sometimes a topographic correction is applied to decouple the SP and topographic data (by subtracting from the SP data a linear, or low order polynomial, regression function between SP and topography). However, caution must be exercised when applying such a "correction" because the topography-related streaming potential that is removed may include the information that was sought by the geophysical survey!

9.5.5 Ground penetrating radar (GPR)

The basic procedure for most GPR methods is to transmit an impulsive electromagnetic wave into the earth, and record the reflections or backscatter produced at interfaces between materials with differing electrical properties (Olhoeft, 2000; Daniels, 1996). In this respect, GPR is similar to seismic reflection, but uses an electromagnetic rather than an acoustic signal, and is sensitive to subsurface electrical properties rather than elastic properties. Because of the very high frequency of GPR signals (50 to 500 MHz for most karst applications) relative to seismic signals, GPR has much higher resolution, although often at the expense of lower depth of investigation.

The propagation of electromagnetic waves is a relatively complicated phenomenon involving coupled electrical and magnetic vector fields (as described by Maxwell's equations) whose behaviour depends upon the electrical conductivity and dielectric constant of the medium of travel. But, most importantly for the geophysical uses of GPR methods, the transmitted electromagnetic pulse may be reflected by interfaces between materials with differing dielectric constant, and scattered by small targets or asperities on electrical interfaces. The reflections and/or backscatter can be recorded, and the travel time and amplitude

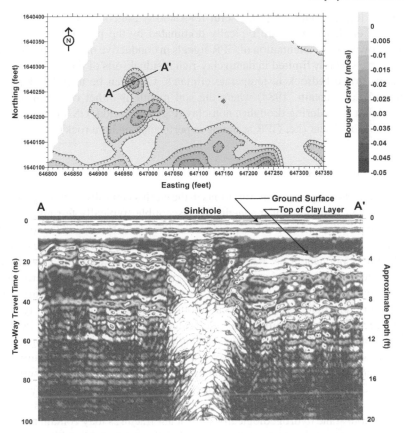

Figure 9.15.   GPR Profile over a Low Gravity Anomaly in Florida, USA. A broken clay layer and underlying anomalously strong reflections (typical of voids) suggest subsurface soil piping into bedrock solution cavities.

(as well as phase shift, and polarization) of the reflected pulses can be measured and used to construct images of the subsurface.

As with seismic reflection, discussion of the processing of GPR reflection data is beyond the scope of this chapter. However, the simplest way to understand the fundamentals of the GPR method is to imagine it as analogous to a sporting "fish finder" that uses an electromagnetic "ping" rather than an acoustic or sonar ping. GPR surveys are performed by towing a transmitter antenna and a receiver antenna along a survey transect. The transmitter rapidly pings the ground, sending out repeated electromagnetic pulses (up to a hundred or more per second), while the receiver records reflections or backscatter. In practice, the transmitter and receiver may be the same antenna toggling rapidly from one function to the other (called a monostatic array), or separate antennae for each function (bistatic). A raw GPR recording plots the reflection/backscatter amplitude as a function of time on an image that can be read as roughly a cross-section of the dielectric properties of the subsurface (Fig. 9.15).

In general, GPR is an excellent, high-resolution method for detecting and imaging any subsurface features whose electrical properties differ from their surroundings. Initially, this suggests that it would be quite useful for karst hydrogeophysics since the electrical properties of earth materials are typically dominated by the presence and chemistry of moisture. However, the attenuation of GPR signals in conductive media leads to penetration depths that are severely limited in damp clay-rich residual soils of the type that commonly mantle karst-prone bedrock in temperate climates. GPR can be useful in karst terranes (Benson and La Fountain, 1984) where the soil mantle is either clay-rich but dry (e.g. temperate karst belts during drought), non-clay (e.g. the Florida, USA coastal plain karst, see Fig. 9.15), or absent (i.e. GPR scanning is performed on bare rock).

9.5.6   Magnetic methods

Magnetic (MAG) surveys measure variations in the earth's naturally occurring DC magnetic field. Where concentrations of magnetically susceptible (typically ferrous) materials are present, the earth's magnetic field magnetizes the ferrous material, and the local total field is anomalous – representing the sum of the induced field in the target and the earth's ambient field. Where permanent or "hard" magnetized materials (i.e. materials with their own magnetic field such as magnetic minerals or magnetized iron or steel or thermoremnantly magnetized or heated materials) are present, the magnetic field of the object also sums with the earth's ambient field, creating a local anomaly. Magnetic surveys are typically performed by recording the magnitude and/or orientation of the local magnetic field vector at gridded stations or along profiles, and examining the data for anomalies. Magnetic surveys are not typically used in karst hydrogeophysics since most hydrogeologically significant features in karst terranes are neither magnetically susceptible, nor hard magnetized. In rare cases, subsurface open voids may create subtle local magnetic anomalies since the magnetic susceptibility of air is zero, while the susceptibility of earth materials is small but measurable. In addition, some hydrogeologically significant fractures may contain mineralization that affects their magnetic susceptibility and makes them detectable with a magnetometer.

9.6   BOREHOLE GEOPHYSICAL LOGGING

Most geophysical methods that are commonly used for surface geophysics or tomography have been adapted in some fashion for borehole geophysical logging. This means recording some type of signal in a borehole using a receiver (or transmitter plus receiver) sonde that is lowered down the bore (Keys, 1990, Paillet, 1994). The energy can be either natural/ambient, or introduced by the sonde. As with surface geophysics, the measured physical parameters can often be related to hydrogeologic conditions.

Table 9.3 lists the commonly available geophysical logging methods, their principles, and their applicability to various tasks. In karst hydrogeology, the typical purpose of geophysical logging is to identify potential water-bearing zones, test for flow, and determine the sense of flow (e.g. in or out of the borehole). While there are many types of geophysical log that can identify potential water-bearing features (e.g. electric, caliper, sonic, televiewer, etc.), there are far fewer that have the capability of detecting actual flow (e.g. fluid temperature, fluid conductivity), and only one type that can determine the sense of flow – impeller, heat pulse, or electromagnetic flow metering. Note that while electromagnetic and heat pulse flow measurements are made at discrete depths, impeller flowmeters can record continuous logs. Chapter 5 of this volume contains further information on borehole flow logging.

Table 9.3. Borehole geophysical logging methods.

Log type	Acceptable construction					Medium in borehole			Resulting data	Uses in karst hydrogeology
	Open hole	Slotted PVC	Solid PVC	Slotted steel	Solid steel	Air	Clear water	Turbid water		
Natural Gamma (nγ)	X	X	X	X	X	X	X	X	Depth vs. natural gamma particle count	Discrimination of carbonate from clay or shale
Short Normal Resistivity	X	X					X	X	Depth vs. apparent resistivity near bore	Detection of electrically conductive clay or shale or fracture zones
Long Normal Resistivity	X	X					X	X	Depth vs. apparent resistivity farther into formation	Detection of electrically conductive clay or shale or fracture zones
Single Point Resistance (SPR)	X	X					X	X	Depth vs. local electrical resistance	Detection of electrically conductive clay or shale or fracture zones
Spontaneous Potential (SP)	X	X					X	X	Depth vs. natural potential	Detection of boundaries of clay or shale, or water chemistry interfaces
Magnetic Susceptibility	X	X	X			X	X	X	Depth vs. magnetic susceptibility	Uncommon
Magnetic Field Vector	X	X	X			X	X	X	Depth vs. magnetic vector magnitude and orientation	Uncommon
Caliper	X					X	X	X	Depth vs. bore diameter	Detection of borehole enlargements at fracture zones or solution cavities
Fluid Temperature	X	X		X	X		X	X	Depth vs. temperature of fluid	Discrimination of water inflow or outflow zones
Fluid Conductivity	X	X		X			X	X	Depth vs. specific conductance of fluid	Discrimination of water inflow or outflow zones
Electromagnetic Induction (Formation Conductivity)	X	X	X			X	X	X	Depth vs. terrain conductivity	Detection of electrically conductive clay or shale or fracture zones
Verticality	X	X	X			X	X	X	Depth vs. bore tilt and azimuth	Uncommon
Full Waveform Sonic	X	X					X	X	Depth vs. seismic V_p and V_s, and seismic wiggle trace (depicts rock quality)	Detection of low velocity/high porosity zones (e.g. fractures)

(Continued)

Table 9.3.   (Continued)

Log type	Acceptable construction					Medium in borehole			Resulting data	Uses in karst hydrogeology
	Open hole	Slotted PVC	Solid PVC	Slotted steel	Solid steel	Air	Clear water	Turbid water		
Digital Video	x	x	x	x	x	x	x	x	Visual record keyed roughly to depth	Visual inspection of borehole walls
Borehole Acoustic Televiewer (BHTV)	x						x	x	Accurately oriented and scaled 360° visual scan	Digitization of borehole walls to locate and measure fractures and solution cavities
Borehole Optical Televiewer (OPTV)	x					x	x		Accurately oriented and scaled 360° acoustic scan	Digitization of borehole walls (or casing) to locate and measure fractures and solution cavities
Heat Pulse Flow Meter	x	x		x			x	x	Measurement of up or down flow rate at discrete depths	Detection and quantification of inflow and outflow zones
Impeller Flow Meter	x	x		x			x	x	Measurement of up or down flow rate at discrete depths	Detection and quantification of inflow and outflow zones
Eletromagnetic Flow Meter	x	x					x	x	Measurement of up or down flow rate at discrete depths	Detection and quantification of inflow and outflow zones
Induced Polarization (IP)	x	x	x			x	x	x	Depth vs. chargeability	Uncommon
Borehole Gravity	x	x	x	x	x	x	x	x	Depth vs. density	Detection of solution zones
Gamma Density (γ-γ)	x	x	x	x	x	x	x	x	Depth vs. density	Detection of fracture or solution zones
Neutron Density	x	x	x	x	x	x	x	x	Depth vs. hydrogen content (porosity)	Quantification of porosity
Water Quality Parameters (pH, O$_2$, etc.)	x	x		x			x	x	Depth vs. parameter	Discrimination of water inflow or outflow zones
Cavity Sonar	x						x	x	Scaled 3-D depiction of cavity dimensions	Quantification of cavity dimensions
Cavity Ultrasound	x					x			Scaled 3-D depiction of cavity dimensions	Quantification of cavity dimensions

Beyond flow data in individual boreholes, hydrogeologists often wish to use borehole geophysical data to correlate features between spaced boreholes. For this task, the advantage of geophysical borehole logging over visual or geologic logging is that geophysical logging provides information on undisturbed, oriented, and in-situ conditions (as opposed to potentially highly disturbed and non-oriented core samples), and provides data even where there was no core recovery. However, a disadvantage is that unlike surface or tomographic geophysics, geophysical logging is typically sensitive only to material in the immediate vicinity of the bore. Therefore, geophysical logging suffers the same limitation as visual logging in that conditions at a borehole may not accurately be represent overall aquifer characteristics – particularly in karst terranes with their extremely heterogeneous porosity. The best use of borehole geophysical logging then is to combine borehole geophysical logs (with their highly accurate but localised information) with surface or tomographic geophysics (with their ambiguous but widely representative information) to constrain possible subsurface hydrogeologic models more tightly than could either type of geophysics alone. Figure 9.16 presents example logs from a karst zone borehole – including acoustic and optical televiewer logs (and inferred feature orientations), geophysical logs, and flow meter logs.

One family of borehole geophysical techniques warrants special mention. While most geophysical logs record a one-dimensional profile of some physical parameter versus depth, the television (video) and televiewer (optical or acoustic) logs allow a geologist to virtually visit and document the inside of a boring. A television log is simply a video recording of the boring walls. These can be forward-looking or side looking, and typically have a depth counter inset in the corner of the frame. The logs can be oriented by placing a compass within the field of view, but quantitative analysis of the video logs is not generally attempted.

Televiewer logs are produced by a transducer that incrementally scans and digitizes the boring walls as the sonde is lowered. This produces a distortion-free, accurately scaled, and geographically oriented depiction of the interior of the boring from which feature dimensions and orientations can be accurately measured. On a log, the view of the interior of the cylindrical borehole is "unwrapped" to provide a two-dimensional image – causing dipping planar features intersecting the borehole to appear as sinusoidal on the televiewer image. An optical televiewer or OPTV (see Fig. 9.16 and Fig. 9.17) creates a visual image (i.e. using reflected visible light), and can only be run in a dry hole or a clear water filled hole.

An acoustic televiewer or ATV performs exactly the same function, but illuminates the borehole with a high frequency sonar pulse. This provides images of both the reflected signal amplitude, and the transit or travel time (see Fig. 9.16). The travel time of the reflected acoustic energy is related to borehole diameter, while the amplitude is related primarily to the roughness of the borehole wall. An acoustic televiewer can only be run in a fluid-filled hole, but the fluid can be entirely turbid and detailed accurate images will still be obtained.

In karst hydrogeological investigations, geophysical logging can be used to locate fractures or other permeable features, and determine their orientation and thickness. Fluid temperature and conductivity logs can determine where water may be entering or leaving a borehole and flow meter logs can quantify vertical water movement in boreholes. In extreme cases, simple video logging can identify water-bearing karst features by visible flow, or movement of sediment or bubbles.

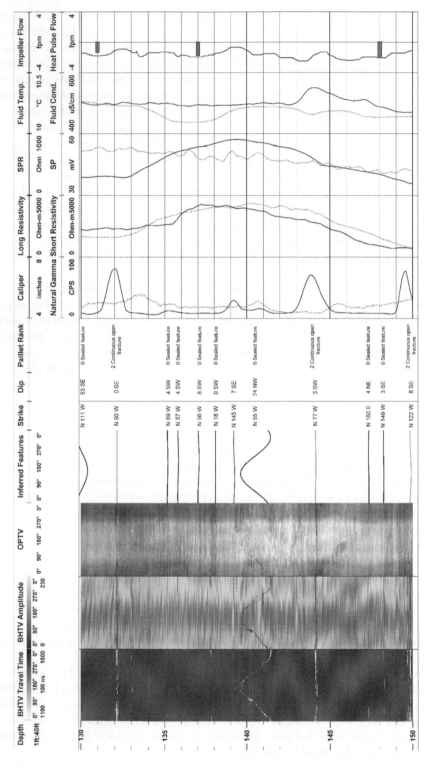

Figure 9.16.   Televiewer, geophysical and flow meter logs from a karst zone borehole.

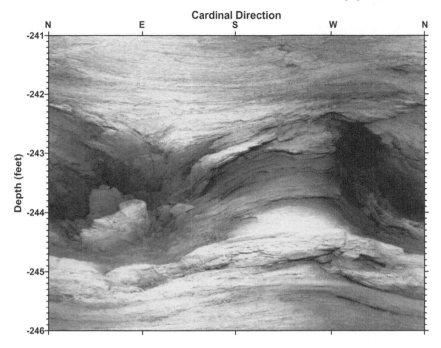

Figure 9.17.   Optical Televiewer Image of a Limestone Solution Cavity Penetrated by a Borehole.

9.7   SUMMARY

Particularly in karst terranes, where the flow and occurrence of water can be laterally and vertically inhomogeneous and anisotropic, geophysical measurements provide critical subsurface measurements that relate to hydrologic parameters, and can suggest or constrain hydrogeologic models. In fact, in karst terranes, geophysics should not be considered an expensive "extra" to be done only if the budget allows. Instead, it should be considered a vital necessity that will reduce the costs of drilling and other direct testing, and therefore the cost of the entire investigation.

Applications of geophysics in karst investigations commonly involve detection and characterization of water-bearing fractures or solution channels (often for well siting), and determination of overburden thickness (e.g. to assess the vulnerability of karst aquifers).

However, for proper application, a method must be selected that is sensitive to the expected physical property contrast between the features to be detected (or imaged) and the surrounding formation. At the same time, the method must be insensitive to local sources of interference. When interpreting the results of even a properly conducted geophysical survey, it is important to remember the unavoidable non-uniqueness of any geophysically based geologic model. Inclusion of all available geologic mapping and borehole data in the model will provide maximum constraint and accuracy. A phased approach is often warranted, in which a geophysical reconnaissance method is first used to optimise drill hole locations. Following drilling, detailed geophysical imaging may be performed to provide accurate interpolation of models between drill holes.

CHAPTER 10

Modelling karst hydrodynamics

Attila Kovács & Martin Sauter

10.1 INTRODUCTION

Mathematical models are exact tools for the quantitative representation of the hydraulic behaviour of aquifer systems. The reconstruction of a groundwater flow field, which is consistent with a given hydraulic conductivity field and given boundary conditions, nearly always requires the use of numerical models (Király 2002). Analytical models of groundwater flow have been developed since the late 1800s. Numerical groundwater flow models have been applied since the 1960s, and their utilisation for granular aquifers has become every-day practice since then. However, the application of numerical methods in karst hydrogeology demands a specially adapted modelling methodology (Palmer et al. 1999). This is because of particular complexities associated with the large heterogeneity of a flow field (Király 1994).

The first step in any modelling study is the schematic representation of a real system. A conceptual model consists of the applied differential equations, aquifer geometry, and of a set of flow parameters, boundary conditions and initial conditions. The hydraulic parameter fields applied in groundwater flow models are usually obtained with an interpolation between discrete observations. Because of the large heterogeneity and high contrast in hydraulic conductivity, interpolation techniques cannot be employed for the characterisation of karst aquifers. The most demanding task in karst modelling is the definition of continuous hydraulic parameter fields.

The second problem is the selection and the development of an appropriate computer code based on numerical methods, which allows the equations defined in the abstract scheme to be solved. Large heterogeneities require the introduction of special features into the numerical models, such as the combination of discrete 1-D elements with a three-dimensional continuum or the double continuum representation of the flow medium.

The third problem is related to the transfer of the simulated results to a real system. Simplifying assumptions made in the conceptual and numerical models must appear as uncertainties in the simulation results. Because of the high degree of heterogeneity, these uncertainties are much larger in karst systems than in most porous unconsolidated aquifers.

This chapter aims to provide a brief overview on karst modelling techniques and related problems. It is intended to assist the modelling hydrogeologist in the selection of appropriate tools as well as the definition of suitable parameterisation techniques.

10.2   CONCEPTUAL MODELS OF KARST SYSTEMS

Every sound conceptual model of karst systems incorporates heterogeneity and accordingly the duality of hydraulic flow processes. These include the duality of infiltration, storage, groundwater flow, and discharge processes. Continuum flow (often termed diffuse flow) processes are active in low permeability matrix blocks or slightly fissured limestone beds, while concentrated flow processes can be observed in a discrete conduit network. Moreover, most conceptual models distinguish three main zones in the vertical direction. These are: soil zone and epikarst, unsaturated or vadose zone, and phreatic zone. Although most conceptual models include similar structural features, the flow and storage processes assigned to them display large variations.

According to the conceptual model of Mangin (1975), the main conduit system transmits infiltration waters towards a karst spring, but is poorly connected to large voids in the adjacent rocks, referred to as the 'annex-to-drain system'. Mangin (1975) associates storage to the 'annex-to-drain system' and also introduced the concept of epikarst, i.e. a shallow, high-permeability karstified zone just below the soil zone (see Chapter 2 of this book). It is believed to act as a temporary storage and distribution system for infiltrating water, similar to a perched aquifer. It is assumed to channel infiltrating water toward enlarged vertical shafts, thereby enhancing concentrated infiltration.

Drogue (1974, 1980) assumes, that the geometric configuration of karst conduit networks follow original rock fracture pattern. Joints constitute a double-fissured porosity system. This network consists of fissured blocks with a size in the order of several hundred metres, separated by high-permeability low storage conduits. Every block is dissected by small scale fissures or fractures with considerably larger storage, with low bulk permeability.

The conceptual model proposed by Király (1975, 2002) and Király et al. (1995) combines the conceptual models of Mangin (1975) and Drogue (1980). This model involves the epikarst and a hierarchical organization of the conduit networks. It also comprises the hydraulic effect of nested groups of different discontinuities. According to Király & Morel (1976a), two classes of hydraulic parameters can adequately reflect the hydraulic behaviour of karst systems. Carbonate aquifers can be considered as interactive units of a high-conductivity hierarchically organised karst channel network with a low-permeability fissured rock matrix. While Mangin (1975) associates storage to the "annex-to-drain system", Drogue (1980) and Király (1975) associate it to the low-permeability matrix. The strength of Király's (1975) conceptual model is based on the fact that it has been tested quantitatively and verified by numerical models (Király & Morel 1976 a, b, Király et al. 1995). A schematic representation of this concept is provided by Doerfliger & Zwahlen (1995) in Fig. 10.1.

10.3   MODELLING APPROACHES

Two, fundamentally different modelling approaches exist for studying and characterizing karst hydrogeological systems.

Global models (lumped parameter models) imply the mathematical analysis of spring discharge time series (hydrographs) that are believed to reflect the overall (global) hydrogeological response of karst aquifers. According to this approach, karst systems can be considered as transducers that transform input signals (recharge) into output signals

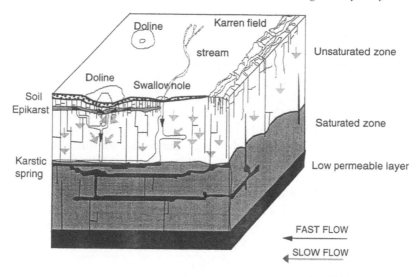

Figure 10.1.   Conceptual model of karst aquifers (Doerfliger & Zwahlen 1995).

(discharge). With the acquisition of spring discharge data being relatively simple, these models have already been employed since the beginning of the last century. Traditional global modelling techniques do not take into account the spatial variations within the aquifer, and cannot provide direct information on the aquifer geometry, hydraulic parameter fields nor physical parameters. However, some of these techniques combined with other exploration methods may be used for the estimation of hydraulic parameters and conduit spacing, information required for distributive flow models. See also Chapter 5, Hydraulics.

For the quantitative spatial simulation of groundwater flow fields, distributive models are needed. The application of distributed parameter methods requires the subdivision of a model domain into homogeneous sub-units, for which groundwater flow can be described by flow equations derived from basic physical laws. Distributive models can consider both spatial and temporal variations of hydraulic parameters and boundary conditions, and thus require detailed information on aquifer geometry, hydraulic parameter fields, and recharge conditions. These parameters can be derived using investigation techniques described in previous chapters, or indirect information obtained with global models.

10.4   GLOBAL MODELS

10.4.1   Introduction

The measurement of discharge with time at the outlet of a karst aquifer provides integral information on the hydraulic behaviour of the entire system. The following two types of spring hydrograph analytical methods can be distinguished (Jeannin & Sauter 1998).

Single Event Models deal with the global hydraulic response of the aquifer to a single storm event. It is widely accepted, that the hydraulic effect of three fundamental factors are reflected in the global response of a karst aquifer: Recharge, storage and transmission (flow). Most of the existing techniques for the analysis of spring hydrographs allow the identification of integral parameters of karst properties in a qualitative but not quantitative sense.

However, most of these methods are based on simple, or sometimes more complex cascades of reservoirs, and involve physical phenomena. Furthermore, some of these methods provide semi-quantitative relationship between the pattern of the global hydraulic response and hydraulic parameters as well as some geometric properties of an aquifer. Therefore, it is more appropriate to speak of "grey box models" rather than "black-box models".

Time Series Analyses relate the global hydraulic response of karst systems to a succession of recharge events. Univariate time series analytical methods can identify cyclic variations. Bivariate time series analyses are designed for the analysis of the relationship between input (recharge) and output (discharge) parameters for different karst systems. These methods are based solely on mathematical operations, and the coefficients obtained cannot directly be related to physical phenomena. Time-series analyses provide limited information concerning the physical properties of the system itself. Consequently, they are "black-box models". On the other hand, the interpretation of the results of such models can sometimes be related to some geometric or hydraulic features of karst systems.

10.4.2   Single event models (grey box models)

An infiltration event over a karst terrain results in a hydrograph peak at the karst spring, which is delayed in time relative to the storm event. This peak can generally be subdivided into three main components: rising limb, flood recession and baseflow recession (Fig. 10.2). The latter is the most stable section of any hydrograph, and also the most characteristic feature of the global response of an aquifer. This is because baseflow recession can be assumed as that segment of a hydrograph least influenced by the temporal and spatial variations of infiltration.

The mathematical description of baseflow recession provided by Maillet (1905) is based on the depletion of a reservoir, and assumes that spring discharge is a function of the volume of water in storage. This behaviour can be described by the following exponential equation:

$$Q_{(t)} = Q_0 e^{-\alpha t} \tag{1}$$

where Q_t is discharge [$L^3 T^{-1}$] at time t, Q_0 is the initial discharge [$L^3 T^{-1}$] at an earlier time, and α is the recession coefficient [T^{-1}] usually expressed in 1/day. Plotting a discharge hydrograph on a semi-logarithmic graph reveals a straight line with the slope $-\alpha$ for favourable conditions. This equation is adequate for the description of the baseflow recession of karst systems.

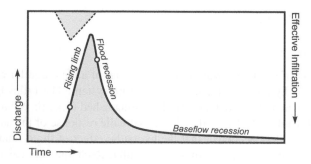

Figure 10.2.   Typical features of a spring hydrograph peak. White dots indicate inflexion points that correspond with maximum infiltration and the end of an infiltration event (Kiraly 1998a, from Kovács 2003).

According to Kovács (2003) and Kovács et al. (2005), the baseflow recession coefficient can be used for the derivation of important information about aquifer hydraulic parameters and conduit network characteristics. The analytical formulae provided by the above authors are based on a simple conceptual model, which involves a regular network of high permeability conduits coupled hydraulically with a low-permeability matrix (Fig. 10.3). The recession coefficient of a karst spring discharging from such a system reflects the structure and hydraulic parameters of the conduit network and the low permeability matrix.

Numerical studies by the above authors showed that the dependence of the recession coefficient on aquifer properties cannot be described using a single formula, but that it follows two significantly different physical principles depending on the overall configuration of the hydraulic and geometric parameter fields (Fig. 10.4, Fig. 10.5).

The baseflow recession of karst systems is controlled by the hydraulic parameters of the low-permeability matrix, and conduit spacing. This flow condition is referred to as matrix-restrained flow regime (MRFR). The baseflow recession coefficient of karst systems can be expressed as follows:

$$\alpha_b = \frac{2\pi^2 T_m f^2}{S_m} \tag{2}$$

The baseflow recession of fissured systems and poorly karstified systems is influenced by the hydraulic parameters of fractures/conduits, low-permeability blocks, fracture spacing, and aquifer extent. This flow condition has been defined as conduit-influenced flow regime (CIFR). The baseflow recession of fractured systems can be expressed as follows:

$$\alpha_h \approx \frac{2}{3} \frac{K_c f}{S_m A} \tag{3}$$

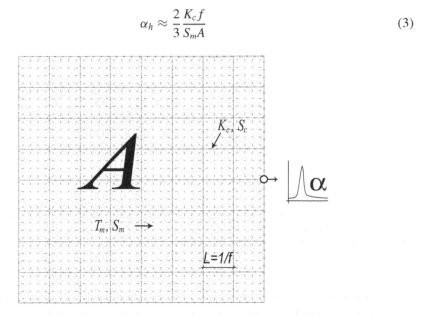

Figure 10.3. Conceptual model of karst systems according to Kovács (2003) and Kovács et al. (2005). Characteristic model parameters are: Transmissivity of the low-permeability matrix (T_m [$L^2 T^{-1}$]), storage coefficient of low-permeability matrix (S_m [-]), conductance of karst conduits or fractures (K_c [$L^3 T^{-1}$]), storage coefficient of karst conduits (S_c [L]), spatial extent of the aquifer (A [L^2]), and the spatial frequency of karst conduits (f [L^{-1}]).

206 *Attila Kovács & Martin Sauter*

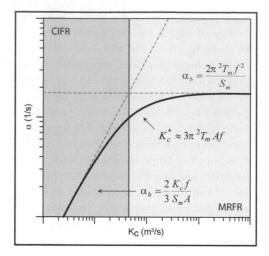

Figure 10.4.   Graphical representation of the relationship between recession coefficient and conduit conductance on log-log graph. K_c^* represents the threshold value of conduit conductance. Higher values entail matrix-restrained baseflow (MRFR), while lower values result in conduit-influenced baseflow (CIFR). Modified after Kovács (2003) and Kovács et al. (2005).

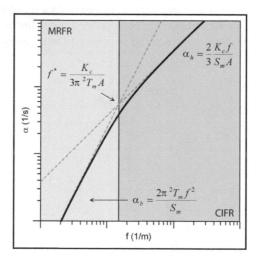

Figure 10.5.   Graphical representation of the relationship between recession coefficient and conduit frequency on log-log graph. f^* represents the threshold value of conduit frequency. Higher values entail conduit-influenced baseflow (CIFR), while lower values result in matrix-restrained baseflow (MRFR). Modified after Kovács (2003) and Kovács et al. (2005).

The importance of these hydrograph analytical formulae lies in their potential for providing crucial input parameters necessary for distributive groundwater flow models. They also reveal some fundamental characteristics of the recession process of strongly heterogeneous hydrogeological systems, and make the estimation of aquifer parameters and conduit spacing possible from spring hydrograph data.

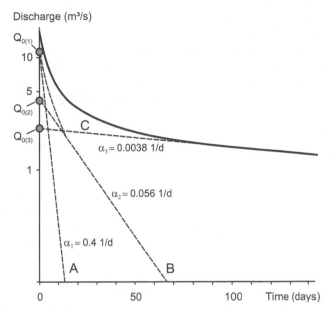

Figure 10.6.   Decomposition of recession curves according to Forkasiewicz & Paloc (1967).

Forkasiewitz & Paloc (1967) extended Maillet's approach to the entire recession process by decomposing hydrograph peak recession limbs into three exponential components (Fig. 10.6). The authors assumed that the components reflect three individual reservoirs, representing a conduit network, an intermediate system of well-integrated karstified fractures, and a low permeability network of pores and narrow fissures. However, the analysis of spring hydrographs simulated by numerical models performed by (Kiraly & Morel 1976b), and later by Eisenlohr et al. (1997a) showed that different exponential hydrograph segments do not necessarily correspond to aquifer volumes with different hydraulic conductivities. Three exponential reservoirs can be fitted to the hydrograph of a karst system consisting of only two classes of hydraulic conductivities. The intermediate exponential could simply be the result of transient phenomena in the vicinity of the high hydraulic conductivity channel network.

Other authors proposed to describe the recession process by employing different mathematical functions. Drogue (1972) described the whole recession process by using one single hyperbolic formula. In contrast, Mangin (1975) distinguished two components on the recession curves, and associated them to the discharge from the unsaturated zone and the saturated zone, respectively.

10.4.3   Time series analysis

Time series analyses imply the mathematical analysis of the response of karst systems to a succession of rainfall events. Because these methods are solely based on mathematical operations, they fail to provide information on physical functioning of karst systems, and can be used mainly for prediction and data compilation purposes. Most of the methods used in time series analysis were principally developed by Jenkins and Watts (1968), and were applied to karst systems by Mangin (1971, 1975, 1981, 1984). Detailed explanation of time-series analytical techniques is provided by Jeannin & Sauter (1998).

Conventional time series analysis uses both univariate (auto-correlation, spectral analysis) and bivariate (cross-correlation, cross-spectral analysis) methods. Univariate methods characterise the structure of an individual time series (hydrograph), while bivariate methods describe the transformation of an input function into an output function (rainfall/discharge).

The Auto-Correlation Method is a tool for the identification of some overall characteristics of a discharge time series, particularly cyclic variations (Mangin 1982, Grasso & Jeannin 1994, Eisenlohr et al. 1997b). Spectral analysis offers considerable potential as a powerful tool for the analysis of periodicities within the time series (Box & Jenkins 1976, Mangin 1984).

Cross correlation methods permit the quantitative comparison of rainfall with spring discharge time series. The technique provides information about strength of the relationship between the two time series and the time lag between them (Jenkins & Watts 1968, Box & Jenkins 1976, Mangin 1981, 1982, 1984, Padilla & Pulido-Bosch 1995, Larocque et al. 1998, Grasso 1998, Grasso & Jeannin 1998).

In the case of linear and steady state systems, a transfer function can be defined, which is a characteristic function of the system that reflects the active processes. The method of identifying the transfer function of an unknown system is known as deconvolution. Assuming the input function to be random, the cross-correlogram can be considered as the transfer function of the system (Neuman & De Marsily 1976, Dreiss 1989, Mangin 1984).

Discharge time series may exhibit non-stationarity in their statistics. While the series may contain dominant periodic signals, these signals can vary both in amplitude and frequency over long periods of time. One possibility for the analysis of non-stationarity of a time series would be to compute a Windowed Fourier Transform using a certain window size, and sliding it along the series. This would provide information on frequency localization, but would still be dependent on the window size used, resulting in an inconsistent treatment of different frequencies (Torrence & Compo 1998).

Wavelet analysis introduced by Grosmann & Morlet (1984) attempts to solve this problem by decomposing a time series into time/frequency space simultaneously. The continuous wavelet transform is defined as the convolution of the time series with a scaled and translated version of a wavelet function (Torrence & Compo 1998). Several types of wavelet functions are in use; each must have zero mean and be localised in both time and frequency space.

Wavelet analysis constitutes an alternative in karst hydrology to spectral and correlation analyses. By varying wavelet scale and sliding the wavelet along a time series, one can construct a picture showing both the amplitude of any periodic signals versus scale and the variation of this amplitude with time. A theoretical explanation of wavelet analysis is provided by Daubechies (1992). The application of wavelets in karst hydrogeology is discussed in Labat et al. (1999 a, b, 2000).

10.5   DISTRIBUTIVE MODELS

10.5.1   Introduction

The spatial heterogeneity of karst aquifers may require the discretisation of a hydrogeological system into homogeneous sub-units, each with its own characteristic hydraulic parameters. The discretisation of a rock volume involves the discretisation of the differential equations describing groundwater flow. The principal formula representing transient groundwater flow in saturated medium is the classical diffusivity equation derived from Darcy's flow law (momentum conservation) and the continuity equation (mass conservation):

$$S\frac{\partial H}{\partial t} = \nabla(K\nabla H) + i \tag{4}$$

where S is the storage coefficient $[L^{-1}]$, K is hydraulic conductivity $[LT^{-1}]$, H is hydraulic head $[L]$, t is time $[T]$, and i is the source term $[T^{-1}]$.

Two types of discretisation methods are used widely in hydrogeology. The Finite Difference Method (FDM) consists of subdividing the model domain into rectangular cells. Partial derivatives are then approximated by simple differences between a given number of adjacent nodes located at the corners or in the centre of each cell. According to the Finite Element Method (FEM), the model domain is subdivided into an irregular network of triangular and/or quadrangular finite elements. The approximation of differential operators is analytical, and it involves integral quantities for each element. The FEM allows the combination of one-, two- and three-dimensional elements of various shapes, thus facilitating proper discretisation and the implementation of discrete features into the model. A detailed explanation of the FDM and FEM discretisation methods can be found in Kinzelbach (1986), Wang & Anderson (1982), and Huyakorn & Pinder (1983).

The equation system obtained by spatial discretisation requires matrix inversion. This can be easily computerised, and thus spatial discretisation allows groundwater flow simulations in complex hydrogeological systems. The solution of an equation system requires the definition of boundary conditions and (for transient problems) initial conditions. Boundary conditions consist of the definition of either hydraulic potential (head boundary) or flux (flux boundary) values along the domain boundaries. Initial conditions consist of the spatial distribution of the unknown function, and are usually taken from an equipotential map or a previous steady state simulation.

Distributed parameter groundwater flow models include two principal concepts. The discrete concept considers the flow within individual fractures or conduits. In contrast, the continuum concept treats heterogeneities in terms of effective model parameters and their spatial distribution. These concepts can be combined into five alternative modelling approaches, according to the geometric nature of the conductive features represented in the model (Teutsch & Sauter 1991, 1998) (Fig. 10.7).

• Discrete Fracture Network Approach (DFN)
• Discrete Channel Network Approach (DCN)
• Equivalent Porous Medium Approach (EPM)
• Double Continuum Approach (DC)
• Combined Discrete-Continuum (Hybrid) Approach (CDC)

The physical parameters of the flow medium can be directly or indirectly derived from real field observations (deterministic models) or can be determined as random variables (stochastic models). Each method has its respective advantages and limitations, and the selection of the appropriate modelling approach may be crucial with respect to the outcome of the simulation.

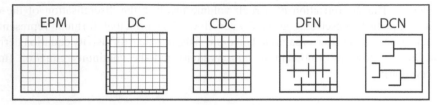

Figure 10.7.   Classification of distributive karst modelling methods.

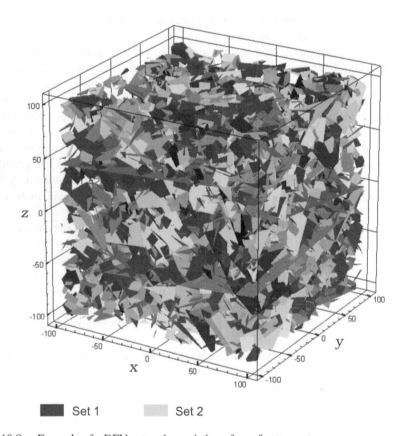

Figure 10.8.   Example of a DFN network consisting of two fracture sets.

10.5.2   Discrete Fracture Network Approach (DFN)

According to the DFN approach, only certain sets of fractures are considered to be per-meable. The matrix medium is assumed to have negligible permeability. This concept simplifies a fissured system into a network of two-dimensional fracture planes (Fig. 10.8). It is mainly applicable to fractured aquifers. However, DFN methods facilitate the representa-tion of karst channels by introducing one-dimensional elements or increased transmissivity zones along individual fractures representing dissolution voids (Dershowitz et al. 2004).

The transmissivity T $[L^2T^{-1}]$ of a single fracture can be expressed by the "cubic law" (Snow 1965) as follows:

$$T = \frac{a^3}{12} \frac{\rho g}{\mu} \qquad (5)$$

where a is the fracture aperture [L], μ is the dynamic viscosity of water $[ML^{-1}T^{-1}]$, ρ is fluid density $[ML^{-3}]$, and g is acceleration due to gravity $[LT^{-2}]$. The "cubic law" is valid for laminar flow in open or closed fractures (Witherspoon et al. 1980). The storativity of a single fracture assuming water compression only, can be expressed as follows:

$$S_f = \frac{\rho g}{E_w} a \qquad (6)$$

The cubic law is based on a parabolic distribution of velocities between smooth parallel plates, which is a condition seldom met in natural fractures. Natural fractures are rough, tortuous and heterogeneous. The cubic law aperture can therefore significantly underestimate the storage and overestimate velocity. For this reason, it is preferable to derive fracture properties directly from field experiments. Packer tests provide information on transmissivity, interference tests estimate storage, and tracer tests determine transport aperture.

The first two-dimensional DFN modelling of synthetic fracture networks using a numerical solver was published by Long et al. (1982). Fracture networks were generated stochastically by applying a Monte Carlo process. Long et al. (1985) constructed a three-dimensional DFN model, and simulated steady-state flow for simple theoretical configurations of orthogonal fractures. Andersson & Dverstorp (1987) presented a modelling case study for conditioning the statistical generation of fracture sets with field data.

According to the Monte Carlo process, probability density functions are fitted to field data describing the spacing, length, direction and aperture of fracture sets identified on stereonets. Parameters are sampled from these statistical distributions, and assigned to individual fractures in the model. Fracture plane centres are generated first, assuming a negative exponential distribution of fracture spacing (Baecher et al. 1977). Fracture sizes are then assigned, usually sampled from lognormal distribution functions. Fracture orientations usually show a Fisher distribution, while fracture transmissivities follow a lognormal distribution.

A major benefit of DFN models is their ability to reflect the compartmentalization phenomenon experienced at several study sites (Shapiro et al. 1994, Sawada et al. 2000). Compartments are adjacent rock volumes that manifest significant differences in hydraulic heads (up to 100 meters difference on a few meters of distance). Robinson (1984) demonstrated that there was a critical number of intersections required to produce percolating pathways. Compartments are the result of insufficient density of the fracture network. The porous medium concept assumes hydraulic continuity between all points in the simulation region, and cannot reflect compartmentalization phenomenon.

One of the most critical shortcomings of the DFN approach is its high data demand. While fracture spacing, orientation and transmissivity distributions can be estimated from downhole measurements, fracture size parameters usually remain uncertain. Computational constraints limit the number of simulated fractures to the order of 10^4 to 10^5. Fracture size distributions are usually truncated, and small size fractures are not represented in the model. This results in an erroneous estimation of the storage and diffusion behaviour, and necessitates the implementation of artificial matrix blocks in the model. As fracture parameters result from a stochastic generation method, model results involve significant spatial uncertainties.

10.5.3   Discrete Conduit Network Approach (DCN)

The DCN approach simulates flow in networks of one-dimensional pipes representing karst conduits or connections between fracture centres. A DCN conduit network can be established deterministically representing a local-scale field situation, or can be derived from stochastic DFN networks by geometric transformations.

The mathematical formulation of the average velocity of laminar flow in one-dimensional conduits may be expressed by the Hagen-Poiseuille law:

$$\vec{v} = -\frac{r^2}{8}\frac{\rho g}{\mu}\vec{I} \tag{7}$$

where r is conduit radius [L]. Conduit conductance is a one-dimensional parameter derived from the Poiseuille law as follows:

$$K_c = \frac{\pi r^4}{8}\frac{\rho g}{\mu} \tag{8}$$

The mathematical formulation of turbulent flow in one-dimensional conduits is given by the Darcy-Weissbach friction law:

$$Q = -K'A_c\sqrt{I} \tag{9}$$

where K' is turbulent flow effective hydraulic conductivity [LT^{-1}], A is cross sectional area [L^2], and $\vec{I}$ is hydraulic gradient [-]. Louis (1968) expressed the effective hydraulic conductivity for fully constricted (phreatic) pipe flow as follows:

$$K' = 2\log\left(1.9\frac{D_h}{\varepsilon}\right)\sqrt{2gD_h} \tag{10}$$

where $D_h = 4R_h$ is the hydraulic diameter [L], and ε is the absolute size of irregularities along conduit walls [L]. Similarly, Strickler (1923) expressed the effective hydraulic conductivity for non-constricted (vadose) pipe flow as follows:

$$K' = K_s R_h^{2/3} \tag{11}$$

where K_s is the Strickler coefficient [$L^{1/3}T^{-1}$] depending on roughness. R_h is the hydraulic radius (cross section divided by wet perimeter) [L].

The storage coefficient of a conduit assuming water compression only, can be formulated as follows (Cornaton & Perrochet, 2002):

$$S_c = \frac{\rho g}{E_w}r^2\pi \tag{12}$$

A summary and comparison of these formulae, along with case studies for their application is provided by Jeannin & Maréchal (1995), and Jeannin (2001). These authors constructed

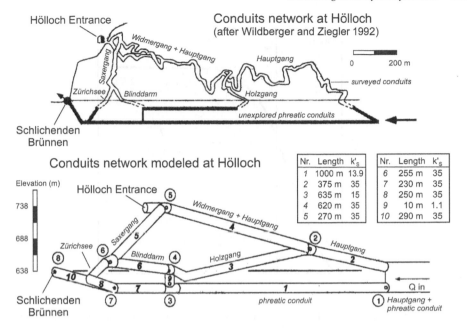

Figure 10.9. The DCN model of the downstream part of the Hölloch cave, Muotatal, Switzerland (Jeannin 2001). The parameter k's is equivalent to K'A.

a two-dimensional fully constricted pipe-flow model of the Hölloch Cave, Muotatal, Switzerland, in order to simulate groundwater flow for various discharge conditions (Fig. 10.9). The geometric properties of the conduits were derived from field observations. The model was calibrated on the basis of several head and discharge measurements in conduits. This study demonstrated how overflow conduits modify aquifer conductivity when they become active. This results in the observed large non-linearity of the system. The typical effective hydraulic conductivity of karst conduits ranges between 1 and 10 m/s and the Louis formula is adequate for calculating head-losses under such conditions.

Another application of the DCN approach is for the representation of fracture networks (Cacas et al. 1990, Dverstorp et al. 1992, Dershowitz et al. 1998). Fractures are usually represented by one-dimensional elements connecting fracture intersections. DCN representation of fracture networks may be useful when no spatial information on fracture flow is necessary. DCN networks derived from DFN models can correctly represent the overall transport behaviour of fissured systems, and can be used for simulating breakthrough curves. The transformation of fractures into one-dimensional pipes requires the estimation of effective pipe width, which remains a calibration parameter. The quality of DCN models in this case strongly depends on the quality of the DFN model from which they originate.

10.5.4 Equivalent Porous Medium Approach (EPM)

The EPM approach utilises discretisation units of similar size. This requires the representative elementary volume to be almost constant all over the model domain, and involves an insignificant change of aquifer hydraulic parameters between adjacent units of discretisation. This is a condition seldom met in karst systems.

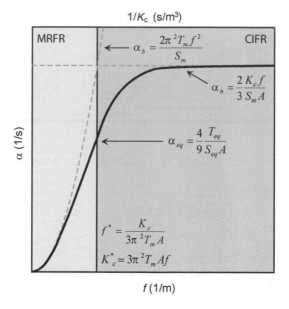

Figure 10.10.   Variations of the recession coefficients for equivalent discrete-continuum models. EPM models correspond to the transition between matrix-restrained and conduit-influenced flow, and designates an inflection point of the recession coefficient curve. Modified after Kovács (2003) and Kovács et al. (2005).

The substitution of strongly heterogeneous rocks with equivalent porous medium (this transformation referred to as upscaling) has long been an interest of hydrogeologists active in the field of fractured rock hydrology. Oda (1985) presented a method for calculating EPM properties from fracture networks. The Oda approach overlays an EPM grid across a fracture network, and derives EPM properties for each grid cell based on the fracture properties contained in that cell. The result is an equivalent permeability tensor, according to a specific grid. Fine discretisation reproduces the hydraulic and transport behaviour of the underlying fractured medium. However, for coarser practical discretisations of tens or hundreds of meters, the Oda approximation produces less accurate results.

The differences in transient hydraulic behaviour between fissured, karstic and equivalent porous systems were demonstrated by Kovács et al. (2005). These authors demonstrated that porous systems manifest fundamentally different temporal hydraulic behaviour from karstified medium (see chapter 4.1.). The global response of EPM models corresponds to the transition between matrix-restrained and conduit-influenced flow (Fig. 10.10).

The application of EPM models for steady-state karst hydrogeological problems is mainly limited by the constraints of discretisation. Transient EPM models fail to reflect karstic hydraulic behaviour. As a consequence, the EPM approach is basically inadequate for modelling karst hydrogeological systems over a large spectrum of applications.

10.5.5   Double Continuum Approach (DC)

The difficulties of obtaining data for constructing DCN or CDC models, and the inability of the EPM approach to take account of the strong heterogeneity of karst aquifers, have

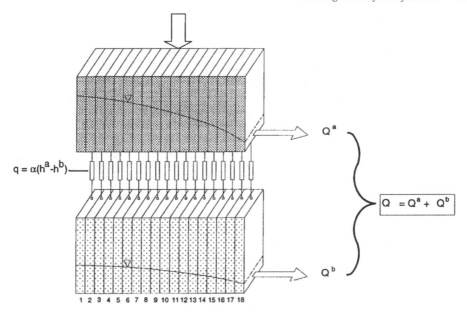

$q = \alpha(h^a\text{-}h^b)$

$Q = Q^a + Q^b$

1 2 3 4 5 6 7 8 9 10 11 12 13 14 15 16 17 18

Figure 10.11. Schematic representation of a one-dimensional DC model (Teutsch 1988).

motivated the development of the DC method, which can simulate specific karst features without requiring detailed knowledge of the conduit network geometry.

The first numerical solution using the DC approach was provided by Teutsch (1988), who used the original concept of Barenblatt et al. (1960) (Fig. 10.11). In a double continuum model the conduit network and the fissured medium are both represented by continuum formulations. The exchange of water and solute between the two continua is calculated based on the hydraulic head difference between them, using a linear exchange term. Flow equations for the two separate media may be formulated as follows:

$$\nabla(K_i \nabla H_i) = S_s^i \frac{\partial H_i}{\partial t} + Q_i \pm \alpha_{ex}(H_1 - H_2) \tag{13}$$

where i is the identifier of the medium considered, the last term is the exchange flux between the two continua, calculated from the hydraulic head difference, multiplied by an empirical steady-state exchange coefficient (α) as follows:

$$q_{ex} = \alpha_{ex}(H_1 - H_2) \tag{14}$$

The α parameter $[L^{-1}T^{-1}]$ characterises the rate of the fluid transfer between the two media, which is composed of the hydraulic conductivity of the exchange zone and geometric factors. The flux value of a node (e.g. spring discharge) can then be evaluated as the sum of the fluxes from the two different media, while head values are observed separately in the two media.

Although the DC concept can adequately describe the dual hydraulic behaviour of karst aquifers, applied hydraulic parameter distributions are volume averaged parameters of the real parameters. As the calibration of the parameters is basically a "trial- and-error" method based on available head data, the quality of the model results strongly depends on the density and location of the observation points. To some extent the parameters of the DC system can be estimated from pumping tests, especially for the matrix system. Sauter (1992) proposed parameter estimation methods for DC models.

The adequacy of DC model results and calibrated parameter fields has been tested by several authors. Mohrlok & Teutsch (1997) demonstrated that simulated spring discharges obtained from calibrated DC models show a good correspondence to field observations. Cornaton (1999) tested the physical meaning of the calibration parameter, and demonstrated a strong dependence of the exchange coefficient on matrix storage and conduit network density. The author also provided a one-dimensional analytical solution for the problem.

Another critical aspect of the DC concept is that it cannot handle the temporal delay of diffuse infiltration, since both subsystems are coupled directly at every node. In order to involve retarded diffuse infiltration in a DC model, a retention function is necessary (Sauter, 1992). Two-dimensional real-world applications of the DC concept were successfully performed by Teutsch (1988), Sauter (1992), and Lang (1995). Three-dimensional real-world applications of the DC method have not yet been performed because of the difficulties in model calibration and the necessity of requiring three-dimensional head data.

The DC approach is an effective tool for modelling karst systems. The spatial distribution of the calibration parameter values provides approximate information on real conduit system configuration. However, the interpretation of applied hydraulic parameters requires further attention.

10.5.6   Combined Discrete-Continuum (hybrid) Approach (CDC)

The CDC approach is capable of handling the discontinuities that exist at all scales in a karst system (fractures, fault zones, karst channels, etc.), by representing them as embedded networks of different orders of magnitude. This nested model explains the duality of karst (duality of the infiltration process, the flow field and discharge conditions), and the scale effect on hydraulic conductivities (Kiraly 1975).

Because the CDC approach uses the FEM discretisation method, it allows the combination of one-, two-, and three-dimensional elements (Király 1979, 1985, 1988, Király et al. 1995, Király & Morel 1976a). According to the CDC method, high conductivity karst channels can be simulated by one-dimensional finite elements, which are set in the low permeability matrix represented by three-dimensional elements. Similarly, the application of two-dimensional elements makes the simulation of fractures and fault zones possible.

The CDC method was developed and first applied by Kiraly & Morel (1976a). This was the first time that the typical temporal behaviour of a karst spring was adequately simulated by a distributive model, using realistic hydraulic parameter distributions (Fig. 10.12). Kiraly (1998b) concluded that the application of EPM models requires artificially high hydraulic conductivities in order to simulate realistic hydraulic heads. However, these discrepancies disappear with the introduction of the high conductivity channel network into the FEM model. Typical spring hydrographs cannot be simulated without applying concentrated infiltration. The rate of this concentrated input should be more than 40% of the total infiltration. The epikarst layer (Mangin 1975) may play an important role in draining and infiltrating water, and channelling it towards the karst network.

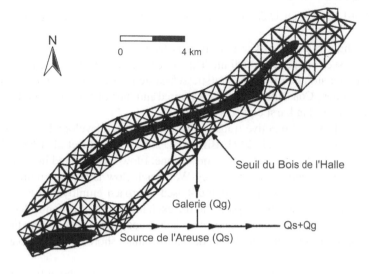

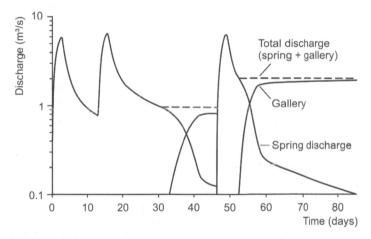

Figure 10.12.   Finite element mesh and simulated spring hydrographs of the first CDC model (Areuse Basin, Switzerland), Kiraly & Morel (1976a).

The CDC approach is the only distributive modelling concept that facilitates the direct application of observed aquifer geometry and measured hydraulic parameters. Consequently, the CDC approach facilitates the testing of conceptual models of karst systems. As demonstrated by Kovács (2003) and Kovács et al. (2005), the correct estimation of conduit network density and hydraulic parameters is crucial for modelling karst systems by the CDC approach. There is only one single parameter configuration that yields appropriate transient model results. These authors also demonstrated, that epikarstic storage can significantly influence the global reactions of a karst system, and thus such systems require the separate estimation of epikarst hydraulic and geometric parameters.

CDC type models are also the basic type of flow models employed the simulation of karst genesis described below (Liedl et al. 2003).

10.6   MODELLING SOLUTE AND HEAT TRANSPORT IN KARST AQUIFERS

Modelling solute transport in karst aquifers is far more complex than the simulation of transport processes in a porous medium. This is because of the large heterogeneity of flow, and the unknown storage in various karstic sub-systems, such as the epikarst, the vadose and phreatic zones. Consequently, only a limited number of transport modelling studies have been carried out in karst aquifers until now.

A CDC type flow and reactive transport model has been developed by various authors (Liedl et al. 2003, Bauer et al. 2003, Birk et al. 2003, Clemens et al. 1996). This model applies the Finite Difference discretisation scheme. Flow is calculated in the pipe network using the Hagen-Poiseuille and the Darcy-Weissbach flow laws, depending on the flow regime and in the matrix by a continuum-based approach employing Darcy's law. The two systems are hydraulically coupled and the exchange of fluid is controlled by the head difference between the two systems and an exchange coefficient. Apart from the simulation of reactive carbonate dissolution, the model can simulate short-term event based heat and solute transport processes.

The simulation of reactive solute transport in conduits of the above model is based on the 1D advection-dispersion equation to which a source term has been added that accounts for the increase of solute concentration from carbonate dissolution:

$$\frac{\partial c}{\partial t} = -v\frac{\partial c}{\partial z} + D\frac{\partial^2 c}{\partial z^2} + \frac{4}{a}F \tag{15}$$

where c denotes the solute concentration ($mol\,m^{-3}$), v the flow velocity ($m\,s^{-1}$), D the dispersion coefficient ($m^2\,s^{-1}$), a the conduit diameter (m), F the dissolution rate ($mol\,m^{-2}\,s^{-1}$), t time (s), and z the spatial coordinate along the conduit (m). The dispersion coefficient is calculated using relationships for turbulent pipe flow (Taylor 1954). The coefficient $4/a$ represents the ratio of surface area and volume of the conduit, required to transform the dissolution rate F into a change in concentration. The dissolution rate F can be calculated by:

$$F = k\left(1 - \frac{c}{c_{eq}}\right)^n \tag{16}$$

where c is the concentration of the dissolved species in the conduit water and c_{eq} is the equilibrium concentration with respect to calcite (Dreybrodt 1996). The rate constant k depends on factors such as rock type, flow conditions, etc. and can be determined by laboratory experiments. The exponent n equals unity if the dissolution process is diffusion controlled. If the dissolution rate is limited by the surface reaction, n is a positive number that has to be determined experimentally.

The model simulates heat convection in the conduit water and heat conduction in the adjacent rock. The heat transport simulation is based on the equation of heat convection expanded by a source term accounting for heat transfer between the conduit water and the conduit wall (Birk et al. 2001, Birk 2002):

$$\frac{\partial T}{\partial t} = -v\frac{\partial T}{\partial z} + \frac{h}{c_w \rho_w}\frac{4}{a}(T_s - T) \tag{17}$$

where h denotes the heat transfer coefficient ($J\,m^{-2}\,s^{-1}\,K^{-1}$), T the temperature of the conduit water (°C), T_s the temperature at the conduit wall (°C), c_w the specific heat of water

$(\mathrm{J\,kg^{-1}\,K^{-1}})$, and ρ_w the water density $(\mathrm{kg\,m^{-3}})$. To calculate the temperature at the conduit wall, the equation of heat conduction in cylindrical coordinates is solved (Carslaw & Jaeger 1959):

$$\frac{\partial T_r}{\partial t} = \kappa_r \left(\frac{\partial^2 T_r}{\partial r^2} + \frac{1}{r} \frac{\partial T_r}{\partial r} \right) \qquad (18)$$

where T_r denotes the rock temperature (°C), κ_r the thermal diffusivity of the rock $(\mathrm{m^2\,s^{-1}})$, and r the radial coordinate (m). Heat and solute transport equations are solved using an explicit finite-difference scheme.

The modelling tool outlined above is employed to simulate flow and transport in a simplified karst catchment. Modelled spring responses can be used to analyse in detail characteristic and complex patterns of discharge and physico-chemical parameters with a process-based approach. It could be shown that the discharge generally responds a lot quicker to recharge events than the physico-chemical parameters (Fig. 10.13). Thus, Ashton (1966) suggested that the water volume discharged during this time lag provides an estimate for the volume of the (phreatic) karst conduit system.

Comparisons between DCN and CDC models demonstrated that the time lag between a spring discharge peak and the subsequent drop in concentration can be much greater in a dual flow system than in an isolated conduit. The water volume discharged during the time lag comprises both pre-event conduit water and water derived from the fissured porous rock. Hence, the method suggested by Ashton (1966) tends to overestimate the volume of the conduit system.

The temperature of spring discharge also represents a physico-chemical property that can potentially be used to estimate conduit volumes. For comparison responses of temperature and solute concentration, heat transport was simulated in the model scenario considered above. Figure 10.14 demonstrates that the temperature response of a CDC system is significantly more dampened than its chemical response, and that both responses exhibit signals that are quite complex (with numerous inflexion points) even for a very simplistic synthetic scenario.

Moreover, the temperature response appears to be delayed in comparison with the chemical response. Both effects are caused by thermal interaction between conduit water and the

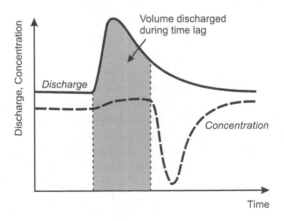

Figure 10.13.   Karst spring response to a recharge event.

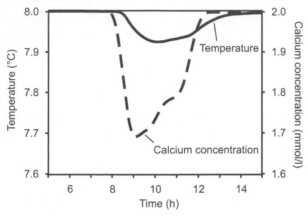

Figure 10.14.   Temperature (solid curve) and concentration (dashed curve) response for the above recharge event. Temperature and concentration axes have been scaled to represent relative parameter changes equivalently.

adjacent rock (Birk 2002) and reveal that water temperature represents a highly 'reactive' tracer. Therefore, it is less suitable as a tracer for the estimation of conduit volumes based on Ashton's (1966) approach. However, temperature responses can provide information about other characteristics of the conduit system, such as the area of the rock-water-interface relative to the conduit volume (Liedl et al. 1998). Thus, solute concentration and water temperature complement each other in aquifer characterisation.

10.7   SUMMARY

This chapter presented a critical review and systematic classification of modelling methods applicable to karst hydrogeological systems. While global models reflect an overall hydraulic response of aquifers, distributive models may provide a complete quantitative representation of spatial and temporal hydrogeological phenomena.

Global models include two fundamentally different approaches: Single event models deal with the global response of karst systems to one particular rainfall event. These models are based on physical laws derived from reservoir models. Consequently, they can provide information on the hydraulic properties of karst systems. Although the mathematical descriptions and interpretations of the fast component of the event hydrograph recession vary, the baseflow recession is generally described by the classical Maillet formula. The recession coefficient, a characteristic parameter of the baseflow recession process, depends on the spatial configuration of the karst conduit network and on the hydraulic parameters of the aquifer. Consequently, single-event hydrograph analysis may be a powerful tool for the estimation of the hydraulic and geometric parameters of an aquifer.

Time series analytical models consist of the mathematical analysis of the long-term response of karst systems to recurrent rainfall events. As these methods are solely based on mathematical operations, they cannot be directly related to any physical parameters. Consequently, they provide only indirect information on the physical functioning of karst hydrogeological systems. Univariate methods deal with discharge time series analysis. These methods are suitable for the identification of cyclic variations of spring hydrographs.

Bivariate statistical methods compare rainfall and discharge time series and analyse their relationship systematically. The drawback of time series analyses is their failure to consider temporal and spatial distributions of rainfall, which may significantly influence statistical parameters.

Distributive groundwater flow models include two principal concepts. While discrete approaches describe flow within networks of fractures or conduits, continuum approaches treat the heterogeneities in terms of effective model parameters and their spatial distribution. The CDC approach represents a combination of these techniques.

The DFN approach simplifies the real system into a network of fractures and fails to consider flow through the low permeability matrix. This method is mainly applicable to fractured aquifers. An advantage of the DFN method is its ability to reflect compartmentalisation phenomenon. Two-dimensional DFN models involve significant discrepancies and are not appropriate for the modelling of real aquifers.

The DCN approach aims to model the flow in a network of one-dimensional karst conduits. The geometry of the conduit network can be derived from DFN networks, or it may represent a local-scale field situation. Local-scale DCN models require detailed knowledge of the conduit network geometry and discharge conditions in every conduit. DCN models facilitate the understanding of the complex global response of karst aquifers but are unlikely to become a standard tool for the modelling of regional groundwater flow.

The EPM approach assumes a low contrast in aquifer parameters. With karst systems typically being highly heterogeneous, the EPM approach is not appropriate for modelling groundwater flow and transport in karst aquifers. Nonetheless, it may be applied to poorly karstified or fissured systems with certain restrictions.

With the DC concept, the conduits and the fissured medium are represented as different continua, and water exchange between the two continua is calculated based on the hydraulic potential difference. Although this method does not incorporate conduit network characteristics, the low data demand makes it a useful tool for several applications. The spatial distribution of calibration parameters can provide approximate information on the real conduit system configuration. This method can be applied for water resources or water budget problems.

The CDC approach can handle discontinuities which exist at all scales in a karst system, as it allows the combination of one-, two-, and three-dimensional elements. The elements can therefore represent karst conduits, fractures and the low permeability matrix. The CDC approach is the only distributive modelling concept, which facilitates the direct application of real aquifer parameters and the simulation of process-orientated reactive transport while still considering computational efficiency. Furthermore, the CDC approach enables the testing of conceptual models of karst systems. Although the practical application of this concept is mainly limited by its large data requirements, spring hydrograph analysis can provide sufficient information for the estimation of conduit network hydraulic and geometric properties, and thus allow for the construction of discrete-continuum models of karst aquifers.

Because of its highly complex nature, solute transport modelling has been mainly performed on synthetic karst systems. CDC-type coupled models have revealed some basic characteristics of solute and heat transport processes in strongly heterogeneous systems. Discharge generally responds a lot quicker to recharge events than the physico-chemical parameters, because hydraulic pressure is almost instantaneously transmitted to the spring, while fluid properties change only after the new recharge water arrives at the spring.

The temperature response of an aquifer is more dampened and delayed in time compared to the chemical response. Therefore, it is less suitable as a tracer for the estimation of conduit volumes. However, temperature responses can provide information about other characteristics of the conduit system, such as the area of the rock-water-interface relative to the conduit volume. Thus, solute concentration and water temperature complement each other in aquifer characterisation.

CHAPTER 11

Combined use of methods

David Drew & Nico Goldscheider

11.1 INTRODUCTION AND SUMMARY OF METHODS

Any full and rigorous groundwater investigations, whether in karstic or non-karstic conditions, requires the application of more than one investigative method. However, the mix of methods used in karst areas may differ significantly from those commonly used where flow is laminar and diffuse and hence where Darcy's law is applicable; in particular the combination of orthodox methods with those appropriate only to karst (speleological investigations for example). This chapter provides a summary of the methods described in this book together with a brief overview of the conjunctive use of methods to address particular karst groundwater problems.

One approach to the use of several techniques in an investigation is to use complementary methods in which data from one technique is used to supplement the inadequacies of data derived from another method. For example, using water tracing to give more accurate information concerning flow paths than can be obtained from piezometric maps. Or using natural tracers (e.g. isotopes, hydrochemistry) to obtain general information regarding all or a part of a groundwater system and then obtaining precise data for a particular locale by using artificial tracers in which the input function is controlled. The methods may also be used to validate results obtained by other means (for example confirming data obtained by indirect means such as geophysics by direct exploration of the aquifer via cave conduits).

Methods may also be use in combination to allow conceptual models of varying degrees of sophistication to be developed. The results of water tracing, groundwater level observations and karst hydrogeomorphological mapping may be combined to build a picture of a karst groundwater system. Spring and swallow hole flow data may be analysed together with hydrochemical data to help to understand the dynamics of flow systems. Finally, some mathematical modelling approaches, as described in this book, may require large amounts of high quality data derived in a variety of ways to produce a holistic view of a groundwater system. However, other analytical modelling approaches, with more limited aims, such as spring hydrograph, chemograph or tracer concentration analysis from springs require comparatively modest data inputs. Some modelling approaches require no real-world input data at all, but are primarily intended to better understand specific processes.

In standard groundwater textbooks such as Fetter (2001) and Brassington (2006) the suite of methods most usually advocated are geological, including lithological interpretation of rocks cores; geophysical, (rock-water properties) and pumping tests allied, if appropriate with water budget computations and modelling. Such a combination may not be the most relevant to a karst groundwater study where an appropriate investigative strategy might be:

- understanding the geological framework
- hydrological mapping (recharge and discharge locations)
- tracer tests to establish hydrological linkages

If practicable and relevant these investigations might be followed by speleological surveys, geophysical studies (either of large areas of the aquifer or to locate particular subsurface features). Depending on the purpose of the study, natural tracer information may be relevant, particularly hydrochemical data if water quality is a concern Hydraulic methods may be employed, but their importance is greatest where access to the aquifer using the other methods listed above is limited.

Table 11.1 summarises the groups of methods described in this book. The methods are grouped according to the chapters of the book and for each the main types of data obtained are listed together with the limitations or drawbacks of the methods.

11.2   COMBINED METHODS FOR SPECIFIC INVESTIGATIONS

In hydrogeological investigations the methods used will be strongly influenced by the questions that need to be answered but also by the available resources. In the case of karst, the type of karst involved will also influence the methods used. For example, borehole based investigations are often of very limited value in upland karsts whilst speleological investigations are often not possible in lowland karsts. Examples of aspects of hydrogeology that are often of particular importance in karst groundwater systems are discussed below.

11.2.1   Determining the catchment areas for springs

Because karst groundwater systems often drain to springs it is often essential to define the area contributing water to that spring, partly to assess potential yields, partly to assist with protecting water quality. Geological information is indispensable to determine the likely limits of a catchment. Geomorphological observations may also be helpful in some cases. It is important to note that groundwater flow paths in karst aquifers may run across valleys and below mountain ridges. Therefore, topographic limits cannot necessarily be translated into groundwater divides, as can often be done in other groundwater environments.

Water balances, though they are often difficult to determine, help in the estimation of the *minimum* surface area of a spring catchment. Karst aquifers are often drained by several springs with overlapping catchments, and this needs to be considered when establishing a water balance for a particular spring. The possible existence of unknown or inaccessible outlets (e.g. submarine springs) represents another difficulty. Mapping of sinking streams and swallow holes and their topographic catchments makes it possible to delineate the allogenic part of the catchment area.

Natural tracers can sometimes resolve ambiguities concerning contributing areas whilst the use of artificial tracers is an essential tool in providing unambiguous information as to whether or not particular areas (tracer input points) lie within a particular spring catchment (e.g. Goldscheider 2005, Lütscher & Perrin 2005). As the delineation of source protection zones usually implies land-use restrictions, which may require financial compensation, it is crucial to have highly reliable data, which can only be provided by tracer tests.

Table 11.1. Summary of methods available for hydrogeological investigations in karst areas.

Method or group of methods	Data obtained/advantages	Disadvantages/limitations
Geological methods	Aquifer framework and geometry information Karstifiability of the rock Orientation, location, type and frequency of potential flow paths Theoretical hydraulic conductivity and porosity	Data not necessarily directly related to groundwater Often not a predictable and unambiguous relation between lithology and hydrogeology
Geomorphological methods	Degree of karstification Types of recharge Historical hydrogeomorphology	Data mainly from the surface (indirect) Static framework rather than hydrodynamics Limited data from covered karsts
Speleological methods	Locating and mapping conduits in 3D Monitoring water quality and quantity within the aquifer Determining the temporal evolution of conduit systems	Access to cave systems may be limited or non-existent Specialist speleological skills required Only a small and perhaps unrepresentative part of the aquifer is likely to be accessible
Hydrological methods	Water budget compilation Characterisation of flow systems by spring hydrograph analysis	Water budget often incomplete, as catchment boundaries are not always clear, and not all inputs and outputs can be monitored Hydrograph alone gives limited information of the origin of the water (needs to be combined with chemograph)
Hydraulic methods	Determination of transmission and storage characteristics Determination of piezometric level Determination of groundwater velocity and flow direction	Many methods not wholly appropriate under non-Darcian conditions Estimates of flow directions and magnitudes may not be accurate Pumping tests may not give representative results
Hydrochemical methods	Hydrochemical characterisation of groundwater bodies Information on water quality and contamination problems May be used as natural tracers for the origin and movement of the water	Difficulties in developing an adequate sampling strategy (high temporal variability) In karst aquifers, microbial contamination is often of greater importance than chemical

(*Continued*)

Table 11.1. (Continued)

Method or group of methods	Data obtained/advantages	Disadvantages/limitations
Isotopic methods	May be used as natural tracers for the origin and movement of the water; this includes: Identifying sources of karst waters and mixing processes Determining residence time/age of karst waters	Input function not known precisely Ambiguities possible in interpreting data
Tracer methods	Determination of flow routes and velocities Determining contributing areas for springs Information on contaminant transport Usually very reliable, precise and unambiguous information	Difficulty in recognising "negative" tracings Usually only gives information for selected points and the hydrological conditions during the tracer test Limited applicability for very deep and large systems with very long transit times Visible colouring and toxicity concerns for some tracers
Geophysical methods	Determining geological structures and overburden thickness Locating conduits, fractures and other preferential flow paths Data can be obtained over wide areas Information on the structure and properties of the underground without drilling, i.e. at relatively low cost	Results may be difficult to interpret without ambiguity (non-uniqueness) Resolution vs. depth of investigation (i.e. the greater the depth, the lower the resolution) Some techniques require very precise location control (gravimetry), others have noise problems or require heavy or expensive equipment
Modelling methods	Conceptualising all or a component of karst aquifer systems May give a better understanding of specific processes, such as speleogenesis, conduit flow and conduit-matrix interactions Simulating groundwater flow and contaminant transport Predicting changes in water quality and quantity	Difficulties in applying conventional models for flow and transport (modelling may lead to significantly erroneous results if the nature of karst is not adequately considered) Exacting data requirements for holistic modelling (distributive models ideally require data on the location and geometry of the entire conduit network, which is never available)

11.2.2 Locating water sources

Springs are the obvious and commonly the preferred source of water supply in karst areas but inevitably there are often requirements for water source in areas that are not close to springs and boreholes must be drilled to obtain a supply.

Boreholes are more uncertain sources of reliable water supply in karst than in other aquifers as groundwater flows are highly localised, often in all three dimensions. Finding optimal locations for supply wells is notoriously difficult but may be assisted by thorough geological and geomorphological investigations (locating inception horizons for conduits, relatively impermeable layers and highly soluble layers within the limestone, faults, fracture zones and lineaments for example).

Geophysical methods are useful in identifying water-bearing zones, particularly in karst aquifers covered by thick soils and sediments, where surface expressions of fracturing and karstification are scarce. Some methods, like microgravity, assist in locating conduits (gravity-lows), particularly if they are relatively large and at shallow depth. Other methods, like seismic, resistivity profiling and various electromagnetic techniques, are powerful tools to locate fracture zones (e.g. Bosch & Müller 2005, Sumanovac & Weisser 2001).

Borehole geophysics and a variety of hydraulic tests can be performed in open boreholes and in screened wells. On one hand, such tests can be used to obtain information on the hydraulic properties of the aquifer. On the other hand, they may also help to optimise the construction and pumping rate of the well, e.g. to locate productive zones, to decide where the well should be screened, and to determine sustainable pumping rates (safe yields).

Exploration and mapping of karst conduits (active cave systems) may be a useful way of locating water supplies in some karsts.

11.2.3 Assessing water quality and contamination problems

Water quality and contamination studies always require hydrochemical and/or microbiological investigations. Microbial pathogens are particularly relevant in karst aquifers, where filtration and inactivation processes are of limited effectiveness. The high variability of karst groundwater flow systems requires adapted, event-based sampling strategies. Automatic samplers and continuous monitoring devices are particularly useful to obtain a better temporal resolution. As such devices are not available for bacteriological parameters, a variety of other parameters, such as turbidity and organic carbon, can be used as indicators for the possible presence of bacterial contamination (Pronk et al. 2006).

A variety of other methods described in this book can also be used within the framework of such investigations. Tracer tests allow the propagation of contaminants to be studied, and specific tracers can be used as surrogates for specific groups of contaminants, e.g. microspheres for *Cryptosporidium* cysts (Renken et al. 2005), and bacteriophages for pathogenic viruses (Rossi et al. 1998). Isotope techniques make it possible to distinguish whether a particular water constituent, such as sulphate or nitrate, is of natural origin or results from human activities (Einsiedl & Mayer 2005).

Geophysical methods can sometimes be used to delineate highly concentrated contaminant plumes (Fels 1999), although the heterogeneity of karst limits this approach. Models that simulate transport processes can also predict the impact of contamination on an aquifer or spring. However, transport modelling in karst is even more problematic than flow modelling and requires input data that are generally not available. Therefore, the use of models for that purpose should be restricted to cases where detailed data are available, and the model must always be validated by independent field data, preferably tracer tests (see the example from Walkerton, Canada, in the Introduction of this book).

11.2.4   Conceptualising karst flow systems

Although all karst aquifers have a considerable degree of similarity in terms of the manner in which they function, there are also considerable differences within and between particular karst aquifers; for example in the spatial distribution of conduits and fissures, in the degree of hierarchical organisation of the drainage system and in the amount and type of storage. Difficulties in reconciling these smaller scale differences with a "global" conceptual model of karst hydrogeology mean that a wide variety of data sources may be required to usefully conceptualise karst hydrogeology. These include detailed information on the location and relative importance of flow routes derived from speleological investigations and detailed information concerning the recharge, storage and transmission characteristics of the aquifer. Quantitative data from water tracing (breakthrough curves for example) and analysis of spring hydrographs and chemographs are particularly relevant in this context. Knowledge of the basic geological framework and values for hydraulic conductivity and hydraulic head, which may suffice to model aspects of many non-karstified aquifers are commonly inadequate in karsts.

11.2.5   Assessing groundwater vulnerability in karst areas

Determining the vulnerability of a groundwater body to contamination is of fundamental importance in groundwater management. The European COST Action 620 provided a conceptual framework for vulnerability and risk mapping for the protection of carbonate (karst) aquifers (Zwahlen 2004). Although some aspects of vulnerability mapping are common to all aquifer types and conditions (assessment of protective cover for example), karst regions require more specific data such as the location of points of concentrated recharge and their local catchments.

Geological, geomorphological, pedological and hydrological mapping provide the basic data required for vulnerability assessment. Speleological observations may also be of great use. Geophysical techniques make it possible to assess the thickness and properties of the overlying layers, and may also help to characterise the epikarst, identify fractures, and localise zones of potential or actual infiltration. However, such techniques are more appropriate for detailed, site-specific vulnerability assessment, while simpler (and less precise) techniques are usually used on the scale of large catchments, regions or countries.

A variety of techniques are available to validate a vulnerability assessment. Releasing a tracer at the land surface and monitoring its breakthrough at the spring is the most direct approach but restricted to selected points (Goldscheider et al. 2001). Other methods, including, hydrochemical and isotopic techniques, also provide information on contaminant pathways, transit times and attenuation processes in the system, and can thus be used for validation. The initial vulnerability map may be modified in the light of these data.

References

Ackers, P., White, W. R., Perkins, J. A., & Harrison, A. J. M. (1978). Weirs and Flumes for Flow Measurement. New York: John Wiley & Sons, 327 p.

Afinowicz, J. D., Munster, C. L. & Wilcox, B. P. (2005). Modelling effects of brush management on the rangeland water budget: Edwards Plateau, Texas. Journal of the American Water Resources Association, 41, 181–193.

Alexander, S. C. (2005) Spectral deconvolution and quantification of natural organic material and fluorescent tracer dyes. In Beck, B. F. (Ed.), Sinkholes and the engineering and environmental impacts of karst, Geotechnical Special Publication 144. American Society of Civil Engineers Reston Virginia, 441–448.

Aley, T. J., Williams, J. H., Massello, J. W. (1972). Groundwater contamination and sinkhole collapse induced by leaky impoundments in soluble rock terrain. Engineering Geology Series No.5, Missouri Geological Survey and Water Resources, 32 p.

Andersson, J., Dverstorp, B. (1987). Conditional simulations of fluid flow in three-dimensional networks of discrete fractures. Water Resources Research, 23, 1876–1886.

Andreo, B. (1997). Hidrogeología de acuíferos carbonatados en las Sierras Blanca y Mijas (Cordillera Bética, Sur de España). Servicio de Publicaciones de la Universidad de Málaga, 489 p.

Andreo, B., Carrasco, F., Sanz de Galdeano, C. (1997). Types of carbonate aquifers according to the fracturation and the karstification in a southern Spanish area. Environmental Geology, 30, 163–173.

Andreo, B., Linan, C., Carrasco, F. De Cisneros, C.J., Caballero, F. & Mudry, J. (2004). Influence of rainfall quantity on the isotopic composition (^{18}O and ^{2}H) of water in mountainous areas. Application for groundwater research in the Yunquera-Nieves karst aquifers (S Spain). Applied Geochemistry, 19, 561–574.

Andrews, J.N., Fontes, J.C., Aranyossy, J.F., Dodo, A., Edmunds, W.M., Joseph, A. & Travi, A.Y. (1994). The evolution of alkaline groundwaters in the continental intercalaire aquifer of the Irhazer-Plain, Niger. Water Resources Research, 30, 45–61.

Appelo, C. A. J. & Postma, D. (2005). Geochemistry, groundwater and pollution. Leiden: Balkema, 649 p.

Arcement, G. J. & Schneider, V. R. (1989). Guide for selecting Manning's roughness coefficients for natural channels and flood plains. U.S. Geological Survey Water-Supply Paper 2339, 38 p.

Ashton, K. (1966). The analysis of flow data from karst drainage systems. Transactions of the Cave Research Group of Great Britain, 7, 161–203.

ASTM (2001). ASTM D5242-92 standard method for open-channel flow measurement of water with thin-plate weirs. West Conshohocken, Pennsylvania: American Society for Testing and Materials, 8 p.

Atkinson, T.C., Smith, D.I., Lavis, J.J. & Whitaker, R.J. (1973). Experiments in tracing underground waters in limestones. Journal of Hydrology, 19, 323–349.

Atteia, O., Kozel, R. (1997). Particle size distributions in waters from a karstic aquifer: from particles to colloids. Journal of Hydrology, 201, 102–119.

Aubert, M. & Atangana, Q.Y. (1996). Self-potential method in hydrogeological exploration of volcanic areas. Ground Water, 34, 1010–1016.

Auckenthaler, A. & Huggenberger, P. (Eds.) (2003) Pathogene Mikroorganismen im Grund- und Trinkwasser. Base: Birkhäuser, 184 p.

Auckenthaler, A., Raso, G. & Huggenberger, P. (2002). Particle transport in a karst aquifer: natural and artificial tracer experiments with bacteria, bacteriophages and microspheres. Water Science and Technology, 46, 131–138.

Audra, P. (1994). Karsts alpins. Genèse de grands réseaux souterrains: Exemples le Tennengebirge (Autriche), l'Ile de Crémieux, la Chartreuse et le Vercors (France). Karstologia Mémoires, 5, 1–280.

Audra, P., Mocochain, L., Camus, H., Gilli, E., Clauzon, G. & Bigot, J. Y. (2004). The effect of the Messinian Deep Stage on karst development around the Mediterranean Sea. Examples from Southern France. Geodinamica Acta, 17, 27–38.

Baecher, G. B., Lanney, N. A. & Einstein, H. H. (1977). Statistical descriptions of rock properties and sampling. Proceedings of the 18th U.S. Rock Mechanics Symposium, 5C1-1-5C1-8.

Bakalowicz, M. (1977). Etude du degré d'organisation des écoulements souterrains dans les aquifères carbonatés par une méthode hydrogéochimique nouvelle. C.R. Acad. Sciences, Paris D, 284, 2463–2466.

Bakalowicz, M. (2005). Karst groundwater: a challenge for new resources. Hydrogeology Journal, 13, 148–160.

Baker, A. & Brundson, C. (2003). Non-linearities in drip water hydrology: an example from Stump Cross Cavern, Yorkshire. Journal of Hydrology, 277, 151–163.

Baker, G. S., Schmeissner, C., Steeples, D. W. & Plumb, R. G. (1999). Seismic reflections from depths of less than two meters. Geophysical Research Letters, 26, 279–282.

Barenblatt, G. I., Zheltov, I. P. & Kochina, I. N. (1960). Basic concepts in the theory of seepage of homogeneous liquids in fissured rocks. Jour. Appl. Math. and Mechanics, 24, 1286–1303.

Bar-Matthews, M., Ayalon, A., Matthews, A., Sass, E., Halicz, L. (1996). Carbon and oxygen isotope study of the active water-carbonate system in a karstic Mediterranean cave: Implications for paleoclimate research in semiarid regions. Geochimica et Cosmochimica Acta, 60, 337–347.

Barnes, C. J., Allison, G. B. (1988). Tracing of water movement in the unsaturated zone using stable isotopes of hydrogen and oxygen. Journal of Hydrology, 100, 143–176.

Barnes, H. H. (1967). Roughness characteristics of natural channels. U.S. Geological Survey Water-Supply Paper 1849, 213 p.

Batiot, C., Emblanch, C. & Blavoux, B., (2003). Total organic carbon (TOC) and magnesium (Mg^{2+}) two complementary tracers of residence time in karstic systems. Comptes Rendus Geosciences, 335, 205–214.

Batsche, H., Bauer, F., Behrens, H., Buchtela, K., Dombrowski, H. J., Geisler, R., Geyh, M. A., Hötzl, H., Hribar, F., Käss, W., Mairhofer, J., Maurin, V., Moser, H., Neumaier, F., Schmitz, J., Schnitzer, W.A., Schreiner, A., Vogg, H. & Zötl, J.

(1970). Kombinierte Karstwasseruntersuchungen im Gebiet der Donauversickerung (Baden-Württemberg) in den Jahren 1967-1969. Steir. Beitr. z. Hydrogeol., 22, 1–165.

Bauer, F. (1989), Die unterirdischen Abflussverhältnisse im Dachsteingebiet und ihre Bedeutung für den Karstwasserschutz. Bericht UBA-89-28, 73 p.

Bauer, S., Liedl, R. & Sauter, M. (2003). Modeling of karst aquifer genesis: Influence of exchange flow. Water Resources Research, 39, Art. No. 1285.

Bäumle, R., Einsiedl, F., Hötzl, H., Käss, W., Witthüser, K. & Wohnlich, S. (2001). Comparative tracer studies in a highly permeable fault zone at the Lindau fractured rock test site, SW Germany. Beitr. z. Hydrogeologie, 52, 136–145.

Bear, J. (1979). Hydraulics of Groundwater. New York: McGraw-Hill, 567 p.

Bear, J., Tsang, C. F. & de Marsily, G. (Eds.) (1993). Flow and contaminant transport in fractured rock. San Diego: Academic Press, 548 p.

Bechtel, T. D., Forsyth, D. W., & Swain, C. J. (1987). Mechanisms of isostatic compensation in the vicinity of the East African Rift, Kenya. Geophysical Journal of the Royal Astronomical Society, 90, 445–465.

Bechtel, T. D., Hojdila, J. I., Baughman, S. H., DeMayo, T. & Doheny, E. (2005). Relost and refound: Detection of a paleontologically, historically, cinematically(?), and environmentally important solution feature in the carbonate belt of Southeastern Pennsylvania. The Leading Edge, 47, 537–540.

Behrens, H. (1970). Zur Messung von Fluoreszenzfarbstoffen. Inst. F. Radiohydrometrie, Jahresbericht 1969, GSF-Bericht R 25, 92–96.

Behrens, H. (1988). Quantitative Bestimmung von Uranin, Eosin und Pyranin in Gemischen mittels Fluoreszenzmessung bei definierten pH-Werten. Steir. Beitr. z. Hydrogeologie, 39, 117–129.

Behrens, H. (1998). Radioactive and activable isotopes. In Käss, W., Tracing Technique in Geohydrology (pp. 167–187). Rotterdam/Brookfield: Balkema.

Behrens, H., Deims, U., Dieter, H., Dietze, G., Eikmann, T., Grummt, T., Hanisch, H., Henseling, H., Käss, W., Kerndorff, H., Leibundgut, C., Müller-Wegener, U., Rönnefahrt, I., Scharenberg, B., Schleyer, R., Schloz, W. & Tilkes, F. (2001). Toxicological and ecotoxicological assessment of water tracers. Hydrogeology Journal, 9, 321–325.

Behrens, H., Benischke, R., Bricelj, M., Harum, T., Käss, W., Kosi, G., Leditzky, H. P., Leibundgut, C., Maloszewski, P., Maurin, V., Rajner, V., Rank, D., Reichert, B., Stadler, H., Stichler, W., Trimborn, P., Zojer, H. & Zupan, M. (1992). Investigations with natural and artificial tracers in the karst aquifer of the Lurbach system (Peggau-Tanneben-Semriach, Austria). Steir. Beitr. z. Hydrogeologie, 43, 9–158.

Behrens, H. & Demuth, N. (1992). Measurement of light input into surface waters by photolysis of fluorescent dye tracers. Proc. of the 6th Int. Symp. on Water Tracing, Karlsruhe, Germany, 21–26 Sept 1992, 49–56.

Bell & Howell (1983). Pressure Transducer Handbook. Pasadena, California: Division of CEC Instruments, 143 p.

Benischke, R. & Leitner, A. (1992). Fiberoptic fluorescence sensors - An advanced concept for tracer hydrology. Proc. of the 6th Int. Symp. on Water Tracing, Karlsruhe, Germany, 21–26 Sept 1992, 41–48.

Benischke, R. & Schmerlaib, H. (1986). Pyranin: a fluorescent dye for tracer hydrology – review of physico-chemical properties, the toxicity and applicability. Proc. of the 5th Int. Symp. on Underground Water Tracing, Athens, 135–148.

Benson, R. C. & La Fountain, L. J. (1984). Evaluation of subsidence or collapse potential due to subsurface cavities. In Sinkholes: Their Geology, Engineering and Environmental Impacts, Proceedings of the 1st Multidisciplinary Conference on Sinkholes, Orlando, Florida, 201–215.

Beres, M. & Haeni, F. P. (1991). Application of ground-penetrating radar methods in hydrogeologic studies. Ground Water, 29, 383–385.

Betts, O. (2004). The Survex project [On-line]. Available: http://www.survex.com/ (5.9.2004).

Birk, S. (2002). Characterization of karst systems by simulating aquifer genesis and spring responses: Model development and application to gypsum karst. Tübinger Geowissenschaftliche Arbeiten C60. Also available at: http://w210.ub.uni-tuebingen.de/dbt/volltexte/2002/558.

Birk, S., Liedl, R., Sauter, M. & Teutsch, G. (2003). Hydraulic boundary conditions as a controlling factor in karst genesis: A numerical modeling study on artesian conduit development in gypsum. Water Resources Research, 39, Art. No. 1004.

Birk, S., Sauter, M. & Liedl, R. (2001). Process-based modeling of concentration and temperature variations at a karst spring. Proc. 7th Conf. on Limestone Hydrology and Fissured Media, Besançon, 41–44.

Blake, R. E. (1992). Stable isotope systematics and ground-water mixing relationships in the Cretaceous Edwards aquifer, Comal, Hays, and Bexar Counties, south-central Texas. MS Thesis, The University of Texas at San Antonio.

Bögli, A. (1980). Karst Hydrology and Physical Speleology. New York: Springer, 270 p.

Bögli, A. & Harum, T. (Eds.) (1981). Hydrogeologische Untersuchungen im Karst des hinteren Muotatales (Schweiz). Steir. Beitr. Hydrogeol., 33, 125–264.

Böhlke, J. K. (2002). Groundwater recharge and agricultural contamination. Hydrogeology Journal, 10, 153–179.

Bolster, C., Groves, C., Meiman, J., Fernández-Cortes, A., & Crockett, C. (2006). Practical limits of high-resolution of carbonate chemistry within karst flow systems. Proceedings of the 8th Conference on Limestone Hydrogeology, Neuchâtel, Switzerland, 79–82.

Bonacci, O. (1987). Karst hydrology with special reference to the Dinaric karst. New York: Springer, 184 p.

Bonacci, O. (1993). Karst springs hydrographs as indicators of karst aquifers. Hydrological Science, 38, 51–62.

Bosch, F. P. & Müller, I. (2001). Continuous gradient VLF measurements: A new possibility for high resolution mapping of karst structures. First Break, 19, 343–350.

Bosch, F. P. & Müller, I. (2005). Improved karst exploration by VLF-EM-gradient survey: comparison with other geophysical methods. Near Surface Geophysics, 3, 299–310.

Boulton, N. S. (1954a). Unsteady radial flow to a pumped well allowing for delayed yield from storage. Int. Assoc. Sci. Hydrology Pub. 37, 472–477.

Boulton, N. S. (1954b). The drawdown of the water table under nonsteady conditions near a pumped well in an unconfined formation. Inst. Civil Engineers Proc. 3, 564–579.

Boulton, N.S. (1963). Analysis of data from non-equilibrium pumping tests allowing for delayed yield from storage: Inst. Civil Engineers Proc., 26, 469–482.

Boulton, N. S. (1970). Analysis of data from pumping tests in unconfined anisotropic aquifers. Journal of Hydrology, 10, 369.

Boulton, N. S. (1973). The influence of delayed drainage on data from pumping tests in unconfined aquifers. Journal of Hydrology, 19, 157–169.

Bouwer, H. (1989). The Bouwer and Rice slug test. Ground Water, 27, 304–309.

Bouwer, H. & Rice, R.C. (1976). A slug test for determining hydraulic conductivity of unconfined aquifers with completely or partially penetrating wells. Water Resources Research, 12, 423–428.

Boving, T. B., Meritt, D. L. & Boothroyd, J. C. (2004). Fingerprinting sources of bacterial input into small residential watersheds: fate of fluorescent whitening agents. Environmental Geology, 46, 228–232.

Box, G. E. P. & Jenkins, G. M. (1976). Time series analysis: forecasting and control. San Francisco: Holden Day, 575 p.

Brassington, R. (2006). Field Hydrogeology, 3rd edition. Chichester: Wiley.

Brown, M. C. & Ford, D. C. (1971). Quantitative tracer methods for investigation of karst hydrologic systems with special reference to the Maligne Basin area. Trans. Cave Res. Group of Great Britain, 13, 37–51.

Brown, M. C., Wigley, T. M. L. & Ford, D. C. (1969). Water budget studies in karst aquifers. Journal of Hydrology, 9, 113–116.

Buchanan, T. J., & Somers, W. P. (1968). Stage Measurements at Gaging Stations. U.S. Geological Survey Techniques of Water-Resources Investigations, book 3, chap. A7, 28 p.

Buchli, R. (1987). Radongehalt des Schweizer Trinkwassers und Bedeutung des Trinkwassers als Radonquelle in Wohnräumen, PSI (EIR), Report TM-81-87-03, Würenlingen, Switzerland.

Butler, D. K. (1984). Microgravimetric and gravity gradient techniques for detection of subsurface cavities. Geophysics, 41, 1084–1096.

Cacas, M. C. (1989). Développement d'un modèle tridimensionnel stochastique discret pour la simulation de l'écoulement et des transports de masse et de chaleur en milieu fracturé. Ph.D. thesis, Ecole des Mines de Paris, Fontainebleau, France.

Cacas, M. C., Ledoux, E., de Marsily, G., Tilie, B., Barbreau, A., Durand, E., Feuga, B. & Peaudecerf, P. (1990). Modeling fracture flow with a stochastic discrete fracture network model: calibration and validation, 1. the flow model. Water Resources Research, 26, 479–489.

Cagniard, L. (1953). Basic theory of the magnetotelluric method of geophysical prospecting. Geophysics, 18, 605–635.

Carslaw, H. S. & Jaeger, J. C. (1959). Conduction of heat in solids. Oxford: Clarendon Press.

Chanson, H. (2004). The hydraulics of open channel flows: an introduction (2nd edition). Oxford: Elsevier, 558 p.

Chilès, J. P. & de Marsily, G. (1993). Stochastic models of fracture systems and their use in flow and transport modeling. In Bear, J. Tsang, C. F. & de Marsily, G. (Eds.) Flow and Contaminant Transport in Fractured Rock (pp. 169–236). San Diego: Academic Press.

Chow, V. T. (1959). Open channel hydraulics. New York: McGraw-Hill, 680 p.

Chow, V. T. (Ed.) (1964). Handbook of Applied Hydrology. New York: McGraw-Hill.

Christensen, T. H., Kjeldsen, P., Bjerg, P. L., Jensen, D. L., Christensen, J. B., Baun, A., Albrechtsen, H. J. & Heron, G. (2001). Biogeochemistry of landfill leachate plumes. Appl. Geochem., 16, 659–718.

Clark, I. D. & Fritz, P. (1997). Environmental isotopes in hydrogeology. Boca Raton: Lewis Publishers, 328 pp.

Clark, J. F., Davisson, M. L., Hudson, G. B. & Macfarlane, P. A. (1998). Noble gases, stable isotopes, and radiocarbon as tracers of flow in the Dakota aquifer, Colorado and Kansas. Journal of Hydrology, 211, 151–167.

Clarke, J. S., Leeth, D. C., Taylor-Harris, D., Painter, J. A. & Labowski, J. L. (2004). Hydraulic properties of the Floridan Aquifer System and equivalent clastic units in coastal Georgia and adjacent parts of South Carolina and Florida. Georgia Geologic Survey Information Circular, 109, 50 p.

Clemens, T., Hückinghaus, D., Sauter, M., Liedl, R. & Teutsch, G (1996). A combined continuum and discrete network reactive transport model for the simulation of karst development. IAHS Publ., 237, 309–318.

Clever, H. L. (1985). Solubility Data Series, Vol. 2, Krypton, Xenon and Radon. Oxford: Pergamon Press.

Close, M. E., Stanton, G. J. & Pang, L. (2002). Use of rhodamine WT with XAD-7 resin for determining groundwater flow paths. Hydrogeology Journal, 10, 368–376.

Cobb, E. D. & Bailey, J. F. (1965) Measurement of Discharge by Dye-Dilution Methods. U.S. Geological Survey Surface-Water Techniques, Book 1, chap. 14, 27 p.

Collier, C. G. (1996). Applications of weather radar systems. a guide to users of radar data in meteorology and hydrology (2nd ed.). Chinchester: Ellis Horwood Limited Publisher, 390 p.

Cook, P. G. (2003). A guide to regional groundwater flow in fractured aquifers. CSIRO Land and Water, Seaview Press, Henley Beach, South Australia, 108 p.

Cooper, H. H. (1963). Type curves for nonsteady radial flow in an infinite leaky artesian aquifer. US Geological Survey Water-Supply Paper, 1545-C, C48-C55.

Cooper, H. H., Bredehoeft, J. D., & Papadopulos, I. S. (1967). Response of a finite-diameter well to an instantaneous charge of water. Water Resources Research, 3, 263–269.

Cooper, H. H., & Jacob, C. E. (1946). A generalized graphical method for evaluating formation constants and summarizing well field history. Am. Geophys. Union Trans., 27, 526–534.

Cornaton, F. (1999). Utilisation de modèles continus discrets et à double continuum pour l'analyse des réponses globales de l'aquifère karstique. Diploma work, CHYN, University of Neuchâtel.

Cornaton, F. & Perrochet, P. (2002). Analytical 1D dual-porosity equivalent solutions to 3D discrete single-continuum models. Application to karstic spring hydrograph modeling. Journal of Hydrology, 262, 165–176.

Costa, J. E., Spicer, K. R., Cheng, R. T., Haeni, F. P., Melcher, N. B., Thurman, E. M., Plant, W. J. & Keller, W.C. (2000). Measuring stream discharge by non-contact methods – a proof-of-concept experiment. Geophysical Research Letters, 27, 553–556.

Craig, D. H. (1988). Caves and older features of Permian karst in San Andres dolomite, Yates Field reservoir, west Texas. In James, N. P. & Choquette, P. W. (Eds.), Paleokarst (p. 345–363). New York: Springer.

Craig, H. (1961). Isotopic variations in meteoric waters. Science, 133, 1702–1703.

Craig, J. D. (1983). Installation and Service Manual for U.S. Geological Survey Manometers. U.S. Geological Survey Techniques of Water-Resources Investigations, Book 8, Chapter A2, 64 p.

Crawford, N. C. (1984). Toxic and explosive fumes rising form carbonate aquifers: a hazard for residents of sinkhole plains. In B. F. Beck (Ed.), First multidisciplinary conference on sinkholes (pp. 297–309). Orlando, FL, USA: A.A. Balkema, Rotterdam.

Crawford, N. C., Lewis, M. A., Winter, S. A. & Webster, J. A. (1999). Microgravity techniques for subsurface investigations of sinkhole collapses and for the detection of groundwater flow paths through karst aquifers. In Beck, B.F., Pettit, A.J. & Herring, J. G.

(Eds.), Hydrology and Engineering Geology of Sinkholes and Karst (pp. 203–218). Rotterdam: A.A. Balkema.

Crawford, N. C. & Ulmer, C.S. (1994). Hydrogeologic investigation of contaminant movement in karst aquifers in the vicinity of a train derailment near Lewisburg, Tennessee. Environmental Geology, 23, 41–52.

Criss, R. E. (1999). Principles of stable isotope distribution. New York: Oxford University Press, 254 p.

Criss, R. E. & Davisson, M. L. (1996). Isotopic imaging of surface water/groundwater interactions, Sacramento Valley, California. Journal of Hydrology, 178, 205–222.

Criss, R. E., Fernandes, S. A. & Winston, W. E. (2001). Isotopic, geochemical and biological tracing of the source of an impacted karst spring, Weldon Spring, Missouri. Environmental Forensics, 2, 99–103.

Criss, R. E. & Winston, W. E. (2003). Hydrograph for small basins following intense storms. Geophysical Research Letters, 30, 1314–1318.

Crowther, J. (1987). Ecological observations in tropical karst terrain, West Malaysia. III. Dynamics of the vegetation-soil-bedrock system. Journal of Biogeography, 14, 157–164.

Cruz Sanjulián, J. J. (1981). Evolución geomorfológica e hidrogeológica reciente en el sector Teba-Cañete la Real (Málaga) a la luz de la datación de formaciones travertínicas. Bol. Geol. y Min., 92, 297–308.

Curl, R. L. (1986). Fractal dimensions and geometries of caves. Mathematical Geology, 18, 765–783.

Currens, J. C. (1999). A sampling plan for conduit-flow karst springs: minimizing sampling cost and maximizing statistical utility. Engineering Geology, 52, 121–128.

Dalrymple, T. & Benson, M. A. (1967). Measurement of peak discharge by the slope-area method. U.S. Geolological Survey Techniques Water-Resources Inv, book 3, chap. A2, 12 p.

Daniels, D. J. (1996). Surface-Penetrating Radar. UK: IEEE, 300 p.

Dansgaard, W. (1964). Stable isotopes in precipitation. Tellus, 16, 436–468.

Dassargues, A. (Ed.) (2000). Tracers and modeling in Hydrogeology. Int. Ass. Sci. Hydrology, 262, 572 p.

Daubechies, I. (1992). The wavelet transform time-frequency localization and signal analysis. IEEE Trans. Inform. Theory, 36, 961–1004.

David, E. (2006). Visual Topo [On-line]. Available: http://vtopo.free.fr/ (8.5.2006).

Davis, G. H., Reynolds, S. J (1996). Structural geology of rocks and regions, 2nd edition. Wiley, 800 p.

Davis, K. J., Dove, P. M., De Yoreo, J. J. (2000). The role of Mg^{2+} as an impurity in calcite growth. Science, 290, 5494, 1134–1137.

Davis, P. M., Atkinson, T. C., Wigley, T. M. L. (2000). Longitudinal dispersion in natural channels: 2. The roles of shear flow dispersion and dead zones in the River Severn, UK. Hydrology and Earth System Sciences, 4, 355–371.

Davisson, M. L. & Criss, R. E. (1996). Stable isotope and groundwater flow dynamics of agricultural irrigation recharge into groundwater resources of the Central Valley, California. International Atomic Energy Agency Symposium on Isotope Hydrology in Water Resources Management, March 1995, Vienna. IAEA-SM-336/14: 405–418.

Davisson, M. L., Smith, D. K., Keneally, J. & Rose, T. P. (1999). Isotope hydrology of southern Nevada groundwater: stable isotopes and radiocarbon. Water Resources Research, 35, 279–294.

Dawson, K. & Istok, J. (1991) Aquifer Testing: Design and Analysis of Pumping and Slug Tests. Boca Raton, Florida: Lewis Publishers, 280 p.

Dean, J. A. (1995). Analytical chemistry handbook. New York: McGraw-Hill, 1168 p.

Dershowitz, W. S., La Pointe, P. R. & Doe, T. W. (2004). Advances in discrete fracture network modeling. Proceedings of the U.S. EPA/NGWA Fractured Rock Conference. Portland.

Dershowitz, W. S., Toxford, T., Sudicky, E., Shuttle, D. A., Eiben, T. & Ahlstrom, E. (1998). PA Works pathways analysis for Discrete Fracture Networks with LTG solute transport. User Documentation, Golder Associates Inc., Redmond, Washington.

Despain, J., Groves, C. & Meiman, J. (2006). Hydrology and rock/water interactions of an Alpine Karst System: Spring Creek, Mineral King, Sequoia National Park, California. Proceedings of the 8th Conference on Limestone Hydrogeology, Neuchâtel, Switzerland, 79–82.

Dincer, T. & Payne, B. R. (1971). An environmental isotope study of the south-western karst region of Turkey. Journal of Hydrology, 14, 233–258.

Doerfliger, N. & Zwahlen, F. (1995). Action COST 65 – Swiss National Report. Bulletin d'Hydrogéologie de l'Université de Neuchâtel, 14, 3–33.

Doesken, N. J. & Judson, A. (1996). The snow booklet. A guide to the science, climatology, and measurement of snow in the United States, 88 p.

Doorenbos, J. (1976). Agro-Meteorological Field Stations. FAO irrigation and Drainage Paper 27. Rome: Food and Agriculture Organisation of the United Nations.

Dorale, J. A., Edwards, R. L., Ito, E. & Gonzales, L. A. (1998). Climate and vegetation history of the mid-continent from 75 to 25 ka: A speoleothem record from Crevice Cave, Missouri, USA. Science, 282, 1871–1874.

Dreiss, S. J. (1983) Linear unit-response functions as indicators of recharge areas for large karst springs. Journal of Hydrology, 61, 31–44.

Dreiss, S. J. (1989). Regional scale transport in karst aquifer. 2: Linear systems and time moment analysis. Water Resources Research, 25, 126–134.

Drew, D. & Hötzl, H. (Eds.) (1999). Karst hydrogeology and human activities. Impacts, consequences and implications. International Contributions to Hydrogeology, 20, 322 pp.

Dreybrodt, W. (1988). Processes in karst systems – physics, chemistry and geology. Heidelberg, New York: Springer, 288 p.

Dreybrodt, W. (1996). Principles of early development of karst conduits. Water Resources Research, 32, 2923–2935.

Dreybrodt, W. (1998). Limestone dissolution in karst environments. Bulletin d'Hydrogéologie, 16, 167–183.

Dreybrodt, W. & Gabrovsek, F. (2000). Dynamics of the evolution of single karst conduits. In A. Klimchouk, D. Ford, C. Palmer, A. & Dreybrodt, W. (Eds.), Speleogenesis: Evolution of karst aquifers (pp. 184–193). Huntsville, Alabama (USA): National Speleological Society.

Dreybrodt, W., Gabrovsek, F. & Romanov, D. (2005). Processes of speleogenesis: a modeling approach. Ljubljana: Zalozba ZRC, 376 p.

Driscoll, F.G. (1986). Groundwater and Wells, Second Edition. St. Paul, Minnesota: Johnson Filtration Systems Inc., 1089 p.

Drogue, C. (1972). Analyse statistique des hydrogrammes de décrues des sources karstiques. Journal of Hydrology, 15, 49–68.

Drogue, C. (1974). Structure de certains aquifères karstiques d'après les résultats de travaux de forage. Comptes Rendus Académie des sciences, Paris, D, 278, 2621–2624.

Drogue, C. (1980). Essai d'identification d'un type de structure de magasins carbonatés fissurés. Application à l'interprétation de certains aspects du fonctionnement hydrogéologique. Mémoires hors série Société Géologique de France, 11, 101–108.

Dunham, R. J. (1962). Classification of carbonate rocks according to depositional texture. Amer. Assoc. Petrol. Geol. Memoir, 1, 108–121.

Dunne, T., & Leopold, L. B. (1978). Water in Environmental Planning (Chapter 13, Snow Hydrology). San Francisco: W. H. Freeman, p. 468–488.

Dverstorp, B., Andersson, J. & Nordqvist, W. (1992). Discrete fracture network interpretation of field tracer migration in sparsely fractured rock. Water Resources Research, 28, 2327–2343.

Eakin, T. E. (1966). A regional interbasin groundwater system in the White River area, southeastern Nevada. Water Resources Research, 2, 251–271.

Eaton, A. D., Clesceri, L. S. & Greenberg, A. E. (Eds.) (1995). Standard methods for the examination of water and wastewater. Washington, DC: American Public Health Association.

Edmunds, W. M., Cook, J. M., Darling, W. G., Kinniburgh, D. G., Miles, D. L., Bath, A. H., Morgan-Jones, M. & Andrews, J. N. (1987). Baseline geochemical conditions in the Chalk aquifer, Berkshire, UK: a basis for groundwater quality management. Applied Geochemistry, 2, 251–274.

Einsiedl, F., Maloszewski, P. & Stichler, W. (2005). Estimation of denitrification potential in a karst aquifer using N-15 and O-18 isotopes of NO_3^-. Biogeochemistry, 72, 67–86.

Einsiedl, F. & Mayer, B. (2005). Sources and processes affecting sulfate in a karstic groundwater system of the Franconian Alb, southern Germany. Environmental Science & Technology, 39, 7118–7125.

Eisenlohr, L., Király, L., Bouzelboudjen, M. & Rossier, I. (1997a). A numerical simulation as a tool for checking the interpretation of karst springs hydrographs. Journal of Hydrology, 193, 306–315.

Eisenlohr, L., Király, L., Bouzelboudjen, M. & Rossier, I. (1997b). Numerical versus statistical modeling of natural response of a karst hydrogeological system. Journal of Hydrology, 202, 244–262.

Eisenlohr, L. & Surbeck, H. (1995). Radon as a natural tracer to study transport processes in a karst system; an example from the Swiss Jura. C.R. Acad. Sci. Paris, 321, IIa, 761–767.

Ekwurzel, B., Schlosser, P., Smethie, W. M., Plummer, L. N., Busenberg, E., Michel, R. L., Weppernig, R. & Stute, M. (1994) Dating of shallow groundwater: Comparison of the transient tracers $^3H/^3He$, chlorofluorocarbons, and ^{85}Kr. Water Resources Research, 30, 1693–1708.

Emblanch, C., Blavoux, B., Puig, J. M. & Mudry, J. (1998). Dissolved organic carbon of infiltration within the autogenic karst hydrosystem. Geophysical research letters, 25, 1459–1462.

Emblanch, C., Zuppi, G. M., Mudry, J., Blavoux, B. & Batiot, C. (2003). Carbon-13 of TDIC to quantify the role of the unsaturated zone: the exemple of the Vaucluse karst systems (Southeastern France). J. Hydrol., 279, 262–274.

Ernston, K. & Scherer, H. U. (1986). Self-Potential variations with time and their relation to hydrogeologic and meteorological parameters. Geophysics, 51, 1967–1977.

Fagerlund, F. & Heinson, G. (2003). Detecting subsurface groundwater flow in fractured rock using self-potential (SP) methods. Environmental Geology, 43, 782–794.

Farrell, D., Mayers, B. L., Moseley, L., Sandberg, S., Mwansa, J., Sutherland, A. & Barnes, H. (2003). Geophysical imaging of seawater intrusion into karst limestones to support groundwater modelling along the west coast of Barbados [On-line]. Available: http://scitec.uwichill.edu.bb/OAS/Reports/Reports.htm

Faybishenko, B., Witherspoon, P. A. & S.M. Benson (Eds.) (2000). Dynamics of Fluids in Fractured Rock. Geophysical Monograph 122, American Geophysical Union, Washington, D.C., 400 p.

Fels, J. B. (1999). Source-identification investigations of petroleum contaminated groundwater in the Missouri Ozarks. Engineering Geology, 52, 3–13.

Felton, G. K. & Currens, J. C. (1994). Peak flow rate and recession-curve characteristics of a karst spring in the Inner Bluegrass, central Kentucky. J. Hydrol., 162, 99–118.

Fernandez-Gibert, E., Calaforra, J. M. & Rossi, C. (2000). Speleogenesis in the Picos de Europa Massif, Northern Spain. In Klimchouk, A., Ford, D.C., Palmer, A.N. & Dreybrodt, W. (Eds.), Speleogenesis. Evolution of karst aquifers (pp. 352–357). Huntsville, Alabama: National Speleological Society.

Ferris, J. G., Knowless, D. B., Brown, R. H., Stallman, R. W. (1962). Theory of aquifer tests. U.S. Geological Survey Water-Supply Paper 1536-E, 69–174.

Fetter, C. W. (1999). Contaminant hydrogeology. Upper Saddle River, NJ: Prentice Hall, 500 p.

Fetter, C.W. (2001). Applied hydrogeology, 4th edition. Upper Saddle River, NJ: Prentice Hall.

Field, M. S. (2002). The QTRACER2 program for tracer-breakthrough curve analysis for tracer tests in karstic aquifers and other hydrologic systems. U.S. Environmental Protection Agency, 600/R-02/001, 179 p.

Field, M. S. (2003). A review of some tracer-test design equations for tracer-mass estimation and sample-collection frequency. Environmental Geology, 43, 867–881.

Field, M. S. & Pinsky, P. F. (2000). A two-region nonequilibrium model for solute transport in solution conduits in karstic aquifers. J. Contam. Hydrol., 44, 329–351.

Fischer, G., Quang, B. V. L. & Müller, I. (1983) VLF ground surveys, a powerful tool for the study of shallow two-dimensional structures. Geophysical Prospecting, 31, 977–991.

Fish, L. (2004). COMPASS [On-line]. Available: http://www.fountainware.com/compass/ (5.9.2004).

Flury, M. & Wai, N.N. (2003). Dyes as tracers for vadose zone hydrology. Reviews of Geophysics, 41, 2.1–2.37.

Flynn, R. M., Schnegg, P. A., Costa, R., Mallen, G. & Zwahlen, F. (2005). Identification of zones of preferential groundwater transport using a mobile downhole fluorometer. Hydrogelogy Journal, 13, 366–377.

Folk, R. L. (1959). Practical petrographic classification of limestones. Bull. Amer. Assoc. Petrol. Geol., 43, 1–38.

Fontes, J. C. & Garnier, J. M. (1979). Determination of the initial ^{14}C activity of the total dissolved carbon: A review of existing models and a new approach. Water Resources Research, 15, 399–413.

Ford, D. C. & Williams, P. W. (1989). Karst geomorphology and hydrology. London: Chapman & Hall, 601 p.

Ford, D. C. & Williams, P. W. (2007). Karst hydrogeology and geomorphology. Wiley & Sons, 448 p.

Forkasiewicz, J. & Paloc, H. (1967). Le régime de tarissement de la Foux-de-la-Vis. Etude préliminaire. Chronique d'Hydrogéologie, BRGM, 3, 61–73.

France, S. (2001). Cave surveying by radio location. Cave radio and electronics group journal, 44, 21–23.

Frederickson, G. C. & Criss, R. E. (1999). Isotope hydrology and time constants of the unimpounded Meramec River basin, Missouri. Chemical Geology, 157, 303–317.

Freeman, L. A., Carpenter, M. C., Rosenberry, D. O., Rousseau, J. P., Unger, R. & McLean, J. S. (2004). Use of Submersible Pressure Transducers in Water-Resources Investigations. U.S. Geological Survey Techniques of Water-Resources Investigations, Book 8, Chapter A3, 50 p.

Frumkin, A. (1994). Hydrology and denudation rates of halite karst. Journal of Hydrology, 162, 171–189.

Füchtbauer, H. (Ed.) (1988). Sedimente und Sedimentgesteine. Stuttgart: Schweizerbart, 1141 p.

Gabrovsek, F., Menne, B. & Dreybrodt, W. (2000). A model of early evolution of karst conduits affected by subterranean CO_2 sources. Environmental Geology, 39, 531–543.

Gaiffe, M. (1987). Processus pédogénetiques dans le karst Jurassien, analyse de la comlexation organo-minérale en ambiance calcique. PhD thesis, University of Besançon, France.

Garnier, J. M. (1985). Retardation of dissolved radiocarbon through a carbonated matrix. Geochimica et Cosmochimica Acta, 49, 683–693.

Gaspar, E. (Ed.) (1987a). Modern trends in tracer hydrology. Vol. 1. Boca Raton: CRC, 145 p.

Gaspar, E. (Ed.) (1987b). Modern trends in tracer hydrology. Vol. 2. Boca Raton: CRC, 137 p.

Genty, D. & Deflandre, G. (1998). Drip flow variations under a stalactite of the Pere Noel cave (Belgium). Evidence of seasonal variations and air pressure constraints. Journal of Hydrology, 211, 208–232.

Gibson, D. (1996). How accurate is radio-location? Cave and Karst Science, 23, 77–80.

Gibson, D. (2001). Cave surveying by radio location. Cave radio and electronics group journal, 43, 24–26.

Glover, R. R. (1976). Cave surveying by electromagnetic induction. In Ellis, B. (Ed.), Surveying Caves. Bridgewater: British Cave Research Association.

Goldscheider, N. (2005). Fold structure and underground drainage pattern in the alpine karst system Hochifen-Gottesacker. Eclogae Geologicae Helvetiae, 98, 1–17.

Goldscheider, N., Göppert, N., Pronk, M. (2006). Comparison of solute and particle transport in shallow and deep karst aquifer systems. 8th Conf. on Limestone Hydrogeology, Neuchâtel, 21–23 Sep. 2006, 133–136.

Goldscheider, N., Hötzl, H., Fries, W. & Jordan, P. (2001). Validation of a vulnerability map (EPIK) with tracer tests. Proc. of the 7th Conf. on Limestone Hydrology, Besançon 20–22 Sep 2001, 167–170.

Goldscheider, N., Hötzl, H. & Käss, W. (2001a). Comparative tracer test in the alpine karst system Hochifen-Gottesacker, German-Austrian Alps. Beitr. z. Hydrogeologie, 52, 145–158.

Goldscheider, N., Hötzl, H., Käss, W. & Ufrecht, W. (2003). Combined tracer tests in the karst aquifer of the artesian mineral springs of Stuttgart, Germany. Environmental Geology, 43, 922–929.

Goldscheider, N., Hötzl, H. & Kottke, K. (2001b). Microbiological decay of Naphthionate in water samples as a source of misinterpretation of tracer tests. Proc. XXXI IAH Congress, Munich 2001, 77–81.

Goldscheider, N., Hunkeler, D. & Rossi, P. (2006). Review: microbial biocenoses in pristine aquifers and an assessment of investigative methods. Hydrogeology Journal, 14, 926–941.

Gonfiantini, R., Zuppi, G. M. (2003). Carbon isotope exchange rate of DIC in karst groundwater. Chemical Geology, 197, 319–336.

Göppert, N., Goldscheider, N. & Hötzl, H. (2005). Transport of colloidal and solute tracers in three different types of alpine karst aquifers, examples from southern Germany and Slovenia. 10th Multidisciplinary Conference on Sinkholes and the Engineering and Environment Impacts of Karst, 24–28 Sept, San Antonio, USA, 385–393.

Göppert, N., Goldscheider, N., Scholz, H. (2002). Karsterscheinungen und Hydrogeologie karbonatischer Konglomerate der Faltenmolasse im Gebiet Hochgrat und Lecknertal (Bayern/Vorarlberg). Beitr. z. Hydrogeologie, 53, 21–44.

Gospodaric, R., Habic, P. (Eds.) (1976). Underground water tracing. Investigations in Slovenia 1972–1975. Ljubljana: Inst. of Karst Research Postojna, 312 p.

Governa, M. E., Lombardi, S., Masciocco, L., Riba, M. & Zuppi, G. M. (1989). Karst and geo-thermal water circulation in the central Apennines (Italy). IAEA-AG-329.2/9, 173–202.

Granger, D., Fabel, D. & Palmer, A. N. (2001). Pliocene-Pleistocene incision of the Green River, Kentucky, determined from radioactive decay of cosmogenic ^{26}Al and ^{10}Be in Mammoth Cave sediments. GSA Bulletin, 113, 825–836.

Grasso, D. A. (1998). Interprétation des réponses hydrauliques et chimiques des sources karstiques. PhD thesis, Centre of Hydrogeology (CHYN), University of Neuchâtel, Switzerland.

Grasso, D. A. & Jeannin, P.-Y. (1994). Etude critique des méthodes d'analyse de la réponse globale des systèmes karstiques. Application au site de Bure (JU, Suisse). Bulletin d'Hydrogéologie, 13, 87–113.

Grasso, D. A. & Jeannin, P.-Y. (1998). Statistical approach to the impact of climatic variations on karst spring chemical response. Bulletin d'Hydrogéologie, 16, 59–74.

Grasso, D. A., Jeannin, P.-Y., & Zwahlen, F. (2003). A deterministic approach to the coupled analysis of karst springs' hydrographs & chemographs. Journal of Hydrology, 271, 65–76.

Gray, D. M., & Male, D. H. (Eds.) (1981). Handbook of snow. Toronto: Pergamon Press, 776 p.

Greene, E. A. (1993). Hydraulic properties of the Madison aquifer system in the western Rapid City area, South Dakota: U.S. Geological Survey Water-Resources Investigations Report 93-4008, Rapid City, South Dakota, 56 p.

Greene, E. A. & Shapiro, A. M. (1995). Methods of conducting air-pressurized slug tests and computation of type curves for estimating transmissivity and storativity. US Geological Survey open-file report 95–424.

Greene, E. A., Shapiro, A. M. & Carter, J. M. (1999). Hydrogeologic characterization of the Minnelusa and Madison Aquifers near Spearfish, South Dakota. U.S. Geological Survey Water-Resources Investigations Report 98-4156, Rapid City, South Dakota, 64 p.

Greswell, R., Yoshida, K., Tellam, J. H. & Lloyd, J. W. (1998). The micro-scale hydrogeological properties of the Lincolnshire Limestone UK. Quarterly J. Eng. Geol., 31, 181–197.

Grillot, J. C. (1979). Structure des systèmes aquifères en milieu fissuré. Contribution méthodologique à cette connaissance. Thèse Doctoral U.S.T.L., Montpellier, 212 p.

Gringarten, A. C., & Ramey, H. J. (1974). Unsteady state pressure distributions created by a well with a single horizontal fracture, partial penetration or restricted entry. Soc. Petrol. Engrs. Journ., 413–426.

Gringarten, A. C., & Whiterspoon, P. A. (1972). A method of analyzing pump test data from fractured aquifers. Int. Soc. Rock Mechanics and Int. Assoc. Eng. Geol., Proc. Symp. Rock Mechanics, Stuttgart, v. 3-B, 1–9.

Grosmann, A. & Morlet, J. (1984). Decomposition of Hardy functions into square integrable wavelets of constant shape. SIAM J. Math. Anal., 15, 723–736.

Groves, C., Bolster, C., & Meiman, J. (2005). Spatial and temporal variations in epikarst storage and flow in South Central Kentucky's Pennyroyal Plateau sinkhole plain. US Geological Scientific Investigations Report 2005-5160, 64–73.

Groves, C. & Howard, A. D. (1994a). Minimum hydrochemical conditions allowing limestone cave development. Water Resources Research, 30, 607–615.

Groves, C. & Howard, A. D. (1994b). Early development of Karst systems, 1. Preferential flow path enlargment under laminar flow. Water Resources Research, 30, 2837–2846.

Groves, C, & Meiman, J. (2001). Inorganic carbon flux and aquifer evolution in the south central Kentucky karst. U.S. Geological Survey Water-Resources Investigations Report 01-4011, 99–105.

Groves, C. & Meiman, J. (2005). Weathering, geomorphic work, and karst landscape evolution in the Cave City groundwater basin, Mammoth Cave, Kentucky. Geomorphology, 67, 115–126.

Grubbs, J. W. (1995). Evaluation of Ground-Water Flow and Hydrologic Budget for Lake Five-O, A Seepage Lake in Northwestern Florida. US Geological Survey Water-Resources Investigations Report 94-4145, 42 p.

Gurk, M. & Bosch, F. (2002). Cave detection using the Self-Potential-Surface (SPS) technique on a karstic terrain in the Jura mountains (Switzerland). Deutsche Geophysikalische Gesellschaft, Kolloquium Elektromagnetische Tiefenforschung, Burg Ludwigstein bei Göttingen, 1–5 Oct 2001.

Halford, K.J. & Kuniansky, E. L. (2002). Documentation of spreadsheets for the analysis of aquifer-test and slug-test data. U.S. Geological Survey Open-File Report 02-197, Carson City, Nevada, 51 p.

Halihan, T., Mace, R. E. & Sharp, J. M. (2000). Flow in the San Antonio segment of the Edwards aquifer: matrix, fractures, or conduits? In Wicks, C. M. & Sasowsky, I. D. (Ed.) Groundwater Flow and Contaminant Transport in Carbonate Aquifers. AA Balkema, pp. 129–146.

Hanshaw, B. B. & Back, W. (1974). Determination of regional hydraulic conductivity through use of ^{14}C dating of groundwater. Memoires, Tome X, 1. Communications. 10th Congress of the International Association of Hydrogeologists, Montpellier, France, 195–198.

Hantush, M. S. (1956). Analysis of data from pumping tests in leaky aquifers. Am. Geophys. Union Trans., 37, 702–714.

Hantush, M. S. (1959). Nonsteady flow to flowing wells in leaky aquifers. Jour. Geophys. Research, 64, 1043–1052.

Hantush, M. S. (1960). Modification of the theory of leaky aquifers. Jour. Geophys. Research, 65, 3713–3725.

Hantush, M. S. (1961a). Drawdown around a partially penetrating well. Jour. Hyd. Div., Proc. Am. Soc. Civil Eng., 87, 83–98.

Hantush, M. S. (1961b). Aquifer tests on partially penetrating wells. Jour. Hyd. Div., Proc. Am. Soc. Civil Eng., 87., 171–194.

Hantush, M. S. (1966a). Wells in homogeneous anisotropic aquifers. Water Resources Research, 2, 273–279.

Hantush, M. S. (1966b). Analysis of data from pumping tests in anisotropic aquifers. Jour. Geophys. Research, 71, 421–426.

Hantush, M. S. & Jacob, C. E. (1955). Non-steady radial flow in an infinite leaky aquifer. Am. Geophys. Union Trans., 36, 95–100.

Hantush, M. S. & Thomas, R.G. (1966). A method for analyzing a drawdown test in anisotropic aquifers. Water Resources Research, 2, 281–285.

Harvey, R. W. (1997). Microorganisms as tracers in groundwater injection and recovery experiments: a review. FEMS Microbiology Reviews, 20, 461–472.

Hauns, M., Jeannin, P.-Y. & Atteia, O. (2001). Dispersion, retardation and scale effect in tracer breakdown curves in karst conduits. Journal of Hydrology, 241, 177–193.

Häuselmann, P. (2002). Cave genesis and its relationship to surface processes: Investigations in the Siebenhengste region (BE, Switzerland). PhD thesis, University of Fribourg, Switzerland, 168 p. Available on-line at: SpeleoProjects.com or at http://ethesis.unifr.ch/theses/index.php

Häuselmann, P. (2005). Cross-formational flow, diffluence and transfluence observed in St. Beatus Cave and Siebenhengste (Switzerland). International Journal of Speleology, 34, 65–70.

Häuselmann, P. & Granger, D. E. (2005). Dating of caves by cosmogenic nuclides: method, possibilities, and the Siebenhengste example (Switzerland). Acta Carsologica, 34, 43–50.

Häuselmann, P., Jeannin, P.-Y., Bitterli, T. (1999). Relationships between karst and tectonics: case-study of the cave system north of Lake Thun (Bern, Switzerland). Geodinamica Acta, 12, 377–387.

Heaton, T. H. E. & Vogel, J. C. (1981). Excess air in groundwater. Journal of Hydrology, 50, 210–216.

Heller, M. (1983). Toporobot – Höhlenkartographie mit Hilfe des Computers. Stalactite, 33, 9–27.

Henderson, F. M. (1966). Open channel flow. New York: MacMillan, 522 p.

Henne, P., Krauthausen, B., Stummer, G. (1994). Höhlen im Dachstein. Derzeitiger Forschungsstand, Anlage der Riesenhöhlensysteme am Dachstein-Nordrand und Bewertung der unterirdischen Abflußverhältnisse. Zeitschrift für Karst- und Höhlenkunde, 45, 48–67.

Herlicska, H., Lorbeer, G., Humer, G., Boroviczeny, F., Mandl, G. W., Trimborn, W. (1995). Pilot Project "Karst Water Dachstein". In COST Action 65, Hydrogeological aspects of groundwater protection in karstic areas (Final Report). Report EUR 16547 EN, 21–34.

Herold, T., Jordan, P. & Zwahlen, F. (2000). The influence of tectonic structures on karst flow patterns in karstified limestones and aquitards in the Jura Mountains, Switzerland. Eclogae Geologicae Helvetiae, 93, 349–362.

Hess, J. W. & White, W. B. (1988). Storm response of the karstic carbonate aquifer of southcentral Kentucky. J. Hydrol., 99, 235–252.

Hess, J. W. & White, W. B. (1989). Water budget and physical hydrology. In White, W.B. & White, E. L. (Eds.), Karst hydrology: concepts from the Mammoth Cave area (pp. 105–126). New York: Van Nostrand-Reinhold.

Hess, J. W. & White, W. B. (1993). Groundwater geochemistry of the carbonate karst aquifer, southcentral Kentucky, USA. Appl. Geochem., 8, 189–204.

Hjulström, F. (1939). Transportation of detritus by running water. In Trask, P.D. (Ed.), Recent marine sediments (pp. 5–31). American Association of Petroleum Geologists.

Hoehn, E. & von Gunten, H. R. (1989). Radon in groundwater: a tool to assess infiltration from surface waters to aquifers. Water Resources Research, 25, 1795–1803.

Holland, H. D., Kirsipu, T. V., Huebner, J. S. & Oxburgh, U. M. (1964). On some aspects of the chemical evolution of cave waters. Journal of Geology, 72, 36–67.

Hoover, R. A. (2003). Geophysical choices for karst investigations. National Groundwater Association Geotechnical Special Publications, 122, 529–538.

Hornberger, G. M., Raffensperger, J. P., Wiberg, P. & Eshleman, K. N. (1998). Elements of physical hydrology. Baltimore: Johns Hopkins University Press, 314 p.

Hötzl, H. (1998). Karst groundwater. In Käss, W., Tracing Technique in Geohydrology (pp. 398–426). Rotterdam, Brookfield: Balkema.

Hötzl, H., Käss, W. & Reichert, B. (1991). Application of microbial tracers in groundwater studies. Water Science and Technology, 24, 295–300.

Howard, A. D. & Groves, C. G. (1995). Early development of karst systems. 2. Turbulent flow. Water Resources Research, 31, 19–26.

Hunkeler, D., Chollet, N., Pittet, X., Aravena, R., Cherry, J. A. & Parker, B. L. (2004). Effect of source variability and transport processes on carbon isotope ratios of TCE and PCE in two sandy aquifers. J. Contam. Hydrol., 74, 265–282.

Hurst, C. J., Crawford, R. L., Knudsen, G. R., McInerney, M. J., Stetzenbach, L. D. (2002). Manual of Environmental Microbiology, 2nd ed. ASM Press, 1158 p.

Huyakorn, P. S. & Pinder, G. F. (1983). Computational methods in subsurface flow. London: Academic Press.

Hwang, N. H. C. & R. J. Houghtalen (1996). Fundamentals of hydraulic engineering systems, 3rd edition. Englewood Cliffs, New Jersey: Prentice Hall, 416 p.

HydroSolve, Inc. (2002). Aqtesolve for Windows, User's Guide. Hydrosolve, Inc., Reston, Virginia, 185 p.

IAEA (2004). Isotope Hydrology Information System. The ISOHIS Database [On-line]. Available: http://isohis.iaea.org

Ingraham, N., Chapman, J., Hess, J. (1990). Stable isotopes in cave pool systems: Carlsbad Cavern, New Mexico, USA. Chemical Geology, 86, 65–74.

Ishido, T. & Pritchett, J. W. (1999). Numerical simulation of electrokinetic potentials associated with subsurface fluid flow. Journal of Geophysical Research, 104(B7), 15247–15259.

ISO (1980). ISO 1438/1-1980(E), Water flow measurement in open channels using weirs and venturi flumes, part 1: thin plate weirs. Geneva: International Organization of Standards, 27 p.

Jacob, C. E. (1963a). Determining the permeability of water-table aquifers. US Geological Survey Water-Supply Paper 1536-I, 245–271.

Jacob, C. E. (1963b). Corrections of drawdown caused by a pumped well tapping less than the full thickness of an aquifer. US Geological Survey Water-Supply Paper 1536-I, 272–292.

Jagnow, D. H., Hill, C.A., Davis, D. G., DuChene, H. R., Cunningham, K. I., Northup, D. E. & Queen, J. M. (2000). History of sulfuric acid theory of speleogenesis in the Guadalupe Mountains, New Mexico. Journal of Cave and Karst Studies, 62, 54–59.

Jaquet, O. (1995). Modèle probabiliste de réseaux karstiques: équation de Langevin et gaz sur réseau. Cahiers de géostatistique, 5, 69–80.

Jaquet, O. & Jeannin, P.-Y. (1993). Modelling the karstic medium: a geostatistical approach. In Armstrong & Dowd (Eds.), Geostatistical simulations (pp. 185–295). Kluwer Academic Press.

Jeannin, P.-Y. (1992). Géometrie des réseaux de drainage karstique: approche structurale, statistique et fractale. Ann. Sci. de l'Université de Besançon, 1–8.

Jeannin, P.-Y. (1996). Structure et comportement hydraulique des aquifères karstiques. PhD thesis, Université de Neuchâtel, 237 p. Available at: http://www.unine.ch/biblio/bc/cyber_liste_fac_inst_FS_geolo.html

Jeannin, P.-Y. (2001). Modeling flow in phreatic and epiphreatic karst conduits in the Hölloch cave (Muotatal, Switzerland). Water Resources Research, 37, 191–200.

Jeannin, P.-Y. & Maréchal, J.-C. (1995). Lois de pertes de charge dans les conduits karstiques: base théorique et observations. Bulletin d'Hydrogéolgie, 14, 149–176.

Jeannin, P.-Y. & Sauter, M. (1998). Analysis of karst hydrodynamic behaviour using global approaches: a review. Bulletin d'Hydrogéologie, Neuchatel, 16, 31–48.

Jeannin, P.-Y., Wildberger, A. & Rossi, P. (1995). Multitracing-Versuche 1992 und 1993 im Karstgebiet der Silberen (Muotatal und Klöntal, Zentralschweiz). Beitr. z. Hydrogeologie, 46, 43–88.

Jenkins, G. M. & Watts, D. G. (1968). Spectral analysis and its applications. San Francisco: Holden Days.

Jennings, J. N. (1985). Karst geomorphology. Ed. Basil Blackwell, 293 p.

John, D. E. & Rose, J. B. (2005). Review of factors affecting microbial survival in groundwater. Enviromental Science and Technology, 39, 7345–7356.

Jones, A. L. & Smart, P. L. (2005). Spatial and temporal changes in the structure of ground-water nitrate concentration time series (1935–1999) as demonstrated by autoregressive modelling. J. Hydrol., 310, 201–215.

Jones, W. K. (1977). Karst hydrology atlas of West Virginia. Special Publication 4, Karst Waters Institute, Charles Town, West Virginia, 111 p.

Joss, J., & Lee, R. (1995) The application of radar-gauge comparisons to operational precipitation profile corrections. Journal of Applied Meteorology, 34, 2612–2630.

Kalbitz, K., Solinger, S., Park, J.-H., Michalzik, B. & Matzner, E. (2000). Controls on the dynamics of dissolved organic matter in soils: A review. Soil Science, 165, 277–304.

Käss, W. (1998). Tracing technique in geohydrology. Rotterdam, Brookfield: Balkema, 581 p.

Käss, W. (2004). Geohydrologische Markierungstechnik. Berlin-Stuttgart: Borntraeger, 557 p.

Kattan, Z. (1997). Environmental isotope study of the major karst springs in Damascus limestone aquifer systems: case of the Figeh and Barada springs. Journal of Hydrology, 193, 161–182.

Katz, B. G. (2004). Sources of nitrate contamination and age of water in large karstic springs of Florida. Environmental Geology, 46, 689–706.

Katz, B. G. & Bullen, T. D. (1996). The combined use of $^{87}Sr/^{86}Sr$ and carbon and water isotopes to study the hydrochemical interaction between groundwater and lakewater in mantled karst. Geochimica et Cosmochimica Acta, 60, 5075–5087.

Katz, B. G., Hornsby, H. D., Bohlke, J. K. & Mokray, M. F. (1999). Sources and chronology of nitrate contamination in spring waters, Suwannee River Basin, Florida. US Geological Survey, Water Resource Investigations Report, 99-4252, 54 p.

Kaufmann, G. & Braun, J. (2000). Karst aquifer evolution in fractured porous rocks. Water Resources Research, 36, 1381–1391.

Keller, G. V. (1971). Natural-field and controlled source methods in electromagnetic sounding methods. Geoexploration, 9, 99–147.

Keller, G. V. (1982). Electrical properties of rocks and minerals. In Carmichael, R. S. (Ed.), Handbook of physical properties of rocks and minerals (pp. 217–294). Boca Raton: CRC Press.

Kennedy, K., Niehren, S., Rossi, P., Schnegg, P.-A., Müller, I. & Kinzelbach, W. (2001) Results of bacteriophage, microsphere and solute tracer migration comparison at Wilerwald test field, Switzerland. Beitr. z. Hydrogeologie, 52, 180–210.

Kertz, W. (1969) Einführung in die Geophysik. Mannheim: Hochschultaschenbücher, 156 p.

Keys, W. S. (1990). Borehole geophysics applied to ground-water investigations. US Geological Survey Techniques of Water-Resources Investigation, Book 2, 150 p.

Kilchmann, S., Waber, H. N., Parriaux, A. & Bensimon, M. (2004). Natural tracers in recent groundwaters form different Alpine aquifers. Hydrogeology Journal, 12, 643–661.

Kincaid, T. (1999). Morphologic and fractal characterization of saturated karstic caves. PhD Thesis, University of Wyoming.

Kinzelbach, W. (1986) Groundwater modeling. Elsevier, Int. Edition.

Kiraly, L. (1975). Rapport sur l'état actuel des connaissances dans le domaine des caractères physiques des roches karstiques. In Burger, A. & Dubertret, L. (Eds.), Hydrogeology of karstic terrains (pp. 53–67). IAH, International Union of Geological Sciences, Series B, 3.

Kiraly, L. (1979). Remarques sur la simulation des failles et du réseau karstique par éléments finis dans les modèles d'écoulement. Bulletin d'Hydrogéologie de l'Université de Neuchâtel, 3, 155–167.

Kiraly, L. (1985). FEM-301, A three dimensional model for groundwater flow simulation. NAGRA Technical Report 84–49, 96 p.

Kiraly, L. (1988). Large-scale 3D groundwater flow modeling in highly heterogeneous geologic medium. In Custodio (Ed.), Groundwater flow and quality modeling (pp. 761–775). D. Riedel Publishing Company.

Kiraly, L. (1994). Groundwater flow in fractured rocks: models and reality. 14th Mintrop Seminar über Interpretationsstrategien in Exploration und Produktion, Ruhr Universität Bochum, 159, 1–21.

Kiraly, L. (1998a). Introduction à l'hydrogéologie des roches fissurées et karstiques. Bases théoriques à l'intention des hydrogéologues. Manuscrit, Université de Neuchâtel.

Kiraly, L. (1998b). Modeling karst aquifers by the combined discrete channel and continuum approach. Bulletin du Centre d'Hydrogeologie, Neuchatel, 16, 77–98.

Kiraly, L. (2002). Karstification and Groundwater Flow. Proceedings of the Conference on Evolution of Karst: From Prekarst to Cessation. Postojna-Ljubljana, 155–190.

Kiraly, L. & Morel, G. (1976a). Etude de régularisation de l'Areuse par modèle mathématique. Bulletin du Centre d'Hydrogéologie, Neuchâtel, 1, 19–36.

Kiraly, L. & Morel, G. (1976b). Remarques sur l'hydrogramme des sources karstiques simule par modèles mathématiques. Bulletin du Centre d'Hydrogéologie, Neuchâtel, 1, 37–60.

Kiraly, L., Perrochet, P. & Rossier, Y. (1995). Effect of epikarst on the hydrograph of karst springs: a numerical approach. Bulletin d'hydrogéologie, 14, 199–220.

Klimchouk, A. (1996). Hydrogeology of gypsum formations. Gypsum Karst of the World: International Journal of Speleology, 25, 83–89.

Klimchouk, A. (1997). The nature and principal characteristics of epikarst. Proceedings of the 12th International Congress of Speleology, La Chaux-de-Fonds, Switzerland, 10–17. Aug 97, 1, 306.

Klimchouk, A. B., Ford, D. C., Palmer, A. N. & Dreybrodt, W. (Eds.) (2000). Speleogenesis, evolution of karst aquifers. Huntsville, Alabama, USA: National Speleological Society, Inc., 527 p.

Knödel, K., Krummel, H. & Lange, G. (1997). Handbuch zur Erkundung des Untergrundes von Deponien und Altlasten. Band 3: Geophysik. Berlin: Springer, 1063 p.

Knop, A. (1878). Über die hydrographischen Beziehungen zwischen der Donau und der Aachquelle im Badischen Oberlande. N. Jb. Min., Geol. u. Palaeont, 1878, 350–363.

Koistinen, J. (1986). The effect of some measurement errors on radar-derived Z-R relationships. 23rd Conference on Radar Meteorology, Boston, American Meteorological Society, 3, 50–53.

Kovács, A. (2003). Geometry and hydraulic parameters of karst aquifers: A hydrodynamic modeling approach. PhD thesis, University of Neuchâtel, Switzerland. 131 p.

Kovács, A., Perrochet, P., Király, L. & Jeannin, P.-Y. (2005). A quantitative method for the characterization of karst aquifers based on spring hydrograph analysis. Journal of Hydrology, 303, 152–164.

Kranjc, A. (Ed.) (1997). Kras: Slovene classical karst. Ljubljana: Karst Research Institute, UNESCO, 254 p.

Kreft, A. & Zuber, A. (1978). On the physical meaning of the dispersion equation and its solution for different initial and boundary conditions. Chem. Eng. Sci., 33, 1471–1480.

Kresic, N. (1991). Kvantitativna hidrogeologija karsta sa elementima zaštite podzemnih voda [in Serbo-Croatian; Quantitative karst hydrogeology with elements of groundwater protection; English abstract and figure captions]. Belgrade: Naučna knjiga, 192 p.

Kresic, N. (2007). Hydrogeology and Groundwater Modeling, Second Edition. Boca Raton/London/New York: CRC Press, Taylor & Francis Group, 806 p.

Kruseman, G. P., & de Ridder, N. A. (1990). Analysis and Evaluation of Pumping Test Data (2nd edition). Wageningen, the Netherlands: International Institute for Land Reclamation and Improvement, 377 p.

Kupusovic, T. (1989). Measurements of piezometric pressures along deep boreholes in karst area and their assessment. Nas Krs, Vol. XV, No. 26–27, 21–30.

Kurz, C. L., Chauvet, S., Andres, E., Aurouze, M., Vallet, I., Michel, G. P. F., Uh, M., Celli, J., Filloux, A., de Bentzmann, S., Steinmetz, I., Hoffmann, J. A., Finlay, B. B., Gorvel, J. P., Ferrandon, D. & Ewbank, J. J. (2003). Virulence factors of the human opportunistic pathogen Serratia marcescens identified by in vivo screening. EMBO Journal, 22, 1451–1460.

Labat, D., Ababou, R. & Mangin, A. (1999a). Analyse en ondelettes en hydrogéologie karstique. 1re partie: analyse univariée de pluies et débits de sources karstiques. Earth and Planetary Sciences, 329, 873–879.

Labat, D., Ababou, R. & Mangin, A. (1999b). Analyse en ondelettes en hydrogéologie karstique. 2e partie: analyse en ondelettes croisée pluie-debit. Earth and Planetary Sciences, 329, 881–887.

Labat, D., Ababou, R. & Mangin, A. (2000a). Rainfall-runoff relations for karstic springs. Part I: convolution and spectral analyses. J. Hydrol., 238, 123–148.

Labat, D., Ababou, R. & Mangin, A. (2000b). Rainfall-runoff relations for karstic springs. Part II: continuous wavelet and discrete orthogonal multiresolution analyses. Journal of Hydrology, 238, 149–178.

Lakey, B. & Krothe, N.C. (1996). Stable isotopic variation of storm discharge from a perenial karst spring, Indiana. Water Resour. Res., 32, 721–731.

LaMoreaux, P., Assaad, F. & McCarley, A. (Eds.) (1993). Annotated Bibliography of Karst Terranes, Vol 5. International Contributions to Hydrogeology (IAH), 14. Hannover: Heise, 425 p.

LaMoreaux, P., Prohic, E., Zötl, J., Tanner, J. M. & Roche, B. N. (Eds.) (1989). Hydrology of Limestone Terranes. Annotated Bibliography of Carbonate Rocks, Vol 4. International Contributions to Hydrogeology (IAH), 10. Hannover: Heise, 267 p.

LaMoreaux, P. E. & Wilson, B. M. (1984). Guide to the Hydrology of Carbonate Rocks. Bernan Assoc, 345 p.

Lang, U. (1995). Simulation regionaler Strömungs- und Transportvorgänge in Karstaquiferen mit Hilfe des Doppelkontinuum-Ansatzes: Methodenentwicklung und Parameteridentifikation. PhD thesis, University of Stuttgart.

Langmuir, D. (1997). Aqueous environmental geochemistry. Upper Saddle River, NJ: Prentice Hall, 600 p.

Lankston, R. W. (1990). High-Resolution Refraction Seismic Data Acquisition and Interpretation. In Ward, S. H. (Ed.), Geotechnical and Environmental Geophysics (pp. 45–74). Society of Exploration Geophysicists.

Larocque, M., Mangin, A., Razack, M. & Banton, O. (1998). Characterization of the La Rochefoucauld karst aquifer (Charente, France) using correlation and spectral analysis. Bulletin d'Hydrogéologie, 16, 49–57.

Larsson, I. (1982). Ground water in hard rocks. Project 8.6 of the International Hydrological Programme, UNESCO, Paris, 228 p.

Lastennet, R. (1994). Rôle de la zone non-saturée dans le fonctionnement des aquifères karstiques. Approche par l'étude physico-chimique et isotopique du signal d'entrée et des exutoires du massif du Ventoux (Vaucluse). PhD thesis, University of Avignon, France.

Lastennet, R. & Mudry, J. (1997). Role of karstification and rainfall in the behavior of a heterogeneous karst system. Environmental Geology, 32, 114–123.

Lee, E. S. & Krothe, N. C. (2001). A four-component mixing model for water in karst terrain in south-central Indiana, USA, using solute concentration and stable isotope as tracers. Chem. Geol., 179, 129–143.

Lee, T. M. (1996). Hydrogeologic controls on the groundwater interactions with an acidic lake in karst terrain. Water Resources Research, 32, 831–844.

Lee, T. M. (2000). Effects of nearshore recharge on groundwater interactions with a lake in mantled karst terrain. Water Resources Research, 36, 2167–2182.

Lee, T. M. & Swancar, A. (1997). The influence of evaporation, groundwater, and uncertainty in the hydrologic budget of Lake Lucerne, a seepage lake in Polk County, Florida. U.S. Geological Survey Water-Supply Paper 2439.

Legates, D. R. (2000). Real-time calibration of radar precipitation estimates. The Professional Geographer, 52, 235–246.

Legates, D. R., Nixon, K. R., Stockdale, T. D. & Quelch, G. E. (1999). Real-time and historical calibration of WSR-88D precipitation estimates. Proceedings, 11th Conference on Applied Climatology. Dallas, Texas: American Meteorological Society, 76–77.

Leibundgut, C. (1998). Surface waters. In Käss, W. (1998), Tracing Technique in Geohydrology (pp. 493–510). Rotterdam, Brookfield: Balkema.

Leibundgut, C., McDonnel, J. & Schulz, G. (1995). Tracer technologies for hydrological systems. Int. Ass. Sci. Hydrology, 229, 320 p.

Li, X. & Götze, H. J. (2001). Tutorial, Ellipsoid, Geoid, Gravity, Geodesy, and Geophysics. Geophysics, 66, 1660–1668.

Liedl, R., Renner, S. & Sauter, M. (1998). Obtaining information about fracture geometry from heat flow data in karst systems. Bulletin d'Hydrogéologie, 16, 143–153.

Liedl, R., Sauter, M., Hückinghaus, D., Clemens, T. & Teutsch, G. (2003) Simulation of the development of karst aquifers using a coupled continuum pipe flow model. Water Resources Research, 39, Art. No. 1057.

Limerinos, J.T. (1970). Determination of the Manning coefficient from measured bed roughness in natural channels. US Geological Survey Water-Supply Paper 1898-B, 47 p.

Linsley, R. K., Kohler, M. A., & Paulhus, J. L. H. (1982). Hydrology for engineers, 3rd edition. New York: McGraw Hill, 508 p.

Liu, Z., Groves, C., Yuan, D., Meiman, J. (2004). South China karst aquifer storm-scale hydrochemistry. Ground Water, 42, 491–499.

Llopis, N. (1970). Fundamentos de Hidrogeología kárstica. Madrid: Editorial Blume, 269 p.

Lohman, S.W. (1972). Ground-water hydraulics. US Geological Survey Professional Paper 708, Washigton, D.C., 70 p.

Loke, M. H. (1999). Electrical imaging surveys for environmental and engineering studies. Sunnyvale, CA: Geometrics, Inc.

Long, J. C. S., Gilmour, P. & Witherspoon, P. A. (1985). A model for steady fluid flow in random three-dimensional networks of disc-shaped fractures. Water Resources Reasarch, 21, 1105–1115.

Long, J. C. S., Remer, J. S., Wilson, C. R., Witherspoon, P. A. (1982) Porous media equivalents for networks of discontinuous fractures. Water Resources Reasarch, 18, 645–658.

Loop, C. M. & White, W. B. (2001). A conceptual model for DNAPL transport in karst ground water basins. Ground Water, 39, 119–127.

López Chicano, M. (1992). Contribución al conocimiento del sistema hidroigeológico kárstico de Sierra Gorda (Granada y Málaga). PhD thesis Univ. of Granada, 387 p.

Louis, C. (1968). Etude des écoulements d'eau dans les roches fissurées et de leurs influences sur la stabilité des massifs rocheux. Bull. Dir. Étud. Rech. Electr. Fr., A3, 5–132.

Lugeon, M. (1933). Barrages et géologie: Méthodes de recherches, terrassement et imperméabilisation. Paris: Dunod, 138 p.

Lütscher, M. & Perrin, J. (2005). The Aubonne karst aquifer (Swiss Jura). Eclogae Geologicae Helvetiae, 98, 237–248.

Madigan, M. T., Martinko, J. M., Parker, J. (2000). Brock biology of microorganisms, 9th edition. Prentice-Hall, 991 p.

Mahler, B. J. & Lynch, F. L. (1999). Muddy waters: temporal variation in sediment discharging from a karst spring. J. Hydrol., 214, 165–178.

Mahler, B. J., Lynch, L. & Bennett, P. C. (1999). Mobile sediment in an urbanizing karst aquifer: implications for contaminant transport. Environmental Geology, 39, 25–38.

Mahler, B. J., Personné, J. C., Lods, G. F. & Drogue, C. (2000). Transport of free and particule-associated bacteria in karst. J. Contam. Hydrol., 238, 179–193.

Maillet, E. (1905). Essais d'hydraulique souterraine et fluviale. Paris: Hermann.

Maloszewski, P., Benischke, R. & Harum, T. (1992). Mathematical modelling of tracer experiments in the karst of Lurbach-System. Steir. Beitr. Z Hydrogeologie, 43, 116–136.

Maloszewski, P., Stichler, W., Zuber, A. & Rank, D. (2002). Identifying the flow systems in a karstic-fissured-porous aquifer, the Schneealpe, Austria, by modeling of environmental ^{18}O and ^{3}H isotopes. Journal of Hydrology, 256, 48–59.

Mangin, A. (1971) Etude des débits classés d'exutoires karstiques portant sur un cycle hydrologique. Annales de spéléologie, 28, 21–40.

Mangin, A. (1975). Contribution a l'étude hydrodynamique des aquifères karstiques. Thèse, Institut des Sciences de la Terre de l'Université de Dijon

Mangin, A. (1981). Utilisation des analyses corrélatoire et spectrale dans l'approche des systèmes hydrologiques. Comptes Rendus de l'Académie des Sciences, Série III, 293, 401–404.

Mangin, A. (1982). L'approche systémique du karst, conséquences conceptuelles et méthodologiques. Proc. Réunion Monographica sobre el karst, Larra., 141–157.

Mangin, A. (1984). Pour une meilleure connaissance des systèmes hydrologiques à partir des analyses corrélatoire et spectrale. Journal of Hydrology, 67, 25–43.

Martin, J. B. & Dean, R. W. (2001). Exchange of water between conduits and matrix in the Floridan aquifer. Chemical Geology, 179, 145–165.

Martin, J. B., Wicks, C. M. & Sasowsky, I.D. (Eds.) (2002). Hydrogeology and biology of post-Paleozoic carbonate aquifers. Proceedings of the Symposium on Karst Frontiers: Florida and Related Environments. KWI Special Publication No 7. Karst Waters Institute, Charles Town, WV.

Martín-Algarra, A., Martín-Martín, M., Andreo, B., Julià, R., González-Gómez, C. (2003). Sedimentary patterns in perched spring travertines near Granada (Spain) as indicators of the paleohydrological and paleoclimatological evolution of a karst massif. Sedimentary Geology, 161, 217–228.

Maslia, M. L. & Randolph, R. B. (1986). Methods and computer program documentation for determining anisotropic transmissivity tensor components of two-dimensional ground-water flow. US Geological Survey Open-File Report 86–227, 64 p.

Matthess, G. (1994). Die Beschaffenheit des Grundwassers. Stuttgart, Berlin: Borntraeger, 499 p.

Mayer de Stadelhofen, C. (1991). Application de la Géophysique aux Recherches d'Eau. Technique et Documentation, Lavoisier.

Mazor, E. (1991). Applied chemical and isotopic groundwater hydrology. New York: Halsted Press, 274 p.

McGrath, R. J., Styles, P., Thomas, E. & Neale, S. (2001). Integrated high-resolution geophysical investigations as potential tools for water resource investigations in karst terrain. Proc. multidisciplinary conference on sinkholes and the engineering and environmental impacts of karsts, 8, 377–382.

McKay, A. J. (2003). Geoelectric fields and geomagnetically induced currents in the United Kingdom. PhD thesis, Univ. of Edinburgh, 238 p.

McKenzie, D. (2004). WALLS Tools for cave survey data management [On-line]. Available: http://www.utexas.edu/depts/tnhc/.www/tss/Walls/tsswalls.htm (5.9.2004).

McNeill, J. D. (1980). Electromagnetic terrain conductivity measurement at low induction numbers. Technical Note TN-6, Mississauga, Ontario, Canada: Geonics, Ltd.

McNeill, J. D. (1990). Use of electromagnetic methods for groundwater studies. In Ward, S. H. (Ed), Geotechnical and environmental geophysics (pp. 191–218). Society of Exploration Geophysicists.

McNeill, J. D. & Labson, V. F. (1991). Geological mapping using VLF radio fields. In Nabighian, M. N. (Ed.), Electromagnetic methods in applied geophysics, Vol. 2 (pp. 521–640). Tulsa, OK: Society of Exploration Geophysicists.

Meiman, J., Groves, C. & Herstein, S. (2001). In-cave dye tracing and drainage basin divides in the Mammoth Cave karst aquifer, Kentucky. US Geological Survey Water-Resources Investigations Report, 01-4011, 179–185.

Meiman, J. & Ryan, M. T. (1993). The Echo River/Turnhole Bend overflow route. Cave Research Foundation Newsletter, 21/1, 16–18.

Milanović, P. (1979). Hidrogeologija karsta i metode istraživanja [in Serbo-Croatian; Karst hydrogeology and methods of investigations]. HE Trebišnjica, Institut za korištenje i zaštitu voda na kršu, Trebinje, 302 p.

Milanovic, P.T. (1981). Karst hydrogeology. Littleton, CO: Water Resources Publications, 434 p.

Milanović, P. T. (2000). Geological engineering in karst. Belgrade: Ed. Zebra, 347 pp.

Militzer, H. & Weber, F. (1985). Angewandte Geophysik, Band 2: Geoelektrik-Geothermik-Radiometrie-Aerogeophysik. Wien: Springer.

Misstear, B., Banks, D. & Clark, L. (2006). Water wells and boreholes. Chichester: Wiley.

Moench, A. F. (1984). Double-porosity models for a fissured groundwater reservoir with fracture skin. Water Resources Research, 21, 1121–1131.

Moench, A. F. (1985). Transient flow to a large-diameter well in an aquifer with storative semiconfining layers. Water Resources Research, 8, 1031–1045.

Moench, A. F. (1993). Computation of type curves for flow to partially penetrating wells in water-table aquifers. Ground Water, 31, 996–971.

Moench, A. F. (1996). Flow to a well in a water-table aquifer: an improved Laplace transform solution. Ground Water, 34, 593–596.

Mohrlok, U., Kienle, J. & Teutsch, G. (1997a). Parameter identification in double-continuum models applied to karst aquifers. Proc. 12th Int. Congress of Speleology, La Chaux-de-Fonds, Switzerland, 2, 163–166.

Mohrlok, U. & Sauter, M. (1997b). Modelling groundwater flow in a karst terrane using discrete and double-continuum approaches – importance of spatial and temporal distribution of recharge. Proc. 12th Int. Congress of Speleology, La Chaux-de-Fonds, Switzerland, 2, 167–170.

Mohrlok, U. & Teutsch, G. (1997). Double continuum porous equivalent (DCPE) versus discrete modeling in karst terranes. Proc. 5th Int. Symp. and Field Seminar on Karst Waters & Environmental Impacts, 10–20 Sep 1995, Antalya, Turkey, 319–326.

Monnin, M., Morin, J. P., Pane, M. B. & Seidel, J. L. (1994). Radon-222 measurements in a fractured karst aquifer. Proc. of the Montpellier-Millau workshop, May 5–8, European Comission, DC Science, Research and Development, Brussels, 81–91.

Mook, W. G. (1980). Carbon-14 in hydrological studies. In Fritz, P. & Fontes, J. C. (Eds.), Handbook of environmental isotope geochemistry (pp. 49–74). New York: Elsevier.

Mooney, H. M. (1980). Handbook of engineering geophysics, volume 2: electrical resistivity. Minneapolis MN: Bison Instruments, Inc.

Mooney, H. M. (1984). Handbook of engineering geophysics, volume 1: seismic. Minneapolis MN: Bison Instruments, Inc.

Moore, C. H. (1989). Carbonate diagenesis and porosity. Development in sedimentology 46. Amsterdam: Elsevier, 338 p.

Morfis, A. & Zojer, H. (Eds) (1986). Karst hydrogeology of the Central and Eastern Peloponnesus (Greece). Steir Beitr z Hydrogeologie, 37/38, 301 p.

Morse, J. W. & MacKenzie, F. T. (1990). Geochemistry of sedimentary carbonates. Elsevier, 696 p.

Mozeto, A. A., Fritz, P. & Reardon, E. J. (1984). Experimental observations on carbon isotope exchange in carbonate-water systems. Geochimica et Cosmochimica Acta, 48, 495–504.

Mudry, J. (1987). Apport du traçage physico-chimique naturel à la connaissance hydrociné-matique des aquifères carbonatés. PhD University of Franche-Comté, Besançon, France (in French), 381 p.

Mudry, J., Charmoille, A., Robbe, N., Bertrand, C., Batiot, C., Emblanch, C. & Mettetal, J. P. (2002). Use of hydrogeochemistry to display a present recharge of confined karst aquifers. Case study of the Doubs valley, Jura mountains, Eastern France. In Carrasco, F., Duran, J. J., Andreo, B. (Eds.). Karst and Environment, 123–129.

Muldoon, M. A., Simo, J. A. & Bradbury, K. R. (2001). Correlation of hydraulic conductivity with stratigraphy in a fractured-dolomite aquifer, northeastern Wisconsin, USA. Hydrogeology Journal, 9, 570–583.

Munnich, K. O., Roether, W. & Thilo, L. (1967). Dating of groundwater with tritium and ^{14}C. Proceedings of an IAEA Symposium on Isotopes in Hydrology, Vienna, 306–320.

Munson, B. R., Young, D. F. & Okiishi, T. H. (1998). Fundamentals of fluid mechanics (3rd edition). New York: John Wiley and Sons, Inc., 844 p.

Musgrove, M. & Banner, J. L. (2004). Controls on the spatial and temporal variability of vadose dripwater geochemistry: Edwards Aquifer, central Texas. Geochimica et Cosmochimica Acta, 68, 1007–1020.

Nabighian, M. N. & Macnae, J. C. (1991). Time-domain electromagnetic prospecting methods. In Nabighian, M. N. (Ed.), Electromagnetic methods in applied geophysics, Vol. 2 (pp. 427–520). Tulsa, OK: Society of Exploration Geophysicists.

Neff, J. C. & Asner, G. P. (2001). Dissolved organic carbon in terrestrial ecoystems: synthesis and a model. Ecosystems, 4, 29–48.

Neuman, S.P. (1972). Theory of flow in unconfined aquifers considering delayed response to the water table. Water Resources Research, 8, 1031–1045.

Neuman, S. P. (1974). Effects of partial penetration on flow in unconfined aquifers considering delayed gravity response. Water Resources Research, 10, 303–312.

Neuman, S. P. (1975). Analysis of pumping test data from anisotropic unconfined aquifers considering delayed gravity response. Water Resources Research, 11, 329–342.

Neuman, S. P. & de Marsily, G. (1976). Identification of linear system response by parametric programming. Water Resources Research, 12, 253–262.

Newman, G. A., Recher, S., Tezkan, B. & Neubauer, F. M. (2003). 3D inversion of scalar radio magnetotelluric field data set. Geophysics, 68, 791–802.

Neuman, S. P. & Witherspoon, P. A. (1969). Theory of flow in a confined two aquifer system. Water Resources Research, 5, 803–816.

Newton, I. (1686). Philosophiae Naturalis Principia Mathematica. London: S. Pepys, Reg. Soc. Press.

Nguyet, V. T. M. & Goldscheider, N. (2006). Tracer tests, hydrochemical and microbiological investigations as a basis for groundwater protection in a remote tropical mountainous karst area, Vietnam. Hydrogeology Journal, 14, 1147–1159.

Niehren, S. & Kinzelbach, W. (1998). Artificial colloid tracer tests: development of a compact on-line microsphere counter and application to soil column experiments. Journal of Contaminant Hydrology, 35, 249–295.

Nixon, K. R., Legates, D. R., Quelch, G. E. & Stockdale, T. D. (1999) A high-resolution weather data system for environmental modeling and monitoring of meteorological conditions. Proc. Conf. American Meteorological Society, Dallas, Texas, 521–524.

Nyquist, J. & Corry, C. E. (2002). Self-Potential: The ugly duckling of environmental geophysics. The Leading Edge, May 2002, 446–451.

O'Neil, J. R. (1986b). Terminology and standards. Reviews of Mineralogy, 16, 561–570.

Oda, M. (1985). Permeability tensor for discontinuous rock masses. Geotechnique, 35, 483–495.

Olhoeft, G. R. (2000). Maximizing the information return from ground penetrating radar. Journal of Applied Geophysics, 43, 175–187.

Osborne, P. S. (1993). Suggested operating procedures for aquifer pumping tests. Ground Water Issue, United States Environmental Protection Agency, EPA/540/S-93/503, 23p.

Padilla, A. & Pulido-Bosch, A. (1995). Study of hydrographs of karstic aquifers by means of correlation and cross-spectral analysis. Journal of Hydrology, 168, 73–89.

Paillet, F. L. (1994). Application of borehole geophysiscs in the characterization of flow in fractured rocks. US Geological Survey Water-Resources Investigations Report 93-4214, Denver, Colorado, 36 p.

Paillet, F. L. (2001). Borehole geophysical applications in karst hydrogeology. US Geological Survey Water-Resources Investigations Report 01-4011, 116–123.

Paillet, F. L. & Reese, R. S. (2000). Integrating borehole logs and aquifer tests in aquifer characterization. Ground Water, 38, 713–725.

Palmer, A. N. (1991). Origin and morphology of limestone caves. GSA Bulletin, 103, 1–21.

Palmer, A. N. (2000). Digital modeling of individual solution conduits. In Klimchouk, A., Ford, D. C., Palmer, A. N. & Dreybrodt, W. (Eds.), Speleogenesis: Evolution of Karst Aquifers (pp. 194–200). Huntsville, Alabama, USA: National Speleological Society.

Palmer, A., Palmer, V. & Sasowsky, I. (1999). Karst Modeling. SP 5, Karst Water Institute, Akron Ohio, 265 p.

Paloc, H. & Back, W. (Eds.) (1992). Hydrogeology of Selected Karst Regions. Int. Contr. Hydrogeology (IAH), 13, Hannover: Heise. 494 p.

Panno, S. V., Hackley, K. C., Hwang, H. H. & Kelly, W.R. (2001). Determination of the sources of nitrate contamination in karst springs using isotopic and chemical indicators. Chem. Geol., 179, 113–128.

Panno, S. V. & Kelly, W. R. (2004). Nitrate and herbicide loading in two groundwater basins of Illinois's sinkhole plain. J. Hydrol., 290, 229–242.

Papadopulos, I. S. (1965). Nonsteady flow to a well in an infinite anisotropic aquifer. Proc. Dubrovnik Symposium on the Hydrology of Fractured Rocks. IAHS, 21–31.

Papadopulos, I. S. & Cooper, H. H. (1967). Drawdown in a well of large diameter. Water Resources Research, 3, 241–244.

Park, C. B., Miller, R. D. & Xia, J. (1999). Multi-channel analysis of surface waves. Geophysics, 64, 800–808.

Patra, H. P. & Mallik, K. (1980). Geosounding principles, 2. Time-varying geoelectric soundings. Amsterdam, Oxford, New York: Elsevier.

Payot, R. (1953). Distribution de la radioactivité en Suisse, PhD thesis, Univ. of Neuchâtel, Switzerland.

Pearce, B. R. (1982). Fractured rock aquifers in central Queensland. Papers of the Groundwater in Fractured Rock Conference, Canberra, Australia, 161–172.

Pearson, F. J., Hanshaw, B. B. (1970). Sources of dissolved carbonate species in groundwater and their effects on carbon-14 dating. In Isotope Hydrology 1970 (p. 271–286). Vienna: IAEA.

Perret, H. (1918). Radioactivité des eaux neuchâteloises et Seelandaises. PhD thesis, Univ. of Neuchâtel, Switzerland.

Perrin, J. (2003). A conceptual model of flow and transport in a karst aquifer based on spatial and temporal variations of natural tracers. PhD thesis, University of Neuchâtel, 227 p. Available at http://www.unine.ch/biblio/bc/cyber_liste_fac_inst_FS_geolo.html

Perrin, J., Jeannin, P.-Y. & Zwahlen, F. (2003a). Epikarst storage in a karst aquifer: a conceptual model based on isotopic data, Milandre test site, Switzerland. Journal of Hydrology, 279, 106–124.

Perrin, J., Jeannin, P.-Y. & Zwahlen, F. (2003b). Implications of the spatial variability of infiltration-water chemistry for the investigation of karst aquifer: a field study at the Milandre test site, Swiss Jura. Hydrogeology Journal, 11, 673–686.

Perrin, J. & Kopp, L. (2005). Hétérogénéité des écoulements dans la zone non saturée d'un aquifère karstique (site de Milandre, Jura suisse). Bulletin d'hydrogéologie, 21, 33–58.

Perrier, F. E., Petiau, G., Clerc, G., Bogorodsky, V., Erkul, E., Jouniaux, L., Lesmes, D., Mcnae, J., Meunier, J. M., Morgan, D., Nascimento, D., Oettinger, G., Schwarz, G., Toh, H., Valiant, M. J., Vozoff, K. & Yazici-Cakin, O. (1997). One year systematic study of electrodes for long period measurement of the electric field in geophysical environments. Journal of Geomagnetism and Geoelectrics, 49, 1677–1696.

Peters, N. E., Hoehn, E., Leibundgut, C., Tase, N. & Walling, D. E. (Eds.) (1993). Tracers in Hydrology. Int. Ass. Sci. Hydrolog., 215, 350 p.

Peterson, E. W., Davis, R. K., Brahama, J. V. & Orndorff, H. A. (2002). Movement of nitrate through regolith covered karst terrane, northwest Arkansas. J. Hydrol., 256, 35–47.

Pinder, G. F. & Jones, J. F. (1969). Determination of the ground water component of peak discharge form the chemistry of natural runoff. Water Resources Research, 5, 438–445.

Plagnes, V. (2000). Structure et fonctionnement des aquifères karstiques. Caractérisation par la géochimie des eaux. Documents du BRGM 294, 352 p.

Pollard, D. D. & Fletcher, R. C. (2005). Fundamentals of structural geology. Cambridge University Press.

Price, R. M., Top, Z., Happell, J. D. & Swart, P. K. (2003). Use of tritium and helium to define groundwater flow conditions in Everglades National Park. Water Resources Research, 39, Art. No. 1267.

Pronk, M., Goldscheider, N. & Zopfi, J. (2006). Dynamics and interaction of organic carbon, turbidity and bacteria in a karst aquifer system. Hydrogeology Journal, 14, 473–484.

Quarto, R. & Shiavone, D. (1996). Detection of cavities by the self-potential method. First Break, 14, 419–431.

Raeisi, E., Groves, C. & Meiman, J. (in press). Effects of partial and full pipe flow on hydrochemographs of Logsdon River, Mammoth Cave Kentucky USA. Journal of Hydrology.

Rank, D., Völkl, G., Maloszewski, G. P. & Stichler, W. (1991). Flow dynamics in an alpine karst massif studied by means of environmental isotopes. Proceedings of a Symposium on the Isotopes Techniques in Water Resources Development, IAEA, Vienna, 327–343.

Rantz, S. E., and others (1982). Measurement and computation of streamflow, volume 1, measurement of stage and discharge. US Geological Survey Water Supply Paper 2175, 313 p. (note: the 16 other major contributors to this publication are listed in its preface)

Ray, J. A. & Blair, R. J. (2005). Large perennial springs of Kentucky: Their identification, base flow, catchment, and classification. 10th Multidisciplinary Conference on Sinkholes and the Engineering and Environment Impacts of Karst, 24–28 Sept, San Antonio, USA, 410–422.

Razack, M. (1984). Application de méthodes numériques et statistiques a l'identification des réservoirs fissurés carbonatés en hydrogéologie. These Doctoral U.S.T.L., Montpellier, 257 p.

Recker, S. A. (1992). Petroleum hydrocarbon remediation of the subcutaneous zone of a karst aquifer, Lexington, Kentucky. Proceedings of the Third Conference on Hydrogeology, Ecology, Monitoring, and Management of Ground Water in Karst Terranes: Ground Water Management, 10, 447–474.

Redpath, B. B. (1973). Seismic refraction exploration for engineering site investigations. US Army Corps of Engineers, Technical Report E-73-4.

Reichert, B. (1991). Anwendung natürlicher und künstlicher Tracer zur Abschätzung des Gefährdungspotentials bei der Wassergewinnung durch Uferfiltration. PhD thesis, Schr. Angew. Geol. Karlsruhe, 13, 226 p.

Renken, R. A., Cunningham, K. J., Zygnerski, M. R., Wacker, M. A., Shapiro, A. M., Harvey, R. W., Metge, D. W., Osborn, C. L. & Ryan, J. N. (2005). Assessing the vulnerability of a municipal well field to contamination in a karst aquifer. Environmental & Engineering Geoscience, 11, 319–331.

Revil, A. & Pezard, P. A. (1999). Streaming potential in porous media 1: Theory of the zeta potential. Journal of Geophysical Research, 104(B9), 20021–20031.

Revil, A. & Schwaeger, H. (1999). Streaming potential in porous media 2: Theory and application to geothermal systems. Journal of Geophysical Research, 104(B9), 20033–20048.

Roberson, J. A., Cassidy, J. J. & Chaudhry, M. H. (1998). Hydraulic Engineering (2nd edition). New York: John Wiley and Sons, Inc., 672 p.

Robinson, P. C. (1984). Connectivity, flow and transport in network models of fractured media. PhD Thesis, St. Catherine's College, Oxford University.

Rodríguez Estrella, T. (2002). Acuíferos kársticos profundos. In Andreo, B. & Durán, J. J. (Eds.), Investigaciones en sistemas kársticos españoles. Hidrogeología y Aguas Subterráneas, 12. IGME.

Rorabaugh, M. I. (1953). Graphical and theoretical analysis of step-drawdown test of artesian well. Proc. Am. Soc. Civi Eng., 79, separate no. 362, 23 p.

Roscoe Moss Company (1990). Handbook of Ground Water Development. New York: John Wiley & Sons, 493 p.

Rose, T. P., Davisson, M. L. (1996). Radiocarbon in hydrologic systems containing dissolved magmatic carbon dioxide. Science, 273, 1367–1370.

Rose, T. P. & Davisson, M. L. (2003). Delineating Pleistocene- versus Holocene-age groundwater in the Great Basin using environmental isotopes. In Enzel, Y., Wells, S. G. & Lancaster, N. (Eds.), Paleoenvironments and paleohydrology of the Mojave and Southern Great Basin Deserts. Geological Society of America Special Paper, 368.

Rose, T. P., Kenneally, J. M., Smith, D. K., Davisson, M. L. & Hudson, G. B. (1997). Chemical and isotopic data for groundwater in Southern Nevada. Lawrence Livermore National Laboratory UCRL-ID-128000, 39 p.

Rossi, C., Cortel, A. & Arcenegui, R. (1997). Multiple paleo-water tables in Agujas Cave System (Sierra de Penalabra, Cantabrian Mountains, N Spain): Criteria for recognition and model for vertical evolution. Proc. 12th Int. Congress of Speleology, La Chaux-de-Fonds, Switzerland, 1, 183–187.

Rossi, P., Doerfliger, N., Kennedy, K., Müller, I. & Aragno, M. (1998). Bacteriophages as surface and ground water tracers. Hydrology and Earth System Sciences, 2, 101–110.

Rossi, P. & Käss, W. (1998). Phages. In Käss, W. (1998), Tracing Technique in Geohydrology (pp. 244–271). Rotterdam, Brookfield: Balkema.

Rozanski, K. (1979). Krypton-85 in the atmosphere 1950-1977: a data review. Environment International, 2, 139–143.

Ryan, M. & Meiman, J. (1996). An examination of short-term variations in water quality at a karst spring in Kentucky. Ground Water, 34, 23–30.

Sacks, L. A., Swancar, A. & Lee, T.M. (1998). Estimating groundwater exchange with lakes using water-budget and chemical mass balance approaches for ten lakes in ridge areas of polk and highlands counties, Florida. U.S. Geological Survey Water Resource Investigations 98-4133, 51 p.

Sauer, V. B., & Meyer, R. W. (1992). Determination of error in individual discharge measurements. U.S. Geological Survey Open-File Report 92-144, 4–7.

Sauter, M. (1992). Quantification and forecasting of regional groundwater flow and transport in a karst aquifer (Gallusquelle, Malm, SW. Germany). Tübinger Geowissenschaftliche Arbeiten, C13.

Sauter, M. & Liedl, R. (2000). Modelling karst aquifer genesis using a coupled continuum-Pipe flow model. In A. Klimchouk, D. Ford, C. Palmer, A. & Dreybrodt, W. (Eds.), Speleogenesis: Evolution of karst aquifers (pp. 212–219). Huntsville, Alabama (USA): National Speleological Society.

Sauvageot, H. (1992). Radar meteorology. Boston: Artech House, Inc., 366 p.

Savoy, L. (2002). Caractérisation du temps de transit et de stockage de l'eau dans la zone non saturée des systèmes karstiques - Utilisation des gaz du sol (Radon et CO_2) comme traceurs naturels. Diploma work, CHYN, University of Neuchatel, 89 p.

Savoy, L. (2007). Storage, transport and biodegradation of solute contaminants in the unsaturated zone of karst systems. PhD thesis, University of Neuchâtel, Switzerland.

Savoy, L. & Surbeck, H. (2003). Radon and CO_2 as natural tracers in a karst system. Proc. 7th Int. Conf. on Gas Geochemistry, ICGG-7, Sept. 22–26, 2003, Freiberg, Germany.

Sawada, A, Uchida, M., Shimo, M., Yamamoto, H., Takahara, H. & Doe, T. W. (2000). Non-sorbing tracer migration experiments in fractured rock at the Kamaishi Mine, Northeast Japan. Engineering Geology, 56, 75–96.

Sayer, C. (1991). Mikrobiologischer Uraninabbau. Jh. geol. Landesamt Baden-Württemberg, 33, 263–286.

Schaer, J. P., Stettler, R., Aragno, P. O., Burkhard, M. & Meia, J. (1998). Géologie du Creux du Van et des Gorges de l'Areuse. In Nature au Creux du Van, Editions du Club Jurassien: 143–215.

Schlosser, P., Stute, M., Dorr, H., Sonntag, C. & Munnich, O. (1988). Tritium/^{3}He dating of shallow groundwater. Earth and Planetary Science Letters, 89, 353–362.

Schlosser, P., Stute, M., Sonntag, C. & Munnich, O. (1989). Tritiogenic ^{3}He in shallow groundwater. Earth and Planetary Science Letters, 94, 245–256.

Schlulz, H. D. (1998). Evaluation and interpretation of tracing tests. In Käss, W., Tracing Technique in Geohydrology (pp. 341–375). Rotterdam/Brookfield: Balkema.

Schmitz, W. (1989). Art und Ausmass des Schadensfalles Sandoz im Rhein. Wasserkalender, 1989, 48–63.

Schmoker, J. W. & Halley, R. B. (1982). Carbonate porosity versus depth: a predictable relation for south Florida. Am. Assoc. Petroleum. Geol. Bull V.66, No.12, 2561–2570.

Schnegg, P. A. (2002). An inexpensive field fluorometer for hydrogeological tracer tests with three tracers and turbidity measurement.; XXXII IAH and ALHSUD Congress Groundwater & Human Development. Bocanegra, E., Martinez, D., Massone, H. (Eds.) ISBN 987-544-063-9.

Schudel, B., Biaggi, D., Dervey, T., Kozel, R., Müller, I., Ross, J. H., Schindler, U. (2003). Application of artificial tracers in hydrogeology – Guideline. Bulletin d'Hydrogéologie, 20, 1–88.

Schwarzenbach, R. P., Gschwend, P. M. & Imboden, D. M. (2003). Environmental organic chemistry. Hoboken NJ: Wiley, 1313 p.

Screaton, E. J., Martin, J. B., Ginn, B. & Smith, L. (2004). Conduit properties and karstification in the unconfined Floridan Aquifer. Ground Water, 42, 338–346.

Seiler, K. P. & Hartmann, A. (1997). Microbiologic activities in karst aquifers with matrix porosity and consequences for groundwater protection in the Franconian Alb, Germany. In Kranjc, A. (Ed.), Tracer Hydrology 97, pp. 339–345.

Seiler, K. P., Maloszewski, P. & Behrens, H. (1989). Hydrodynamic dispersion in karstified limestones and dolomites in the upper Jurassic of the Franconian Alb, F.R.G. Journal of Hydrology, 108, 235–247.

Shapiro, A. M. (2001). Characterizing ground-water chemistry and hydraulic properties of fractured rock aquifers using the multifunction Bedrock-Aquifer Transportable Testing Tool (BAT^3). U.S. Geological Survey Fact Sheet FS-075-01, 4 p.

Shapiro, A. M. & Greene, E. A. (1995). Interpretation of prematurely terminated air-pressurized slug tests. Ground Water, 33, 539–546.

Shapiro, A. M. & Hsieh, P. (1994). Overview of research at the Mirror Lake site: use of hydrologic, geophysical, and geochemical methods to characterize flow and transport in fractured rock. U.S.G.S. Water Resources Investigation Report 94-4015.

Sheppard, S. M. F. (1986). Characterization and isotopic variations in natural waters. Reviews of Mineralogy, 16, 165–183.

Shevenell, L. (1996). Analysis of well hydrographs in a karst aquifer – Estimate of specific yields and continuum transmissivities. Journal of Hydrology, 174, 331–355.

Shuster, E.T. & White, W. B. (1971). Seasonal fluctuations in the chemistry of limestone springs: A possible means for characterizing carbonate aquifers. Journal of Hydrology, 14, 93–128.

Siemers, J. & Dreybrodt, W. (1998). Early development of karst aquifers on percolation networks of fractures in limestone. Water Resources Research, 34, 409–419.

Sklash, M. G. & Farvolder, R. N. (1979). The role of groundwater in storm runoff. Journal of Hydrology, 43, 45–65.

Smart, C. C. (1988a). Artificial tracer techniques for the determination of the structure of conduit aquifers. Ground Water, 26, 445–453.

Smart, C. C. (1988b). Quantitative tracing of the Maligne Karst Aquifer, Alberta, Canada. Journal of Hydrology, 98, 185–204.

Smart, C. C. (1992). Temperature compensation of electrical conductivity measurements in glacial meltwaters. Journal of Glaciology, 38, 9–12.

Smart, C. C. (1997). Hydrogeology of glacial and subglacial karst aquifers: Small River, British Columbia, Canada. Proc. VI Conf. on Limestone Hydrology and Fissured Media, La Chaux de Fonds, Switzerland, 315–319.

Smart, C. C. (2005). Error and technique in fluorescent dye tracing. In Beck, B. F. (Ed.), Sinkholes and the Engineering and Environmental Impacts of Karst, Geotechnical Special Publication 144, American Society of Civil Engineers, 500–509.

Smart, C. C. & Karunaratne, K. C. (2002). Characteristation of fluorescence background in dye tracing Environmental Geology, 42, 492–498.

Smart, C. C. & Simpson, B. (2002). Detection of fluorescent compounds in the environment using granular activated charcoal detectors. Environmental Geology, 42, 538–545.

Smart, C. C. & Worthington, S. R. H. (2003). Electrical conductivity profiling as a means of identifying karst aquifers. 9th multidisciplinary conference on sinkholes and the engineering and environment impacts of karst, Huntsville, Alabama, USA, 265–276.

Smart, C. C., Zabo, L., Alexander, C. A. & Worthington, S. R. H. (1998). Some advances in fluorometric techniques for groundwater tracing. Environmental Monitoring and Assessment, 53, 305–320.

Smart, P .L. & Friederich, H. (1986). Water movement and storage in the unsaturated zone of a maturely karstified carbonate aquifers, Mendip Hills, England. Proc. Conf. Env. problems of karst terranes and their solutions, 59–87.

Smart, P. L. & Laidlaw, I. M. S. (1977). An evaluation of some fluorescent dyes for dye tracing. Water Resources Research, 13, 15–33.

Smethie, W. M., Solomon, D. K., Schiff, S. L. & Mathieu, G. G. (1992). Tracing groundwater flow in the Bordon Aquifer using krypton-85. Journal of Hydrology, 130, 279–297.

Snow, D. T. (1965). A parallel plate model of fractured permeable media. PhD thesis University of California Berkeley.

Solomon, D. K. & Sudicky, E. A. (1991). Tritium and helium 3 isotope ratios for direct estimation of spatial variations in groundwater recharge. Water Resources Research, 27, 2309–2319.

Spicer, K. R., Costa, J. E. & Placzek, G. (1997). Measuring flood discharge in unstable stream channels using ground-penetrating radar. Geology, 25, 423–426.

Stallman, R. W. (1961). The significance of vertical flow components in the vicinity of pumping wells in unconfined aquifers. U.S. Geological Survey Professional Paper 424-B, B41–B43.

Stallman, R. W. (1965). Effects of water-table conditions on water-level changes near pumping wells. Water Resources Research, 1, 295–312.

Stallman, R. W. (1971). Aquifer-test, design, observation and data-analysis. U.S. Geological Survey Techniques of Water-Resources Investigations, book 3, chap. B1, 26 p.

Stanton, W. I. & Smart, P. L. (1981). Repeated dye traces of underground streams in the Mendip Hills, Somerset. Proc. University of Bristol Speleol Soc. 16, 47–58.

Steeples, D. W. & Miller, R. D. (1990). Seismic reflection methods applied to engineering, environmental, and groundwater problems. In Ward, S. H. (Ed.), Geotechnical and Environmental Geophysics (pp. 1–30). Society of Exploration Geophysicists.

Stock, G. M., Anderson, R. S. & Finkel, R. C. (2004). Pace of landscape evolution in the Sierra Nevada, California, revealed by cosmogenic dating of cave sediments. Geology, 32, 193–196.

Stock, G., Granger, D., Sasowsky, I., Anderson, R. & Finkel, R. (2005). Comparison of U-Th, paleomagnetism, and cosmogenic burial methods for dating caves: Implications for landscape evolution studies. Earth and Planetary Science Letters, 236, 388–403.

Streltsova, T. D. (1974). Drawdown in compressible unconfined aquifer. Jour. Hyd. Div., Proc. Am. Soc. Civil Eng., 100, HY11, 1601–1616.

Streltsova, T. D. (1988). Well testing in heterogeneous formations. New York: John Wiley & Sons, 413 p.

Strickler, A. (1923). Beiträge zur Frage der Geschwindigkeitsformel und der Rauhigkeitszahlen für Ströme, Kanäle und geschlossene Leitungen. Mitt. Amt. für Wasserwirt, 16, 21–38.

Stueber, A. M. & Criss, R. E. (2005). Origin and transport of dissolved chemicals in a karst watershed, southwestern Illinois. Journal American Water Resources Association, 41, 267–290.

Sturchio, N. C., Du, X., Purtschert, R., Lehmann, B. E., Sultan, M., Patterson, L. J., Lu, Z.-T., Muller, P., Bigler, T., Bailey, K., O'Connor, T. P., Young, L., Lorenzo, R., Becker, R., El Alfy, Z., El Kaliouby, B., Dawood, Y., Abdallah, A. M. A (2004). One million year old groundwater in the Sahara revealed by krypton-81 and chlorine-36. Geophysical Research Letters, 31, L05503.

Suijlen, J. M., Staal, W., Houpt, P. M. & Draaier, A. (1994). A HPLC-based detection method for fluorescent sea water tracers using on-line solid phase extraction. Continental Shelf Research, 14, 1523–1538.

Sumanovac, F. & Weisser, M. (2001). Evaluation of resistivity and seismic methods for hydrogeological mapping in karst terrains. Journal of Applied Geophysics, 47, 13–28.

Surbeck, H. (1992). Nature and extent of a ^{226}Ra anomaly in the western Swiss Jura Mountains. Proc. 1992 Int. Symp. on Radon and Radon Reduction Technology, EPA-Report EPA-600/R-93-0836, NTIS PB93-296202, U.S. EPA, Washington DC, 8–19.

Surbeck, H. (1993). Radon monitoring in soils and water. Nuclear Tracks Radiation Measurement, 22, 463–468.

Surbeck, H. & Eisenlohr, L. (1993). Radon as a tracer in hydrogeology; a case study. Proc. 2nd Int. Conf. on Rare Gas Geochemistry, July 5–9, Besançon, France.

Surbeck, H. & Medici, F. (1991). Rn-222 transport from soil to karst caves by percolating water. Proc. 22nd Congress of the IAH, 27 Aug–1 Sept. 1991, Lausanne, Switzerland, 348–355.

Taylor, G. I. (1954). The dispersion of matter in turbulent flow through a pipe. Proceedings of the Royal Society of London Series A 223, 446–468.

Telford, W. M., Geldart, L. P., Sheriff, R. E. & Keys, D. A. (1990). Applied geophysics. Cambridge University Press, 770 p.

Teutsch, G. (1988). Grundwassermodelle im Karst: Praktische Ansätze am Beispiel zweier Einzugsgebiete im Tiefen und Seichten Malmkarst der Schwäbischen Alb. Ph.D. thesis, University of Tubingen.

Teutsch, G. & Sauter, M. (1991) Groundwater modeling in karst terrains: Scale effects, data acquisition and field validation. 3rd Conference on hydrology, ecology, monitoring and management of ground water in karst terranes, Nashville, USA.

Teutsch, G. & Sauter, M. (1998) Distributed parameter modeling approaches in karst-hydrological investigations. Bulletin d'Hydrogéologie, 16, 99–109.

Theis, C.V. (1935). The relation between the lowering of the piezometric surface and the rate and duration of discharge of a well using ground-water storage. Am. Geophys. Union Trans., 16, 519–524.

Tissot, G. & Tresse, P. (1978), Les systèmes karstiques du Lison et du Verneau – région de Nans-sous-Sainte-Anne (Doubs). PhD University of Franche-Comté, Besançon, France (in French), pp. 134.

Tooth, A. F. & Fairchild, I. J. (2003). Soil and karst hydrological controls on the chemical evolution of speleothem-forming drip waters, Crag Cave, southwest Ireland. Journal of Hydrology, 273, 51–68.

Toride, N., Leij, F. J., & van Genuchten, M. T. (1999). The CXTFIT code for estimating transport parameters from laboratory or field tracer experiments. US Salinity Laboratory, USDA, ARS, Riversied, CA.

Torrence, C. & Compo, G. (1998). A practical guide to wavelet analysis. Bulletin of the American Meteorological Society, 79, 61–78.

Toussaint, B. (1971). Hydrogeologie und Karstgenese des Tennengebirges (Salzburger Kalkalpen). Steir. Beitr. z. Hydrogeologie, 23, 5–115.

Uccellini, L. W. (1997). Snow measurement guidelines for National Weather Service Cooperative Observers. U.S. Department of Commerce, National Oceanic and Atmospheric Administration, National Weather Service Office of Meteorology, 4 p.

US Army Corps of Engineers (1995). Geophysical exploration for engineering and environmental investigations. Engineering Manual 1110-1-1802, Washington DC.

USACOE (U.S. Army Corps of Engineers) (1999). Groundwater hydrology. Engineer Manual 1110-2-1421, Washington, D.C.

USBR (1977). Ground water manual, a water resources technical publication; a guide for the investigation, development, and management of ground-water resources. Washington DC: US Government Printing Office, 480 p.

USGS (2004). Vertical flowmeter logging [On-line]. Available: http://water.usgs.gov/ogw/bgas/flowmeter/

Van Der Pluijm, B. A. & Marshak, S. (2003). Earth structure: an introduction to structural geology and tectonics, 2nd edition. WW Norton & Company, 656 pp.

Veselic, M., Cencur-Curk, B. & Trcek, B. (2001). Experimental field site Sinji Vrh. Beitr. z. Hydrogeologie, 52, 45–60.

Vesper, D. J., Loop, C. M. & White, W. B. (2000). Contaminant transport in karst aquifers. Theoretical and applied karstology, 13, 101–111.

Vesper, D. J. & White, W. B. (2003). Metal transport to karst springs during storm flow: an example from Fort Campbell, Kentucky/Tennessee, USA. J. Hydrol., 276, 20–36.

Vesper, D. J. & White, W. B. (2004a). Spring and conduit sediments as storage reservoirs for heavy metals in karst aquifers. Environmental Geology, 45, 481–493.

Vesper, D. J. & White, W. B. (2004b). Storm pulse chemographs of saturation index and carbon dioxide pressure: implications for shifting recharge sources during storm events in the karst aquifer at Fort Campbell, Kentucky/Tennessee, USA. Hydrogeology Journal, 12, 135–143.

Vichabian, Y. & Morgan, F. D. (2002). Self potentials in cave detection. The Leading Edge, 866–871.

Vogel, J. C., Thilo, L. & Van Dijken, M. (1974). Determination of groundwater recharge with tritium. Journal of Hydrology, 23, 131–140.

Von Gunten, H. R., Surbeck, H. & Rössler, E. (1996). Uranium series disequilibrium and high thorium and radium enrichments in karst formations. Environment Science & Technology, 30, 1268–1274.

Vortmann, G. & Timeus, G. (1910). L'applicazione di sostanze radioattive nelle richerche d'idrologia sotteranea. Le origine del Timavo. Boll. Soc. Adriatica di Scienze Naturali in Trieste, XXV, 247–260.

Walker, F. W., Parrington, J. R. & Feiner, F. (1989). Nuclides and Isotopes (14th ed.). San Jose, California: General Electric Co.

Walton, W. C. (1987). Groundwater pumping tests, design & analysis. Chelsea, Michigan: Lewis Publishers, 201 p.

Wanfang, Z. & Beck, B. F. (1998). Investigation of groundwater flow in karst areas using component separation of natural potential measurements. Environmental Geology, 37, 19–25.

Wang, H. F. & Anderson, M. P. (1982). Introduction to groundwater modeling. San Francisco: W.H. Freeman & Co.

Wanielista, M., R. Kerstein, & R. Eaglin (1997). Hydrology: water quantity and quality control (2nd edition). New York: Wiley & Sons, 567 p.

Ward, W. C., Cunningham, K. J., Renken, R. A., Wacker, M. A., & Carlson, J. I. (2003). Sequence-stratigraphic analysis of the regional observation monitoring program (ROMP) 29A test corehole and its relation to carbonate porosity and regional transmissivity in the Floridan aquifer system, Highlands County, Florida. U.S. Geological Survey Open-File Report 03-201, Tallahassee, Florida, 34 p.

Warner, D. (1997). Hydrogeologic evaluation of the Upper Floridan aquifer in the southwestern Albany area, Georgia. U.S. Geological Survey Water-Resources Investigations Report 97-4129, Atlanta, GA, 27 p.

Wendt, I., Stahl, W., Geyh, M. & Fauth, F. (1967). Model experiments for 14C water-age determinations. In Isotopes in Hydrology, Proceedings of a Symposium, IAEA, Vienna, 1966, p. 321–337.

Wenzel, L. K. (1936). The Thiem method for determining permeability of water-bearing materials and its application to the determination of specific yield; results of investigations in the Platte River valley, Nebraska. U.S. Geological Survey Water Supply Paper 679-A, Washington, D.C., 57 p.

Werner, A. (1998a). Hydraulische Charakterisierung von Karstsystemen mit künstlichen Tracern. PhD thesis, Schr. Angew Geol. Karlsruhe, 51, 169 p.

Werner, A. (1998b). TRACI – an example for mathematical tracing-interpretation-model. In Käss, W., Tracing Technique in Geohydrology (pp. 376–381). Rotterdam, Brookfield: Balkema.

Wernli, H. R. (1986). Naphthionat – ein neuer Fluoreszenztracer zur Wassermarkierung. Deutsche Gewässerkundl. Mitt. (DGM), 16–19.

White, W. B. (1969). Conceptual models for carbonate aquifers. Ground Water, 7, 15–21.

White, W. B. (1988). Geomorphology and hydrology of karst terrains. New York: Oxford University Press, 464 p.

White, W. B. & White, E. L. (1989). Karst hydrology: Concepts from the Mammoth Cave area. Van Nostrand Reinhold, 346 p.

Wiedemeier, T. H., Newell, C. J., Rifai, H. S. & Wilson, J.T. (1999). Natural attenuation of fuels and chlorinated solvents in the subsurface. New York: John Wiley & Sons, 617 pp.

Wigley, T. M. L. (1975). Carbon-14 dating of groundwater from closed to open systems. Water Resources Research, 11, 324–328.

Wildberger, A. (1996). Zur Geologie und Hydrogeologie des Karstes der Sulzfluhhöhlen (St. Antönien, Graubünden). Stalactite, 46, 112–118.

Williams, J. H., Lane, J. W., Singha, K., & Haeni, F. P. (2001). Application of advanced geophysical logging methods in the characterization of a fractured-sedimentary bedrock aquifer, Ventura County, California. U.S. Geological Survey Water-Resources Investigations Report 00-4083, 28 p.

Williams, P. W. (1983). The role of the subcutaneous zone in karst hydrology. Journal of Hydrology, 61, 45–67.

Williams, P. W. & Dowling, R. K. (1979). Solution of marble in the karst of the Pikikiruna range, Northwest Nelson, New-Zealand. Earth Surface Processes and Landforms, 4, 15–36.

Williams, S. D., Wolfe, W. J. & Farmer, J. J. (2006). Sampling strategies for volatile organic compounds at three karst springs in Tennessee. Ground Water Monitoring and Remediation, 26, 53–62.

Wilt, M. & Stark, M. (1982). A simple method for calculating apparent resistivity from electromagnetic sounding data. Geophysics, 47, 1100–1105.

Winston, W. E. & Criss, R. E. (2002). Geochemical variations during flash flooding, Meramec River basin, May 2000. Journal of Hydrology, 265, 149–163.

Winston, W. E. & Criss, R. E. (2004). Dynamic hydrologic and geochemical response in a perennial karst spring, Water Resources Research, 40, W05106, 11p.

Witherspoon, P. A. (2000). Investigations at Berkeley on fracture flow in rocks: from the parallel plate model to chaotic systems. In Faybishenko, B., Witherspoon, P. A. & Benson, S. M. (Eds.), Dynamics of fluids in fractured rock (pp. 1–58). Geophysical Monograph 122. Washington DC: American Geophysical Union.

Witherspoon, P. A., Wang, J. S. Y., Iwai, K., Gale, J. E. (1980). Validity of cubic law for fluid flow in a deformable rock fracture. Water Resources Reasarch, 16, 1016–1024.

Witthüser, K. (2002). Untersuchungen zum Stofftransport in geklüfteten Festgesteinen unter besonderer Berücksichtigung der Matrixdiffusion. Schr. Angew. Geol. Karlsruhe, 64, 145 p.

Witthüser, K., Reichert, B. & Hötzl, H. (2003). Contaminant transport in fractured chalk: laboratory and field experiments. Ground Water, 41, 806–815.

Wolfe, W. J. & Haugh, C. J. (2001). Preliminary conceptual models of chlorinated- solvent accumulation in karst aquifers. US Geological Survey Water-Resources Investigations Report 01-4011, 157–162.

Wolff, R. G. (1982). Physical properties of rocks – Porosity, permeability, distribution coefficients, and dispersivity. U.S. Geological Survey Open-File Report 82–166, 118 p.

Woo, M. K. & Marsh, P. (1978). Analysis of error in the determination of snow storage for small high Arctic basins. J. Appl. Meteorol., 17, 1537–1541.

Worthington, S. R. H. (1991). Karst hydrogeology of the Canadian Rocky Mountains. PhD thesis, McMaster University, 227 p.

Worthington, S. R. H. (2003). The Walkerton Karst Aquifer. Canadian Caver, 60, 42–43.

Worthington, S. R. H. & Smart, C. C. (2003) Empirical equations for determining tracer mass for sink to spring tracer testing in karst. In Beck, B. F. (Ed), Sinkholes and the Engineering and Environmental Impacts of Karst (pp. 287–295). Geotechnical Special Publication 122, American Society of Civil Engineers.

Worthington, S. R. H., Smart, C. C. & Ruland, W. W. (2003). Assessment of groundwater velocities to the municipal wells at Walkerton. Proceedings of the 2002 joint annual conference of the Canadian Geotechnical Society and the Canadian chapter of the IAH, Niagara Falls, Ontario, 1081–1086.

Yang, D., & Woo, M. K. (1999). Representativeness of local snow data for large-scale hydrological investigations. Hydrolological Processes, 13, 1977–1988.

Yonge, C. J., Ford, D. C., Gray, J. & Schwarcz, H. P. (1985). Stable isotopes studies of cave seepage waters. Chemical Geology, 58, 97–105.

Yuan, D., Zhu, D., Weng, J., Zhu, X., Han, X., Wang, X., Cai, G., Zhu, Y., Cui, G. & Deng Z. (1991). Karst of China. Beijing: Geological Publishing House, 224 p.

Yungul, S. H. (1996). Electrical methods in geophysical exploration of deep sedimentary basins. London, UK: Chapman & Hall.

Zimmerman, R. W. & Yeo, I. W. (2000). Fluid flow in rock fractures: from the Navier-Stokes equations to cubic law. In Faybishenko, B., Witherspoon, P. A. & Benson, S. M. (Eds.), Dynamics of fluids in fractured rock (pp. 213–224). Geophysical Monograph 122. Washington DC: American Geophysical Union.

Zötl, J. (1961). Die Hydrographie des nordostalpinen Karstes. Steir. Beitr. z. Hydrogeol., 1960/61 (2), 183 p.

Zötl, J. (1974). Karsthydrogeologie. Vienna: Springer, 291 p.

Zuber, A. & Motyka, J. (1994). Matrix porosity as the most important parameter of fissured rocks for solute transport at large scale. Journal of Hydrology, 158, 19–46.

Zwahlen, F. (Ed.) (2004). Vulnerability and risk mapping for the protection of carbonate (karst) aquifers, final report COST action 620. EUR 20912, 297 p.

Index